International Conference on Scanning Probe Microscopy (SPM-2018)

IOP Conference Series: Materials Science and Engineering Volume 443

Ekaterinburg, Russia
26 – 29 August 2018

ISBN: 978-1-5108-7658-3
ISSN: 1757-8981

Printed from e-media with permission by:

Curran Associates, Inc.
57 Morehouse Lane
Red Hook, NY 12571

Printed by Curran Associates, Inc. (2019)

For permission requests, please contact the Institute of Physics
at the address below.

Institute of Physics
Dirac House, Temple Back
Bristol BS1 6BE UK

Phone: 44 1 17 929 7481
Fax: 44 1 17 920 0979

techtracking@iop.org

Additional copies of this publication are available from:

Curran Associates, Inc.
57 Morehouse Lane
Red Hook, NY 12571 USA
Phone: 845-758-0400
Fax: 845-758-2633
Email: curran@proceedings.com
Web: www.proceedings.com

TABLE OF CONTENTS

IOP Conf. Series: Materials Science and Engineering **443** (2018) 011001 doi:10.1088/1757-899X/443/1/011001

Guest Editorial

These Proceedings contain a collection of papers based on the invited, oral, and poster presentations at the **International Conference "Scanning probe microscopy" (SPM-2018)** combined with the **International Workshop "Modern Nanotechnologies" (IWMN-2018)**, which was held in Ekaterinburg, Russia, August 26-29, 2018.

International Conference SPM-2018 covered a wide range of scientific topics:

- SPM in materials science
- PFM, MFM, KPFM, SNOM, ESM, SEM, and related techniques
- Tip-enhanced phenomena
- Probe lithography and nano-indentation
- Biocompatible & organic materials
- Multiferroic phenomena and magnetoelectric coupling
- Interface and domain engineering
- Ferroelectrics, piezoelectrics, and ionic conductors
- 1D and 2D nanostructured materials
- Theory, modeling, and data processing

The conference was organized by the School of Natural Sciences and Mathematics and Ural Center for Shared Use "Modern Nanotechnology" of the Ural Federal University.

The objective of SPM-2018 was to provide an international and interdisciplinary exchange of information on a wide range of topics related to modern state-of-the-art and future development of various Scanning Probe Microscopy techniques in different fields of science in order to ensure effective cooperation among scientists and specialists from universities, research institutions and industry.

More than 150 participants from 11 countries: Belarus, Belgium, Brazil, China, France, India, Ireland, Korea, Poland, Portugal, and Russia presented their results. The conference comprised 169 presentations, including 18 invited, 43 oral, and 108 poster ones.

The organizers are very grateful to the conference sponsors and exhibitors: Taylor & Francis Group, Russian Foundation for Basic Research, NT-MDT Spectrum Instruments, Promenergolab, Ostec ArtTool, OPTEC, INTERTECH Corporation, Merck, IMC Systems, and Alexander Lukanin.

Vladimir Shur
Victor Mironov
Chairmen of **SPM-2018**

IOP Conf. Series: Materials Science and Engineering **443** (2018) 011001 doi:10.1088/1757-899X/443/1/011001

SPM-2018 Committees

Conference Chairmen

Prof. Vladimir Shur Prof. Victor Mironov

Local Organizing Committee

Prof. Vladimir Shur Mrs. Alevtina Shur
Mrs. Elena Pelegova Dr. Ekaterina Shishkina
Dr. Dmitry Pelegov Ms. Victoria Pryakhina
Ms. Olga Dyachuk

Program Committee

Prof. A.V. Ankudinov, A.F. Ioffe Physical-Technical Institute RAS, Russia
Prof. A.A. Bukharaev, Kazan E. K. Zavoisky Physical -Technical Institute RAS, Russia
Prof. V.A. Bykov, NT-MDT Spectral Instruments, Russia
Prof. A.V. Latyshev, Rzhanov Institute of Semiconductor Physics SB RAS, Russia
Prof. V.L. Mironov, Institute for Physics of Microstructures RAS, Russia
Prof. G.M. Mikhailov, Inst. of Microelectronics Technology and High Purity Materials RAS, Russia
Prof. A.A. Saranin, Institute of Automation and Control Processes FEB RAS, Russia
Prof. V.Ya. Shur, Ural Federal University, Russia
Prof. I.V. Yaminsky, Moscow State University, Russia
Prof. S.A. Chizhik, A.V. Luikov Heat and Mass Transfer Institute NASB, Belarus
Prof. A.L. Gruverman, University of Nebraska-Lincoln, USA
Dr. A.L. Kholkin, University of Aveiro, Portugal
Prof. O.V. Kolosov, University of Lancaster, UK
Dr. V.V. Shvartsman, University of Duisburg-Essen, Germany
Dr. A.P. Volodin, University of Leuven, Belgium

IOP Conf. Series: Materials Science and Engineering **443** (2018) 011001 doi:10.1088/1757-899X/443/1/011001

SPM-2018 Sponsors and Exhibitors

Taylor & Francis Group
Russian Foundation for Basic Research
NT-MDT Spectrum Instruments
Promenergolab LLC
Ostec-ArtTool Ltd.
OPTEC
INTERTECH Corporation
IMC Group Ltd.
Merck LLC
Alexander Lukanin

IOP Conf. Series: Materials Science and Engineering **443** (2018) 011001 doi:10.1088/1757-899X/443/1/011001

List of SPM-2018 Participants

Belarus
Baraishuk Sergey
Kukharenko Lyudmila
Kuznetsova Tatyana
Lapitskaya Vasilina
Melnikova Galina

Belgium
Volodin Alexander

Brazil
Araujo Eudes

China
Hu Qingyuan
Li Zhenrong
Wei Xiaoyong
Liu Xin
Xu Zhuo
Zhao Ye
Zhou Di

France
Neradovskaia Elizaveta
Neradovskiy Maxim

Ireland
Tofail Syed

India
Khanuja Manika

Korea
Kim Myong Ho
Kim Won Jeong
Kim Dajeong
Kim Junchan
Park Tae-Gone
Song Tae Kwon

Portugal
Kholkin Andrei
Pegah Mirzadeh

Poland
Bauer Joanna

Russia
Abramov Alexander
Adishchev Sergey
Ageev Vladislav
Akbaeva Galina
Akberova Elmara
Akhmatkhanov Andrei
Akovantseva Anastasia
Al-Alwani Ammar
Alekseev Alexander
Alikin Denis
Andreeva Natalia
Ankudinov Alexander
Avilov Vadim
Barabanova Ekaterina
Begletsova Nadejda
Bodnarchuk Yadviga
Borodin Bogdan
Butiaikin Sergey
Bykov Victor
Bykov Andrei
Bystrov Vladimir
Chaika Alexander
Chaplygin Konstantin
Cheburkin Viktor
Chezganov Dmitry
Chouprik Anastasia
Chunikhin Alexander
Chuvakova Maria
Dalidchik Fedor
Dementieva Maria
Demin Alexander
Dunaevskiy Mikhail
Dyachuk Olga
Esin Alexander
Fedotov Alexander
Fomin Lev
Frolova Anastasia
Gavrikov Maksim
Gimadeeva Lubov
Golubev Evgeny
Greshnyakov Evgeny
Gruzdev Dmitry
Gunina Gushchina
Gushchina Gushchina
Il'in Oleg
Ilina Marina
Ivanov Yuri
Jityaev Igor
Kamenev Anton
Khanipov Timur
Kiryakov Arseny
Kiseleva Elina
Klinova Svetlana
Kokatev Alexander
Kolomiytsev Alexei
Komshin Aleksandr
Kosareva Ekaterina
Kosobokov Mikhail
Kots Ivan
Kozodaev Maxim
Kunkel Tatiana
Kuzmenkov Yaroslav
Lashkova Anastasia
Lebedev Vasily
Leontyev Andrey
Linker Eduard
Lisov Evgeny
Lobov Ivan
Lukyanenko Anna
Makaev Andrei
Mamin Rinat
Maslyanaya Kristina
Mingaliev Evgeny
Minigalieva Ilzira
Minin Maxim
Mironov Viktor
Morozova Anna
Mushinsky Sergey
Nasedkin Andrey
Neudachina Vera
Nuraeva Alla
Nurgazizov Niyaz
Pakhomov Alexey
Pavlov Dmitry
Pelegov Dmitry
Pelegova Elena
Polyakov Vyacheslav
Polyakova Viktoria
Privalova Larisa
Pronin Igor
Pryakhina Viktoria
Pugachev Alexey
Pukhova Valentina
Raevskaya Svetlana
Raevski Igor
Rassadin Alexander
Rezvan Alexei
Safiullina Aisylu
Saranin Alexander
Savchenkov Evgeny
Sazanova Tatiana
Senkevich Stanislav
Sergeeva Olga
Sharipov Talgat
Shatalov Alexander
Shavkuta Boris
Scherbakova Olga
Shikhova Vera
Shishkina Ekaterina
Shkalei Ivan
Shubin Andrei
Shumov Ivan
Shur Alevtina
Shur Vladimir
Skorokhodov Evgeny
Slautin Boris
Smirnov Vladimir
Smolyarova Tatyana
Sokolov Denis
Stekleneva Lubov
Syritskii Antoni
Tolstikhina Alla
Tominov Roman
Trusov Mikhail
Turygin Anton
Udalov Artur
Urusov Damir
Ushakov Andrei
Vakulenko Aleksandr
Vakulov Zakhar
Vasilev Semen
Vasileva Daria
Vasil'eva Vera
Vlasov Evgeny
Yaminsky Igor
Yashina Natalia
Yuzhakov Vladimir
Zelenovskiy Pavel
Zhukov Alexey
Zlobina Irina
Zotov Andrey
Zubarev Ilya

IOP Conf. Series: Materials Science and Engineering **443** (2018) 011002 doi:10.1088/1757-899X/443/1/011002

Peer review statement

All papers published in this volume of *IOP Conference Series: Materials Science and Engineering* have been peer reviewed through processes administered by the proceedings Editors. Reviews were conducted by expert referees to the professional and scientific standards expected of a proceedings journal published by IOP Publishing.

Switching current shape analysis in LBGO single crystals

A R Akhmatkhanov[1], K S Plashinnov[1], E V Milov[2], I V Shneidshtein[2], A P Turygin[1] and V Ya Shur[1]

[1] School of Natural Sciences and Mathematics, Ural Federal University, 620000, Ekaterinburg, Russia
[2] Faculty of Physics, Lomonosov Moscow State University, 119991, Moscow, Russia

andrey.akhmatkhanov@urfu.ru

Abstract. The switching current shape during polarization reversal at room temperature by rectangular field pulses was studied. The activation-type external field dependence of switching time was revealed. The domain structure was imaged after partial polarization reversal by piezoresponse force microscopy with high-spatial resolution and the main geometrical parameters were extracted. The switching currents for complete polarization reversal in field range from 15 to 20 kV/mm were fitted by modified Kolmogorov-Avrami formula taking into account the geometrical parameters of the domain structure. The obtained excellent fitting of the switching current confirmed the proposed model of the switching process.

1. Introduction

The borate-based nonlinear-optical crystals have emerged as the most promising candidates for creation of efficient laser sources in UV range based on frequency conversion of coherent light from visible and IR sources. The $LaBGeO_5$ (LBGO) single crystals grown for the first time in 1990s [1,2] with high nonlinear optical coefficients and ferroelectric properties allow realizing quasi-phase matched second harmonic generation (SHG) with the highest nonlinear optical coefficient (d_{33}=1.3 pm/V [3]) and without the light wave walk-off effect. Moreover, the LBGO crystals are non-hygroscopic in contrast to other borate-based crystals (β-BaB_2O_4 and $CsLiB_6O_{10}$) applied for UV light generation. The LBGO single crystals possessing the stillwellite structure have the ferroelectric phase transition at 532°C [4,5]. The outstanding crystal properties have stimulated the strong scientific interest resulting in detail study of thermal, dielectric [4], nonlinear optical [3], structural [5,6], and main ferroelectric properties [7,8].

The first polarization reversal in LBGO at room temperature was realized in Ref. [9]. The frequency and activation-type temperature dependences of coercive field during polarization reversal by triangular field pulses were studied [9,10]. It was shown that the switching time obtained during polarization reversal by rectangular field pulses followed the activation-type temperature dependence with activation energy 0.86 eV [10]. The attempt to analyse and describe the shape of switching currents obtained during polarization reversal at 120°C using the Kolmogorov-Avrami (K-A) model was made [11] with poor fitting of experimental data.

The domain structure imaging in LBGO single crystals was realized by the optical microscopy after selective chemical etching [11] and by SHG interference microscopy [10] with low spatial resolution. It is necessary to point out that the modern high-resolution domain imaging methods were not applied.

The obtained results in studying the polarization reversal process in LBGO single crystals allowed developing the domain engineering methods for creation of periodically poled samples for the first-order SHG at 365 nm [12] with period 5.5 μm. However, the attempts to create the period 2.2 μm for generation at 266 nm ended with creation of period 4.4 μm for the second-order SHG with lower efficiency [13]. Thus, the deeper study of domain structure evolution is required for further development of domain engineering methods in LBGO to obtain the periods below 3 μm.

In the present paper, we provide the detail study of the switching current shapes in LBGO single crystals obtained during polarization reversal at room temperature accompanied by visualization of static domain structure with high spatial resolution.

2. Experimental

The studied c-cut plates of LBGO single crystal grown by Czochralski pulling method (Oxide corp., Japan) had dimensions 5x15x0.5 mm^3 along a, b, and c axes, correspondingly. The liquid electrodes (LiCl saturated aqueous solution) were used for polarization reversal. The sample was glued over the 1-mm-in-diameter round aperture in the supporting glass and sandwiched between two glass plates covered by transparent ITO electrodes [14]. The liquid electrodes provided the contacts between the ITO electrodes and polar surfaces of the sample.

The rectangular voltage pulses were generated by ADC/DAC board (NI PCI 6251, National Instruments, USA) and amplified by high-voltage amplifier (TREK 20/20C, Trek Inc., USA). The applied voltage and switching current were measured by feedback ammeter. The rectangular field pulse with rise-time from 0.1 to 0.01 s, amplitude E_{ex} from 15 to 20 kV/mm and duration long enough to obtain full polarization reversal under the electrode area was used for forward switching. The obtained switching currents were smooth enough, so the measurement rate of 5 KHz could be used. The unipolar pulse of reverse polarity with E_{ex} = 20 kV/mm and duration 3 s was applied after each forward switching. The 3 min interval was maintained before application of the next forward field pulse. The switching charge was calculated by integration of the switching current. The switching time was obtained as the time interval corresponding to 0.95 of the total switching charge.

3. Results and discussion

The typical shape of the switching current in LBGO is presented in Figure 1a. The smooth shape of the switching current was obtained for the whole studied field range. The switching current consisted of two peaks: the first one – at the rise of external field, and the second one – at the constant field. It was demonstrated by applying two subsequent field pulses of the same polarity that the first peak was not related only to the capacitive input of the switching current (charging of the linear sample capacitance) as this input was much smaller (Fig. 1b). In previous works this peak was attributed to

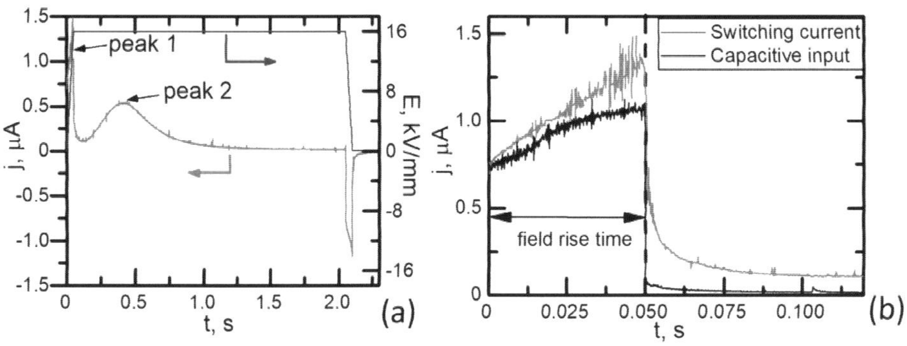

Figure 1. Typical switching current obtained in LBGO during switching with liquid electrodes: (a) switching pulse and applied electric field, (b) the magnified first peak of the switching pulse (red line) and capacitive input to measured switching current (black line) measured during application of the second subsequent forward field pulse. Amplitude of the field pulse 16 kV/mm, rise time 0.05 s.

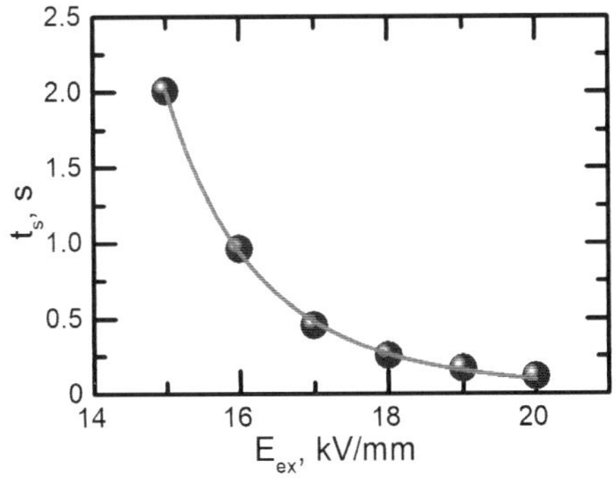

Figure 2. Field dependence of the switching time fitted by Eq. 1.

the nucleation of new domains [11,15]. In our experiments, this peak can be related to the domain nucleation at the electrode edges, while the second peak – to formation and growth of domains under electrode area.

The switching time t_s was measured by analysis of the time dependence of the switching charge corresponding to the second peak only. The activation-type dependence was used for fitting t_s (E_{ex}) data (Fig. 2):

$$t_s(E_{ex}) = t_0 \exp\left(-\frac{E_{ac}}{E_{ex}}\right) \tag{1}$$

where E_{ac} is activation field and t_0 is switching time in the limit of large switching fields. The best-fit parameter values: $E_{ac} = (180 \pm 5)$ kV/mm, $t_0 = (11 \pm 3)$ ms.

The detail study of the switching current requires the preliminary information about the domain structure geometry. The sample was switched partially by electric field pulse with E_{ex} = 15 kV/mm and duration 0.7 s (Fig. 3). The chosen pulse parameters allowed switching about the half of the area covered by electrode.

The imaging of static domain structure was realized using the scanning probe microscope (SPM) Asylum MFP-3D SA (Asylum Research, USA) in Piezoresponse Force Microscopy (PFM) mode. Commercial probes NSC18 with titanium-platinum conductive coating (MikroMash, Estonia) with a radius of curvature R = 35 nm, resonance frequency f = 70 kHz, and spring constant k = 3 N/m were used. Piezoresponse measurements were realized by the application of 5 V AC voltage close to contact resonance frequency to the SPM tip. The static domain structure (Fig. 4) represented the irregularly shaped domains nucleated with density of about $4 \cdot 10^4$ mm^{-2}. Such high values of domain nucleation densities can be attributed to defects of the sample surface. It can be seen from the Figure 4b that new domains nucleate mostly along the linear imperfections of the surface.

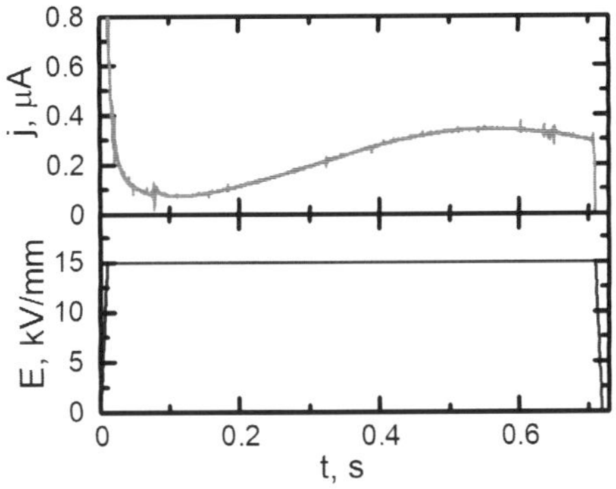

Figure 3. External field pulse and switching current obtained during partial polarization reversal.

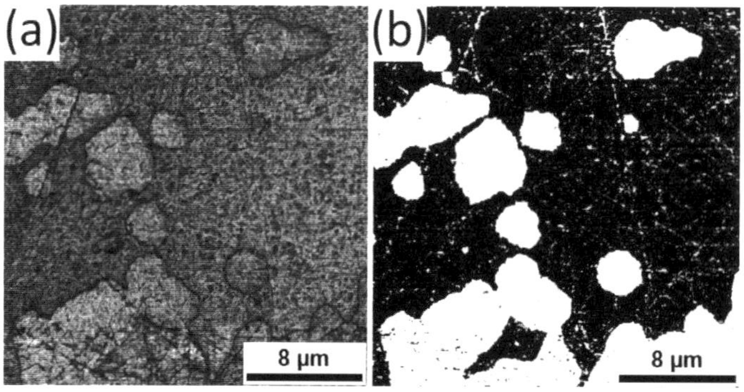

Figure 4. SPM visualization of static domain structure after partial polarization reversal: (a) amplitude signal, (b) phase signal. White contrast at (b) corresponds to switched domains.

The visualized large number of domains with wide variety of sizes ranged from 100 nm to 10 μm at the middle of the polarization reversal process allows to suppose that the main part of the switching process occurs by α(2D) process within the K-A model, while at the end of the switching the β(1D) process is expected. It should be noted that the proposed model is applicable for describing the second (main) peak of the switching current only. Additional study of the domain structure is required for full description of switching current.

The switching currents obtained for complete polarization reversal in field range from 15 to 20 kV/mm were fitted by K-A formula modified for switching in finite volume taking into account the transition from α(2D) to β(1D) processes (geometrical catastrophe) at t_{cat} [16,17]:

$$j(t) = \begin{cases} 2P_S A \frac{3}{t_{0\alpha}} \left(\frac{t-t_{st}}{t_{0\alpha}}\right)^2 exp\left(-\left(\frac{t-t_{st}}{t_{0\alpha}}\right)^3 \left(1-\frac{t}{t_m}\right)\right), \text{ for } t \leq t_{cat} \\ 2P_S A \frac{1}{t_{0\beta}} exp\left(\frac{-(t-t_{st})}{t_{0\beta}}\right), \quad \text{for } t \geq t_{cat} \end{cases} \quad (2)$$

where P_S = 4 μC/cm^2 is the spontaneous polarization, $t_{0\beta}$ and $t_{0\alpha}$ are characteristic times of corresponding processes, t_{st} is the time of switching process start, t_m accounts for impingement of growing domains on the electrode edges (boundaries of switching media) [16,17], and A is the switched area.

The obtained excellent fitting of the switching current (Fig. 5) confirmed the proposed model of the switching process. The fit parameter A was significantly higher than the real electrode area indicating that the switching outside the electrode area occurred.

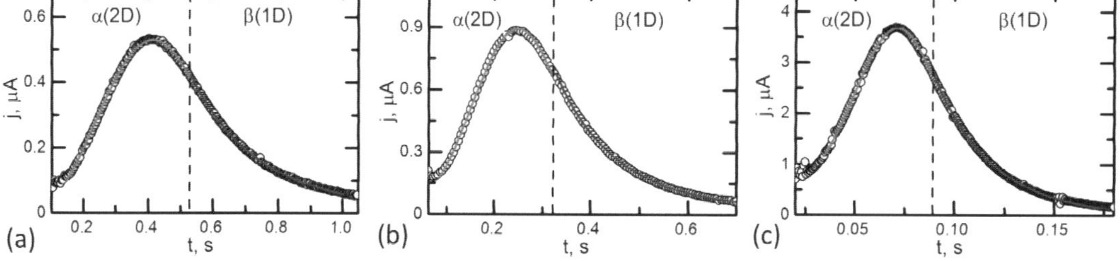

Figure 5. Switching current fitted by K-A formula modified for switching in finite volume taking into account the change of the growth model from α(2D) to β(1D) (Eq. 2). Amplitude of field pulse (a)16 kV/mm, (b) 17 kV/mm, (c) 20 kV/mm. The experimental points number was reduced to improve the visibility.

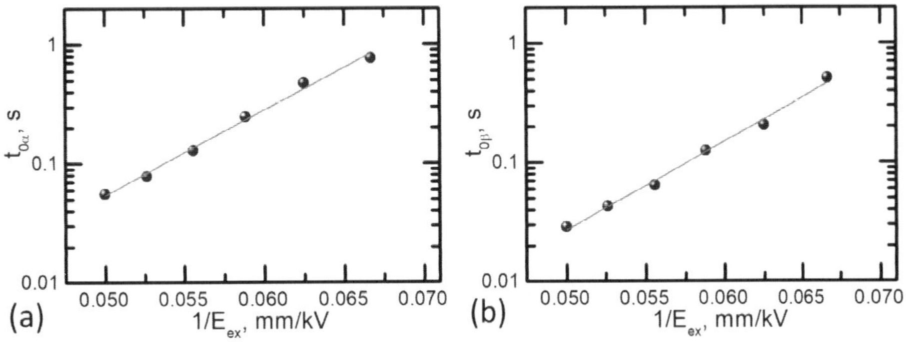

Figure 6. Field dependence of characteristic process times: (a) α(2D) process, (b) β(1D) process.

It was shown that the characteristic times $t_{0\alpha}$ and $t_{0\beta}$ followed the activation type field dependences (see Eq. (1)) with activation fields (165±5) kV/mm and (170±7) kV/mm, respectively. The equal (within the experimental error) activation fields indicate that both processes are governed by the same field-activated mechanism of domain growth.

4. Conclusion

The switching current shape obtained during polarization reversal at room temperature by rectangular field pulses was studied in LBGO single crystals. It was shown that the switching time followed the activation-type field dependence with activation field (180 ± 5) kV/mm. The visualization of static domain structure after partial polarization reversal using the scanning probe microscope in piezoresponse force microscopy mode demonstrated that domain structure represented the irregularly shaped domains nucleated with density of about $4\cdot10^4$ mm^{-2}. The switching currents obtained for complete polarization reversal in the field range from 15 to 20 kV/mm were fitted by Kolmogorov-Avrami formula modified for switching in finite volume taking into account the transition from α(2D) to β(1D) processes (geometrical catastrophe). The model choice was based on the results of domain structure imaging. The obtained excellent fitting of the switching current confirmed the proposed model of the switching process. It was shown that the characteristic times of α(2D) and β(1D) processes followed the activation type field dependences with activation fields (165±5) and (170±7) kV/mm, respectively. The equal (within the experimental error) activation fields indicated that both processes were governed by the same field-activated mechanism of domain growth. Obtained results improved understanding of domain structure evolution in LBGO single crystals and will be used for development of domain engineering for creation of compact and effective laser light sources in UV range.

Acknowledgements

The equipment of the Ural Centre for Shared Use "Modern nanotechnology" UrFU was used. The research was made possible by Russian Science Foundation (Project №14-12-00826).

References

[1] Kaminskii A A, Mill' B V and Butashin A V 1990 *Sov. J. Quantum Electron.* **20** 875-6
[2] Kaminskii A A, Butashin A V, Maslyanizin I A, Mill B V, Mironov V S, Rozov S P, Sarkisov S E and Shigorin V D 1991 *Phys. Status Solidi* **125** 671-96
[3] Horiuchi N, Osakabe E, Uesu Y and Strukov B A 1995 *Ferroelectrics* **169** 273-80
[4] Onodera A, Strukov B A, Belov A A, Taraskin S A, Haga H, Yamashita H and Uesu Y 1993 *J. Phys. Soc. Japan* **62** 4311-5
[5] Uesu Y, Horiuchi N, Osakabe E, Omori S and Strukov B A 1993 *J. Phys. Soc. Japan* **62** 2522-3

IOP Conf. Series: Materials Science and Engineering **443** (2018) 012001 doi:10.1088/1757-899X/443/1/012001

[6] Belokoneva E L, David W I F, Forsyth J B and Knight K S 1997 *J. Phys. Condens. Matter* **9** 3503-19
[7] Strukov B A, Uesu Y, Onodera A, Gorshkov S N and Shnaidshtein I V 1998 *Ferroelectrics* **218** 249-55
[8] Milov E V and Strukov B A 2001 *Phys. Solid State* **43** 513-6
[9] Milov E V., Strukov B A and Milov V N 2002 *Ferroelectrics* **269** 15-20
[10] Strukov B A, Milov E V, Milov V N, Korobtsov A P, Tomida T, Sato K, Fukunaga M and Uesu Y 2005 *Ferroelectrics* **314** 105-13
[11] Milov E, Milov V, Strukov B, Ymazaki K and Uesu Y 2006 *Ferroelectrics* **341** 39-48
[12] Hirohashi J, Hatori M, Sakairi M, Miyazawa S, Takekawa S, Taniuchi T and Furukawa Y 2013 *Advanced Solid-State Lasers Congress* (Washington, D.C.: OSA) p AM3A.2
[13] Hirohashi J, Taniuchi T, Imai K and Furukawa Y 2015 *Cleo 2015* STh3H.5
[14] Shur V Ya, Esin A A, Alam M A and Akhmatkhanov A R 2017 *Appl. Phys. Lett.* **111** 152907
[15] Shur V Ya, Nikolaeva E V, Shishkin E I, Kozhevnikov V L, Chernykh A P, Terabe K and Kitamura K 2001 *Appl. Phys. Lett.* **79** 3146-8
[16] Shur V Ya, Rumyantsev E L and Makarov S D 1998 *J. Appl. Phys.* **84** 445-51
[17] Shur V Ya, Rumyantsev E L, Makarov S D and Volegov V V 1994 *Integr. Ferroelectr.* **5** 293-301

IOP Conf. Series: Materials Science and Engineering **443** (2018) 012002 doi:10.1088/1757-899X/443/1/012002

Reconstruction of volume structure of carbon-based conductive polymer composites

A M Alekseev[1,2], A E Efimov[3] and G de With[4]

[1] National Research University "MIET", Moscow, 124498, Russia
[2] National Laboratory Astana, Nazarbayev University, Astana, 010000, Kazakhstan
[3] Shumakov National Medical Research Center for Transplantology and Artificial Organs, Moscow, 123182, Russia
[4] Eindhoven University of Technology, Eindhoven, 5612 AZ, The Netherlands

e-mail: alalrus@gmail.com

Abstract. Conductive polymer composites with different carbon fillers are under intensive investigation nowadays due to their unique properties and possible applications in different areas of science and technology. The percolation character of the conductivity behaviour in these materials and influence of nanoscale structure of conductive filler on macroscopic characteristics of composites make it important to determine the 3D structure of the components. Various electrical methods of atomic force microscopy (AFM) are efficient tools for visualization of a conductive network on a sample surface. In this work, we use a combination of AFM methods and ultramicrotomy for the 3D reconstruction of the filler structure in graphene-polystyrene and carbon black-epoxy-amine composites. Common features of the various composites, such as the formation of clusters with similar conductivity and limited number of filler particles forming a conductive network, are discussed.

1. Introduction

During the last two decades a new class of the conductive materials based on carbon fillers and nonconductive polymers was discovered and intensively investigated [1-4]. Various carbon materials were used as a conductive filler: carbon nanotubes (CNT), graphene (Gr), and carbon black (CB). The percolation character of the electrical conductivity in such composite materials provides a low value of resistance at very small amount of filler (usually parts of percent). Such composites are promising for using them as conductive coatings, paintings, transparent electrodes, sensors, etc. [5]. Knowledge of the nanoscale filler structure and its influence on the macroscopic properties is necessary for the development of new functional composite materials and improvement of their properties. The conductive filler inside a polymer forms a three-dimensional network. The structure of such a network can be visualized by various microscopy methods, including the various forms of atomic force microscopy (AFM) [4, 6] and electron microscopy [7]. The advantage of AFM is that it is possible to detect a range of electrical properties, such as conductivity, surface potential, and local charges, with nanometer resolution, which is important for reconstruction of the conductive network. Since AFM is a surface sensitive technique, additional sample treatment is needed for reconstruction of the 3D filler structure [8]. Controllable removal of a thin layer of material from the surface by methods, like ultramicrotomy, etching, and AFM probe scratching, is required for volume structure reconstruction by AFM. A combination of conductive-AFM (C-AFM) and ultramicrotomy was used for the 3D reconstruction of nanoscale conductivity in CNT-polystyrene (CNT-PS) composites [9]. Clusters with a similar conductivity level were observed inside the sample volume. It was concluded that these clusters were linked by high resistance connections, which determined the overall current in a single cluster. The possibility to distinguish between the conductive network of graphene and electrically isolated flakes using different modes of AFM was demonstrated in Ref. [10]. In this article, we use a combined AFM-ultramicrotome device for the 3D reconstruction of the filler network in conductive Gr-polystyrene (Gr-PS) and CB-epoxy-amine (CB-EA) composites and compare results obtained for different carbon-based composites.

2. Experimental

Samples of a PS-Gr composite with 2% Gr were prepared using latex technology. Details of preparation route are described elsewhere [4]. CB-EA composites with 2.5% CB were prepared by mixing CB particles into epoxy matrix. All composites had filler concentration well above the percolation threshold.

The device combining AFM and ultramicrotomy (Ntegra Tomo, NT-MDT) was used for measurements and 3D reconstruction [11]. The probes were Au-coated CSG10 (NT-MDT) and NSC12 (Micromash). C-AFM was performed in contact mode, in which the voltage applied to the probe typically was in the range 0.1-1 V, while the samples were grounded. Electrostatic force microscopy (EFM) was performed in tapping mode using the two-pass technique. 3D reconstruction was performed with ImagePro software.

3. Results and discussion

The procedure of data acquisition and subsequent 3D reconstruction was the same as described before [9-11]. The topography and current distribution of a PS-Gr sample after microtoming are shown in Figure 1. The surface is rather flat (typical rms is 3-4 nm on 10×10 μm^2 area), except some pores and nearly vertical scratches due to the ultramicrotome knife. Since the topography is nearly flat, it is approximated by a plane for 3D reconstruction. This approximation is valid as we use a relatively large z-step of 180 nm. Combining 9 images separated by z-step of 180 nm leads to a reconstructed volume element of $15.7\times13.6\times1.5$ μm^3 (Fig. 1c). Figure 1 shows the presence of conductive domains of filler with different current levels, similar to results obtained on carbon nanotubes-PS composites [9]. Such clustering of nanoscale current pathways inside the composite is confirmed by Figure 1c. Unfortunately, it is not possible to see connection between the domains inside sample volume, since the size of graphene flakes is too large in comparison with the depth available with the method used for measurements. The z-range available for reconstruction using the AFM-ultramicrotome approach with reasonable z-resolution is usually limited by a maximum of 20-30 slices, leading to a thickness of ~20% from the x- and y-size [11]. Penetration of domains with similar current level inside the sample volume is additional proof for the clustering of current pathways in Gr-PS.

Figure 1 demonstrates the distribution of graphene flakes forming a conductive network, since the current is going through the sample volume between the conductive probe and the grounded sample holder with a total distance between the electrodes in the range of ~0.2-1 mm. However, as was shown before [10], there are electrically isolated flakes, which can be visualized by EFM. It is not obvious that EFM can be used for 3D reconstruction, since there are different sources of EFM contrast: tip-sample capacitance and Coulomb interaction. As shown before [10], the penetration depth of EFM signal in Gr-PS composite is within 50-100 nm, which roughly corresponds to d_{air}/ε, where d_{air} is the distance from the surface, when the EFM contrast disappears, and ε is the permittivity of the polymer. From this observation, it can be concluded that the main source of contrast in EFM imaging of Gr-PS composites is the Coulomb interaction between the tip and the charged Gr flakes. Thus, relatively short range Coulomb interaction makes it possible to reconstruct the volume structure of electrically isolated graphene flakes inside composites by EFM. Figure 2 shows a 3D reconstruction of the graphene network inside a composite basing on 7 EFM images in a volume of $3\times2.2\times0.2$ μm^3 with a z-step of 30 nm. It is clear from Figure 2c that z-step of 30 nm is sufficient to get information on volume shape of graphene flakes. At the same time, both lateral resolution and z-resolution of 3D EFM reconstruction are worse than those of C-AFM shown in Figure 1. Thus, both C-AFM and EFM can be used for 3D graphene network reconstruction, giving complementary information on the conductive network and the electrically isolated Gr flakes distribution.

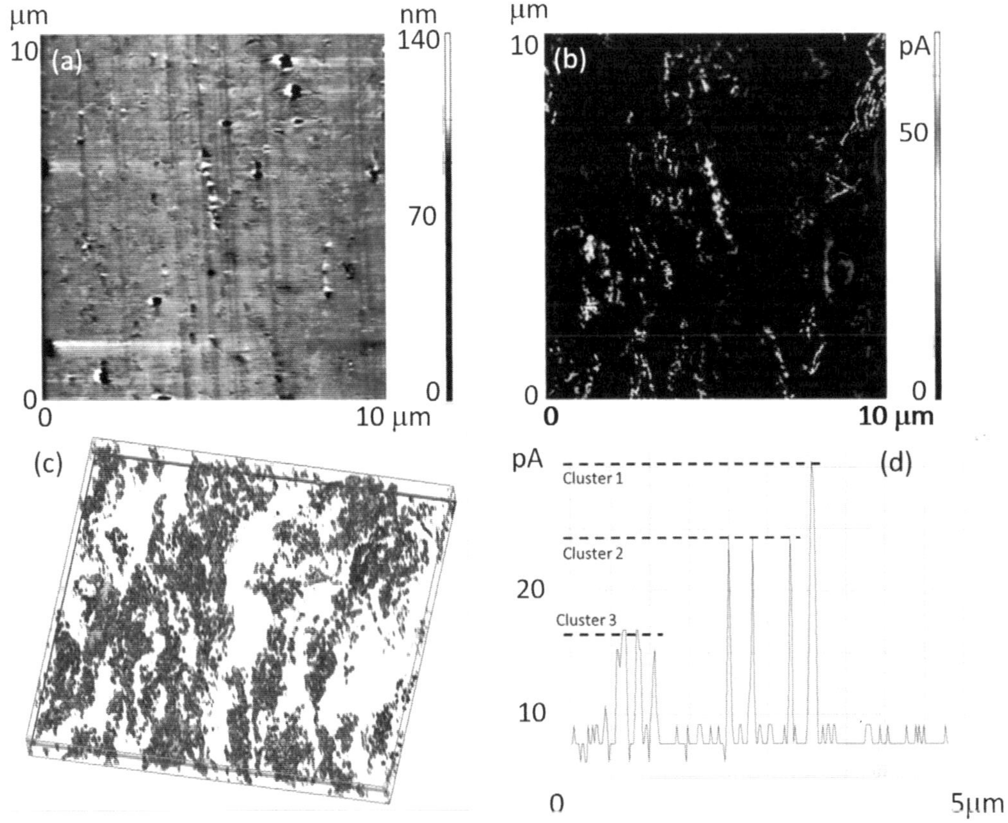

Figure 1. Gr-PS composite: (a) topography, (b) current distribution on the same area, (c) reconstructed volume structure of current inside a composite volume, (d) section of the current image.

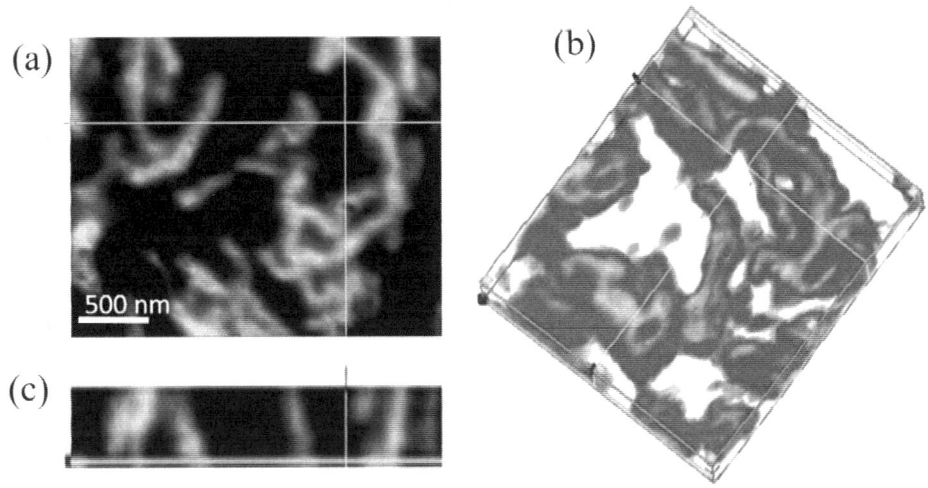

Figure 2. 3D reconstruction of Gr-PS based on EFM images: (a) top view (one of EFM images), (b) 3D EFM image, (c) vertical cross-section of (b) scaled for better visibility.

Another composite material studied by 3D AFM imaging is CB-EA. We used C-AFM for reconstruction of the nanoscale current inside a volume of this composite. The results of a reconstruction based on 28 slices and a z-step of 20 nm resulting in a volume of 4.5×4.5×0.54 μm^3 are shown in Figure 3. Several conclusions can be drawn from these results. First, from a comparison of the topography and the current distribution measured on the same area (Figs. 3a,b), it is clearly seen that only part of the CB particles is connected to the conductive filler network, while many particles are electrically isolated. The particles are clearly visible in the topography image due to the different mechanical properties of the matrix and the filler, which leads to a different surface deformation during cutting. Second, again clustering of the current pathways inside the composite is observed. Since the *z*-step used for 3D reconstruction was comparable with the size of the CB particles, several connections between particles were obtained within the *z*-range used. Analysis of the obtained 3D data revealed the locations, where changes of current level occur (Figure 3d). In Figure 3d, the transition from a low current cluster (blue color) to a cluster with a higher current level can be observed. The current level is indicated by the color bar in Figure 3b. Thus, these locations are interpreted as "bottle necks" with a high resistance between two neighboring clusters inside the sample volume. The size of the conductive clusters is directly related to the size of the filler particles: clusters of current pathways in CB-based composite are much smaller than those in Gr- or CNT-filled materials and more connections between particles are observed in the same volume. The probability to find a transition between clusters is higher just below the surface, since the number of conductive branches increases with increasing distance from tip and then a "bottle neck" should be expected after a few connections between the particles.

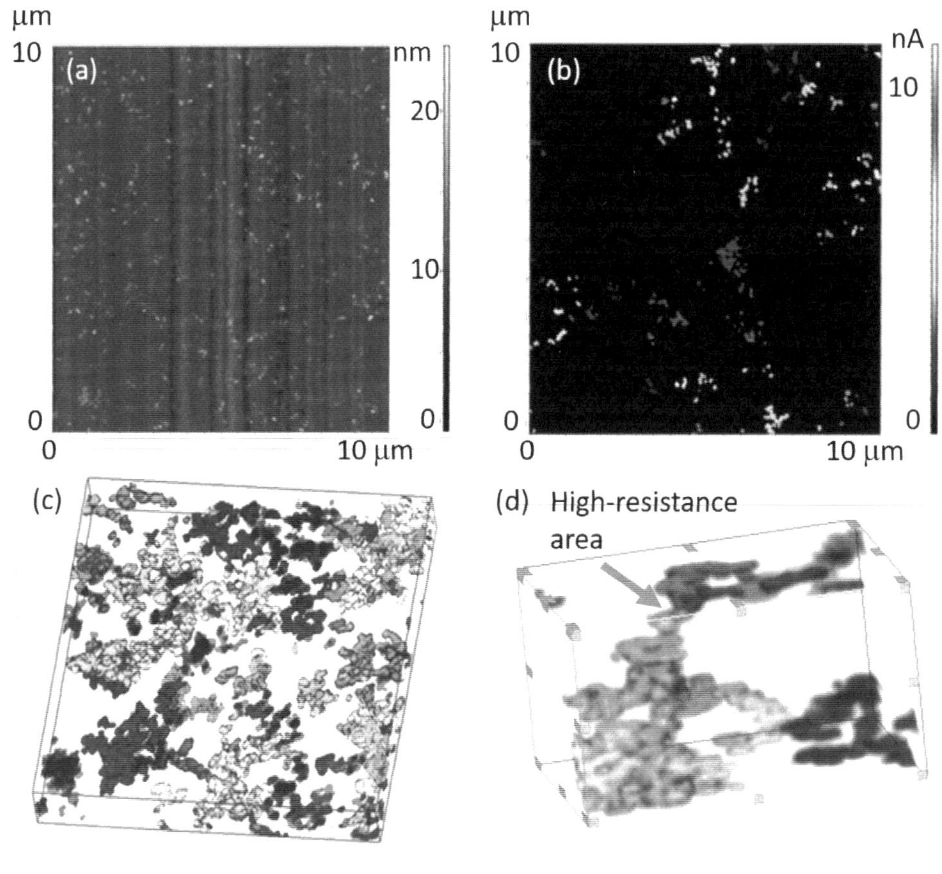

Figure 3. CB-EA composite: (a) topography, (b) current distribution on the same area, (c) reconstructed volume structure of current inside the composite volume, (d) zoomed part of (c).

The results obtained for three conductive composites (CNT-PS, Gr-PS, and CB-EA) made by different technologies with three types of fillers and different matrixes are very similar in general and have common features. This similarity is also present in the local I-V curves measured by C-AFM: in all composites both linear and nonlinear I-V characteristics were observed, possibly reflecting the existence of a tunneling conductivity mechanism between the filler particles. It may be thought that the existence of "bottle necks" is due to the dispersion methods used for realizing these composites. However, conductive composites realized by infiltration of CNT networks by a polymer [12] demonstrate very similar characteristics.

4. Conclusion

The nanoscale volume structure of conductive fillers in conductive Gr-PS and CB-EA composites was reconstructed using a combination of AFM and ultramicrotomy. The similar character of conductivity in all measured samples as well as in previously investigated CNT-PS composites shows the universal character of the revealed peculiarities. In all investigated carbon-based conducted composites, the filler particles form clusters with similar conductivity level, separated by connections with high resistivity. Only a part of filler particles forms a conductive network, while others remain electrically isolated. The size of the domains in CB-based composites is smaller than that for Gr- and CNT-based composites, which makes it possible to directly visualise highly resistive connections between clusters in a volume. 3D reconstruction of Gr-PS structure by EFM is possible due to strong influence of Coulomb interaction on the measured contrast between the AFM probe and the charged graphene flakes.

Acknowledgments

This work was supported by Federal Target Program of MES of Russian Federation, contract 14.575.21.0149 (RFMEFI57517X0149). The authors thank Prof. C. Koning, Dr. J. Loos, T. Wu, Dr. M. Ghislandi, and Dr. E. Tkalya (TU/e) for discussions and providing samples.

References

[1] Zhang W, Dehghani-Sanij A A and Blackburn R S 2007 *J. Mater. Sci.* **42** 3408-18
[2] Kim H, Abdala A A and Macosko C W 2010 *Macromolecules* **43** 6515-30
[3] Wang M, Duan X, Xu Y, Duan X 2016 *ACS Nano* **10** 7231-47
[4] Tkalya E, Ghislandi M, Alekseev A, Koning C and Loos J 2010 *J. Mater. Chem.* **20** 3035-9
[5] Li C, Thostenson E T and Chou T-W 2008 *Compos. Sci. Technol.* **68** 1227-49
[6] Balberg I, Azulay D, Toker D and Millo O 2004 *Int. J. Mod. Phys.* B **18** 2091-122
[7] Loos J, Alexeev A, Grossiord N, Koning C E and Regev O 2005 *Ultramicroscopy* **104** 160-7
[8] Alekseev A, Efimov A, Loos J, Matsko N and Syurik J 2014 *Eur. Pol. J.* **52** 154-65
[9] Alekseev A, Efimov A, Lu K and Loos J 2009 *Adv. Mater.* **21** 4915-9
[10] Alekseev A, Chen D, Tkalya E E, Ghislandi M G, Syurik Y, Ageev O, Loos J and de With G 2012 *Adv. Func. Mater.* **22** 1311-8
[11] Efimov A E, Tonevitsky A G, Dittrich M and Matsko N B 2007 *J. Microsc.* **226** 207-16
[12] Bârsan O A, Hoffmann G G, van der Ven L G J and de With G 2016 *Adv. Funct. Mater.* **26** 4377-85

IOP Conf. Series: Materials Science and Engineering **443** (2018) 012003 doi:10.1088/1757-899X/443/1/012003

Reversible modification of electrical properties at the nanoscale level in bilayer oxide systems

N V Andreeva, D A Chigirev, A S Kunitsyn and A A Petrov

Saint Petersburg Electrotechnical University "LETI", Saint-Petersburg, Russia

nvandr@gmail.com

Abstract. We report the results of experimental study of resistive switching effects in heterogeneous oxide systems based on bilayers with different sequences of two oxide layers (TiO_2 and Al_2O_3) by means of tunnelling atomic-force microscopy (AFM). These systems demonstrate a bipolar resistive switching. Moreover, the resistance state of metal-insulator-metal capacitors with TiO_2/Al_2O_3 bilayers can be electrically tuned over seven orders of magnitude. To elucidate a possible influence of nanoscale characteristics of these bilayers on the parameters of both bipolar resistive switching and electrical tuning of the resistance state, electrical properties of TiO_2/Al_2O_3 bilayers with different thicknesses of the top Al_2O_3 layer and Al_2O_3/TiO_2 bilayers were investigated. The presence of more conductive anatase crystallites in the amorphous phase of TiO_2 layer was experimentally observed only for TiO_2/Al_2O_3 bilayers. Measurements of the current distribution justify the reversible formation of conductive areas in the top Al_2O_3 layer located directly above the anatase crystallites, under the voltage application between the AFM tip and the bottom electrode of the structure. These conductive areas could be switched back into the high resistive state by application of bias voltage of reverse polarity.

1. Introduction

We have recently reported multilevel resistive switching in thin-film bilayer metal-insulator-metal (MIM) structures based on two oxide layers (TiO_2 and Al_2O_3) combined to form a bilayer structure [1,2]. It has been shown that at certain technological parameters, the resistance of MIM capacitors with TiO_2/Al_2O_3 bilayer structure can be electrically tuned in the range from 10^{12}·to 10^5 Ω. To achieve a certain resistance state (resistance level) of the structure in the given range, a pulse of bias voltage (in the range of 2-4 V) is applied between the top and bottom electrodes. The value of applied voltage determines the level of the structure's resistance: the greater the value of the voltage applied, the smaller the resistance of the TiO_2/Al_2O_3 bilayer after its application. Relative to the given resistance level (set by pulse voltage application) in MIM capacitors based on TiO_2/Al_2O_3 structures, bipolar resistive switching occurs. At the SET process, the resistance of this MIM structure switches from a high resistive state (HRS) to a low resistive state (LRS), whereas the RESET process implies the resistance switches back from a LRS to a HRS. The on-resistance/off-resistance ratio for a certain resistance level of this structure could vary from one to two orders of magnitude depending on the resistance level (Fig. 1).

Unfortunately, until now a stable tuning of the resistance level in MIM capacitors based on TiO_2/Al_2O_3 structures back from low resistance level (10^5 Ω) to high resistance level (10^{12} Ω) has not been achieved. The issue of controllable switching between different resistance levels in these structures also remains open. Answers on these questions will largely determine the prospects for application of these bilayers with multilevel resistive switching in the field of non-volatile memory devices.

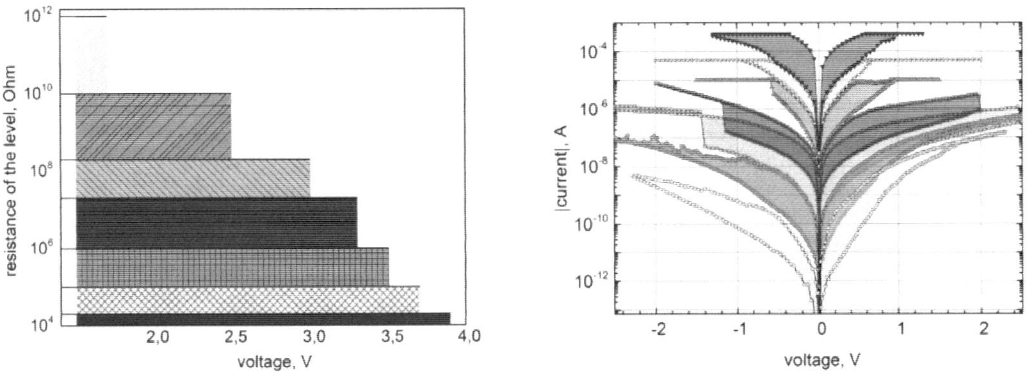

Figure 1. (a) Tuning the resistance of the $Pt/TiO_2/Al_2O_3/Pt$ structure by pulse voltage application; (b) *I-V* curves typical for bipolar resistive switching of the $Pt/TiO_2/Al_2O_3/Pt$ structure, seven resistance levels depicted in (a).

The goal of this work is to clarify the specificity of resistive switching effects in MIM structures based on two oxide layers (TiO_2 and Al_2O_3) by means of tunnelling atomic force microscopy (AFM) at the nanoscale level. This investigation was taken with a view to elucidate possible influence of nanoscale characteristics of these heterogeneous oxide systems on the parameters of both bipolar resistive switching and electrical tuning of the resistance level in the given bilayers. We suppose that deeper understanding the nature and mechanism of multilevel resistive switching will allow us to develop a proper algorithm of controllable switching between intermediate levels of the resistance in these structures. Moreover, peculiarities of resistive switching effects studied at the nanoscale level could give ideas on the deviation of electrical behaviour of the resistive random-access memory (characterized by storing data in cells through changing the resistance of thin metal oxide layer), among which is an uncertainty in the value of the RESET voltage [3].

2. Materials and methods

MIM capacitors with Pt electrodes were prepared with TiO_2/Al_2O_3 and Al_2O_3/TiO_2 bilayer structures as follows. First, a 100-nm-thick SiO_2 layer was formed by thermal oxidation on a p-type Si (100) substrate. A 10-nm-thick titanium adhesion layer was then deposited on the SiO_2 layer, followed by the deposition of a 100-nm-thick platinum bottom electrode (Pt-BE) by DC sputtering. Next, the TiO_2/Al_2O_3 or Al_2O_3/TiO_2 bilayer structures were deposited by atomic layer deposition at 200°C by using trimethylaluminium or tetrakis(dimethylamino)titanium and H_2O gases. Post-deposition annealing was done at 200°C for 30 s under ambient oxygen.

The thickness of the titanium oxide layer in both types of bilayer structures was 30 nm, while the thickness of the top aluminium oxide layer in TiO_2/Al_2O_3 bilayer was varied and was equal to 1, 3, and 5 nm for different structures. In case of Al_2O_3/TiO_2 bilayers, the thickness of aluminium oxide layer was 5 nm (Fig. 2).

The Pt top electrode (Pt-TE) layers were formed by electron-beam evaporation by using a metal shadow mask. The area of the top electrode was 10^{-4} cm^2. Schematically, the structure of formed bilayers is shown in Figure 2.

The bilayer structures thus fabricated were characterized by ellipsometry, electron microscopy, Auger electron spectroscopy, high-energy electron diffraction (HEED), and optical techniques. The ellipsometry results indicate that the thickness irregularity of the thin-deposited oxide films was less than 1% over the 100 mm substrate. Auger spectroscopy was used for stoichiometry and depth profiling element analyses of the TiO_2 thin film. The results justify the in-depth TiO_2 film homogeneity, the absence of uncontrollable impurity in the film, and the relatively narrow interface regions. The results of optical measurements and HEED indicate that TiO_2 in the film is in the anatase phase with presence of amorphous phase in a polycrystalline TiO_2 film [2]. The Al_2O_3 layers less than 5 nm thick have an amorphous structure independent of the technological parameters of film deposition.

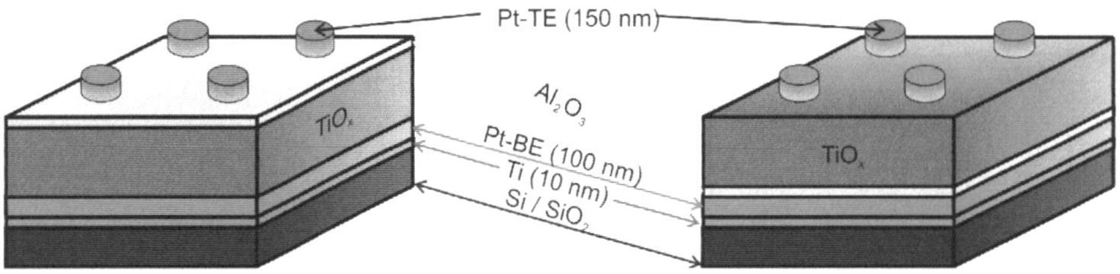

Figure 2. Schematics of Pt-BE/TiO_2/Al_2O_3/Pt-TE (on the left) and Pt-BE/Al_2O_3/TiO_2/Pt-TE (on the right) capacitors.

The tunnelling AFM measurements were performed at room temperature using Veeco scanning probe microscope. Commercially available Pt-coated silicon tips were employed for mapping the current distribution over the surface of the thin oxide films. To investigate resistive switching in the fabricated capacitors, *I-V* curves were measured under ambient condition using Keithley 4200-SCS semiconductor characterization system. The operating voltage was applied to the Pt-TE with the Pt-BE grounded.

3. Results and discussion

Earlier it was shown that the sequencing of titanium oxide and aluminium oxide layers could significantly influence the electrical properties of MIM structures [1]. Pt-BE/Al_2O_3/TiO_2/Pt-TE capacitors demonstrated a stable bipolar resistance switching with on-resistance (at low resistance state)/off-resistance (at high resistance state) ratio of ~10^2 driven by a relatively small voltage ±0.8 V after forming process. The resistance in HRS was 2×10^3 Ω, while in LRS it was 2×10^1 Ω. Pt-BE/TiO_2/Al_2O_3/Pt-TE structures do not demand a forming process, its resistance could be electrically tuned, starting from ~10^{12} Ω and up to ~10^5 Ω, by negative bias application to the top electrode of the structure. Moreover, compared to the "prepared" (by voltage application) level of the structure's resistance, a stable bipolar resistance switching was observed.

Conductive AFM measurements aim to estimate the possible influence of nanoscale characteristics of bilayers on the parameters of both bipolar resistive switching and electrical tuning of the resistance (in case of Pt-BE/TiO_2/Al_2O_3/Pt-TE capacitors). For this purpose, we scan the surface of bilayers with voltage applied between the conductive AFM tip and the bottom electrode of the structure, registering the topography and the current distribution over the sample's surface simultaneously. We distinguish two range of voltages applied between an AFM tip and the bottom electrode of structures. First, a reading voltage, the value of which was small enough to ensure that the resistance states remained unchanged during and after scanning the samples. The second one is so-called writing voltage, the value of which could provoke the certain changes in the current distribution (resistance state) over the scanned surface. The experiment was organized and pursued along the lines indicated below:

1) an area of sample's surface of 3×3 μm² in size was scanned with reading bias applied between an AFM tip and the bottom electrode of the structure to visualize the initial distribution of the current over the sample's surface;

2) the same area was then scanned with writing bias, the value of which sweeping in the range of -4V - +6V, to induce the changes in current distribution (resistance state of the structure) could be associated with resistive switching effects;

3) after each scan with writing voltage application, the same surface area was scanned with reading voltage application to register possible changes in current distribution induced by writing voltage application.

3.1. Topography of TiO_2/Al_2O_3 and Al_2O_3/TiO_2 bilayers

The results of topography measurements of TiO_2/Al_2O_3 and Al_2O_3/TiO_2 are shown in Figure 3.

The surface topography of bilayers with top aluminium oxide layer is qualitatively different from the samples with top titanium oxide. Measurements of the TiO_2/Al_2O_3 bilayers visualize the single grains on the topography distributed uniformly over the sample's surface (Fig. 3a,c). The character of topography is similar for all TiO_2/Al_2O_3 bilayers independently of the thickness of the top aluminium oxide layer. Statistical analysis of grains made for these type of bilayers justifies that the fraction of grains occupies from 9 to 15 % of the total scan area and may vary slightly from scan to scan. This variation is rather due to the location of the measurements than due to the effect of Al_2O_3 top layer thickness in different bilayers. An interesting fact is that these grains are distributed almost uniformly both over their height (from 8 to 14 nm) and lateral sizes (mean diameter is approximately 80 nm) for all bilayers (Fig. 3e,f). By supplementing the results of optical and HEED measurements, we consider that these grains are crystallites of anatase phase alternating with amorphous phase (seen as flat areas on the surface topography) in TiO_2 layer. The specificity of atomic layer deposition technique implies that the top aluminium oxide layer is an atomically flat, conformal layer. At all given thicknesses, Al_2O_3 layers have an amorphous structure and should be visualized as a flat surface on the topography. Thus, we can conclude that the only reason for appearance grains on the topography of TiO_2/Al_2O_3 bilayers is an inclusion of anatase phase in TiO_2 layer. Further confirmation of this fact follows from the results of measurements of single TiO_2 layers (of the same thickness) grown on the bottom platinum electrode and annealed at different temperatures. Increasing the temperature of postdeposition led to decrease in amount of amorphous phase in TiO_2 film, according to the results of HEED and AFM measurements of the topography (not shown here). The topography of a single TiO_2 layer deposited at the same technological conditions on a bottom platinum electrode was similar to those obtained on TiO_2/Al_2O_3 bilayers.

Meanwhile, the surface topography of bilayers with top titanium oxide layer is atomically flat (Fig. 3b), without grains. This result allows us to conclude that deposited on aluminium oxide layer, TiO_2 layer more likely has an amorphous structure in Pt-BE/Al_2O_3/TiO_2/Pt-TE capacitors.

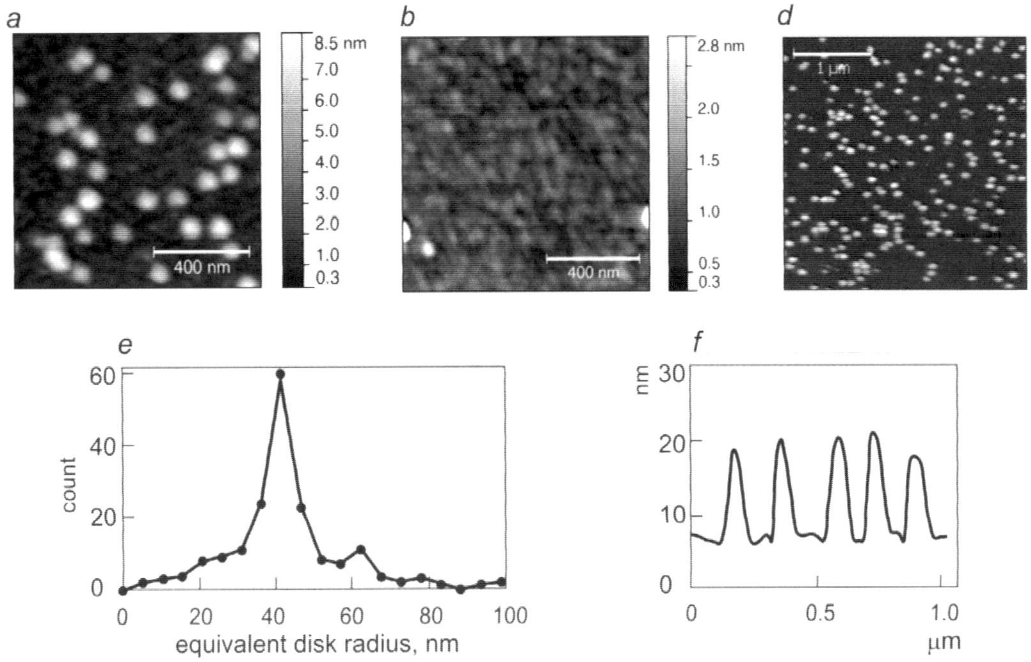

Figure 3. (a) topography of TiO_2/Al_2O_3 (30 nm/5 nm) bilayer's surface; (b) topography of Al_2O_3/TiO_2 (5 nm/30 nm) bilayer's surface; (c) topography of TiO_2/Al_2O_3 (30 nm/1 nm) bilayer's surface (3×3 μm^2); (d) statistical grain's analysis, performed over the 9 μm^2 surface area from (b); (e) height profile taken above the area of the topography containing several grains.

3.2. Current distribution over the surface of TiO_2/Al_2O_3 and Al_2O_3/TiO_2 bilayers

The results of measurements of the current distribution over the surface of TiO_2/Al_2O_3 bilayers with the thickness of top aluminium oxide layer equal to 1 nm (Fig. 4a,b) suggest that areas of the samples associated with grains are more conductive. Taking into account the conformity of Al_2O_3 layer and the fact that at the given thickness this layer has a high transparency for tunnelling current, we can conclude that the contrast in current distribution (observed at scanning with reading voltage) is related to difference in conductivity values of anatase crystallites and amorphous phases of titanium oxide. This is consistent with the fact that the resistivity of undoped titanium oxide is 10^4–10^7 Ω cm, but with the formation of oxygen vacancies related to Ti^{3+}, its value reduces to 10 Ω cm for anatase and 100 Ω cm for rutile [4].

With the increase in the thickness of top aluminium oxide layer, the value of measured current through TiO_2/Al_2O_3 bilayers decreases due to the negligible impact of tunnelling current. In this case, the resistance of TiO_2/Al_2O_3 bilayers will be provided by the Al_2O_3 layer (the resistivity of Al_2O_3 is 10^{13}-10^{15} Ω cm as opposed to 10^4-10^7 Ω cm for TiO_2). In the initial state, the resistance of the $Pt/TiO_2/Al_2O_3/Pt$ structure is ~10^{12} Ω. For the system with the thickness of top aluminium oxide layer equal to 3 nm, there was no contrast on current distribution taken at reading values of voltages applied between the AFM tip and bottom electrode. Increasing the value of voltage to -2V leads to a situation, when all areas of aluminium oxide above the crystallites of anatase in TiO_2 layer become conductive.

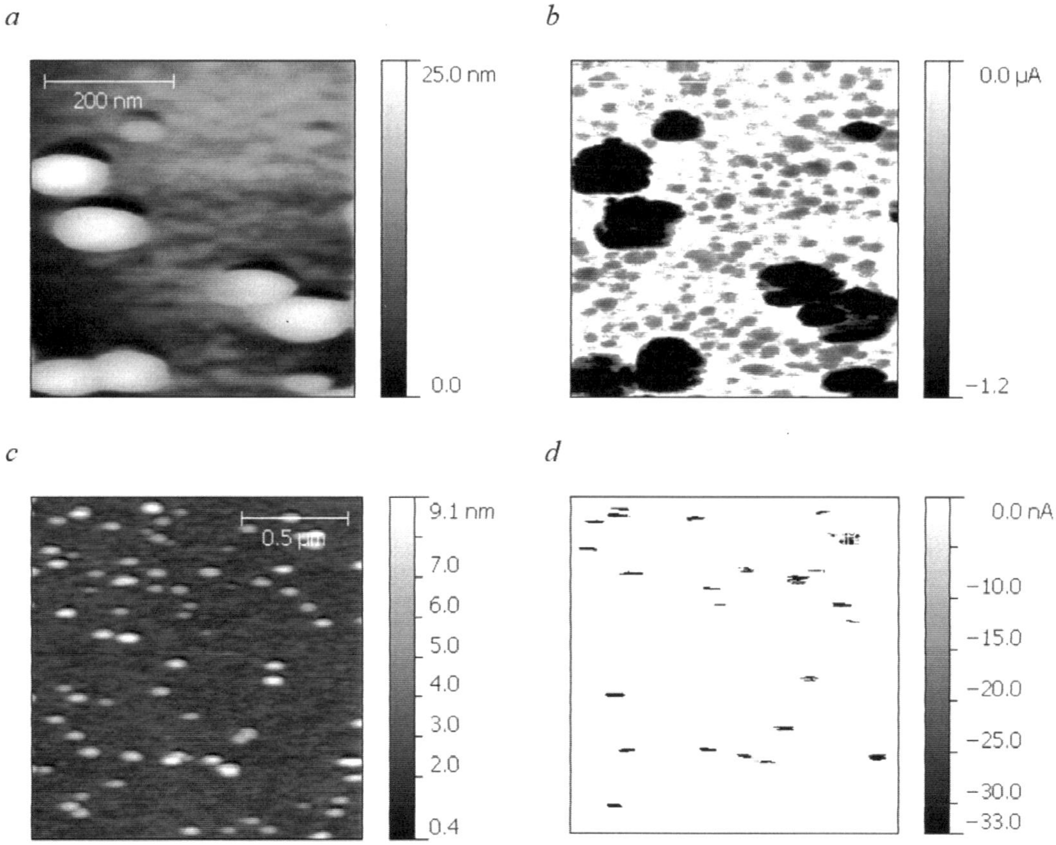

Figure 4. (a) Topography and (b) current distribution over the surface of TiO_2/Al_2O_3 bilayer (30 nm/1 nm) obtained at -1V of bias voltage applied between the conductive AFM tip and the bottom electrode of the structure. (c) Topography and (d) current distribution over the surface of TiO_2/Al_2O_3 bilayer (30 nm/5 nm) obtained at -4V of bias voltage applied between the conductive AFM tip and the bottom electrode of the structure.

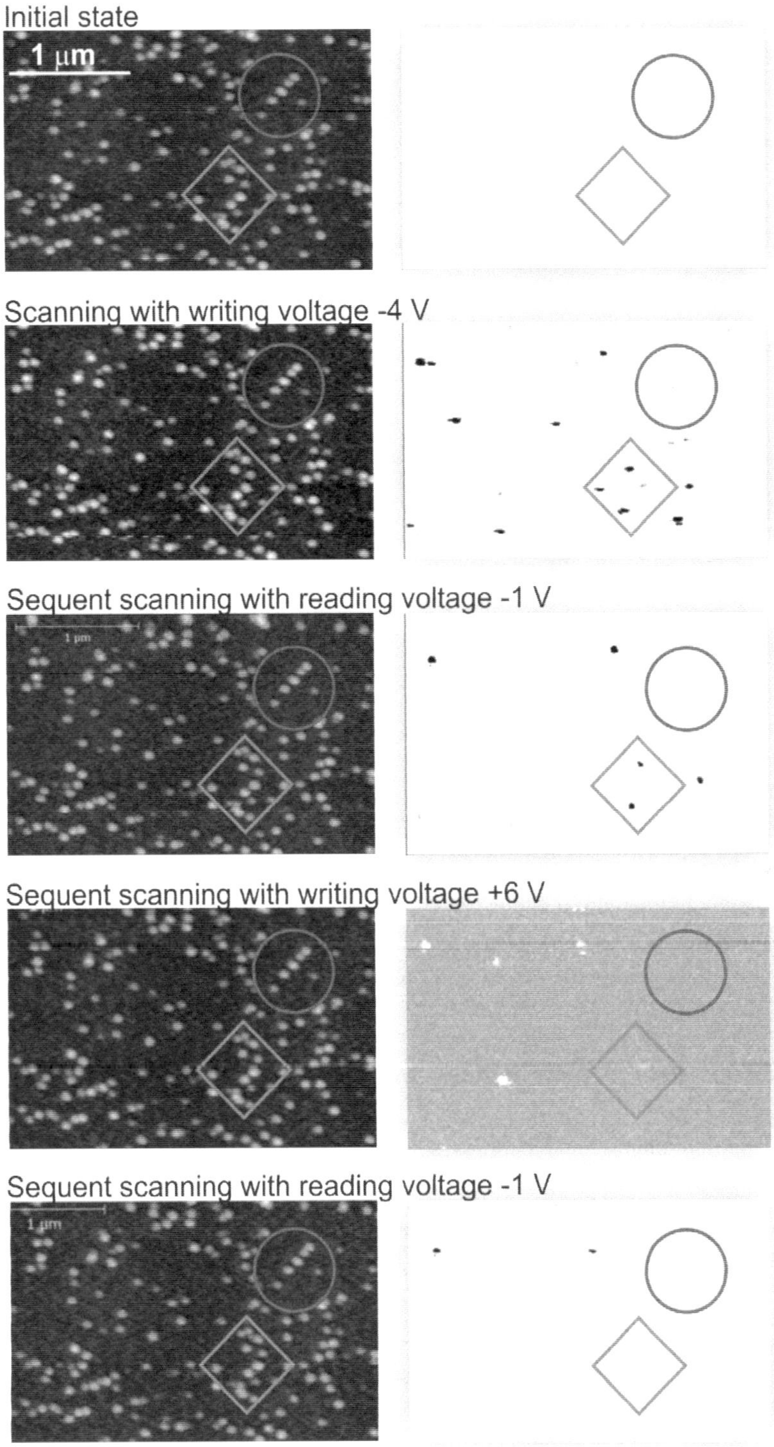

Figure 5. Topography (on the left) and corresponding current distribution (on the right) for TiO_2/Al_2O_3 bilayer (30 nm/5 nm) showing the dynamics of reversible formation of stable conductive areas in the top aluminium oxide layer after scanning with voltage applied between the AFM tip and the bottom electrode of the structure.

The most interesting results were obtained on the system with 5-nm-thick top aluminium oxide layer. In this case, the value of bias voltage applied between the AFM tip and the bottom electrode of the structure correlates with the number of conductive areas appeared above the crystallites of TiO_2 layer at scanning (Fig. 5). The higher the voltage, the larger number of conductive areas appears at writing voltages. Moreover, a number of conductive areas visualized on a current distribution at scanning with writing voltages also remain conductive at reading voltage. Application of writing voltage of reverse polarity leads to switching previously formed conductive areas back into the high resistance state. Here, it should be mentioned, that *I-V* characteristics with bipolar resistive switching could be measured with AFM tip only in location of previously formed conductive areas. Whereas, in Al_2O_3/TiO_2 (5 nm/30 nm) bilayer, observing resistive switching with *I-V* curves measurements is possible independently on the location of the AFM tip on the surface of the sample.

We consider that the varying number of conductive areas in the top aluminium oxide layer is a reason of instability of the value of both SET and RESET voltages for TiO_2/Al_2O_3 (30 nm/5 nm) bilayers with multilevel resistive switching effects. Here, we would like to clarify why we do not relate an appearance of these conductive areas with formation conductive filaments, a popular model to explain the resistive switching effect in metal oxide structures [5-6]. We explain the appearance of multilevel resistance states in $Pt/TiO_2/Al_2O_3/Pt$ structures by modification of properties of the Al_2O_3 layer due to oxygen vacancies drifting under a bias voltage. In this case, the TiO_2 layer acts as a reservoir of oxygen vacancies. We assume (according to the results of low temperature measurements [2]) that a bipolar resistive switching relatively the given level of resistance happens due to purely electronic processes and is not related to thermal or thermochemical processes. Thus, forming (by one polarity of bias) and erasing (by another polarity) of the conductive areas at scanning the surface of TiO_2/Al_2O_3 with AFM tip highlights once again the possibility to reversible change the conductivity of the Al_2O_3 layer, which we relate to multilevel resistive states in this structure. An appearance of a certain number of these areas could be explained by the specificity of measuring technique. We obtain a current distribution by scanning (moving) the conductive AFM tip along the surface, i.e. apply voltage locally. Taking into account that the tip-surface contact area does not exceed tens of nanometres in diameter and strongly influencing by topography, it seems reasonable to suspect that conductive areas should appear consequently in places associated with an existence of anatase crystallites in TiO_2 layer. If the structure is covered with top Pt electrode, the bias is applied to the whole area underneath the electrode. In this case, conductivity of the Al_2O_3 layer changes simultaneously and more or less uniformly. Presence of anatase crystallites in TiO_2 layer will introduce the instability in the values of SET and RESET voltages.

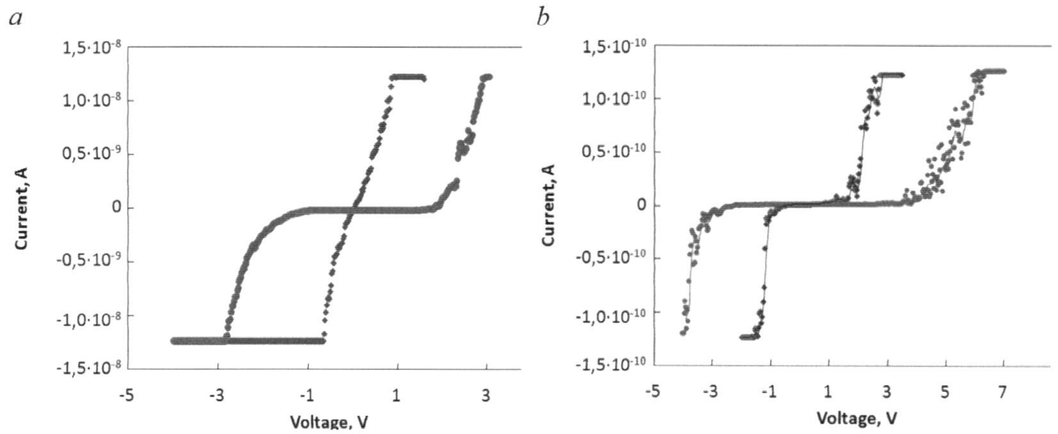

Figure 6. (a) *I-V* curves of Al_2O_3/TiO_2 bilayer and (b) *I-V* curves of TiO_2/Al_2O_3 bilayer measured with AFM tip.

One important remark should also be done here to explain why we do not connect an appearance of conductive areas with common filamentary approach, explaining the bipolar resistive switching in metal oxides. In case of Al_2O_3/TiO_2 (5 nm/30 nm) bilayer we were able to measure the *I-V* characteristics with AFM tip in any place on the surface of top TiO_2 layer (Fig. 6). This bilayer does not exhibit any crystallites (Fig. 3b) in TiO_2 layer, but demonstrates a stable bipolar resistive switching [7]. Thus, we assume that in our case reversible formation of conductive areas in TiO_2/Al_2O_3 bilayers are related to changes in the conductivity of the top aluminium oxide layer of this structure.

4. Conclusion

This paper presents an investigation of $Pt/TiO_2/Al_2O_3/Pt$ and $Pt/Al_2O_3/TiO_2/Pt$ bilayers by means of tunnelling atomic force microscopy. The presence of more conductive anatase crystallites in the amorphous phase of TiO_2 layer was experimentally observed only in $Pt/TiO_2/Al_2O_3/Pt$ bilayers. Scanning with bias applying between the AFM tip and the bottom electrode of these structures leads to reversible formation of conductive areas in the top aluminium oxide layer located directly above the anatase crystallites. These conductive areas could be switched back into the high resistance state by application of bias voltage of reverse polarity. Fraction of the conductive area of the sample varies with the value of applied bias between the tip and the bottom electrode of the sample. We consider an appearance of conductive areas in TiO_2/Al_2O_3 bilayers with the possibility to reversibly change the conductivity of the Al_2O_3 layer, which we relate to multilevel resistive states in these structures. Nevertheless, the presence of heterogeneity at the nanoscale level (owing to the anatase crystallites) in $Pt/TiO_2/Al_2O_3/Pt$ structures could influence on the stability of the parameters of both bipolar resistive switching and electrical tuning of the resistance in these heterogeneous oxide systems.

In $Pt/Al_2O_3/TiO_2/Pt$ capacitors, TiO_2 is amorphous (grown on the Al_2O_3 layer). Both $Pt/TiO_2/Al_2O_3/Pt$ and $Pt/Al_2O_3/TiO_2/Pt$ structures demonstrate a bipolar resistive switching on *I-V* characteristics measured with AFM tip located on the surface of the top layer of given bilayers.

Acknowledgments
The work was funded by the Russian Ministry of Education and Science under Grant No. 16.2873.2017/4.6.

References

[1] Alekseeva L, Nabatame T, Chikyow T and Petrov A 2016 *Jpn. J. Appl. Phys.* **55** 08PB02
[2] Andreeva N, Ivanov A and Petrov A 2018 *AIP Advances* **8** 025208
[3] Niu D, Xiao Y and Xie Y 2012 *17th Asia and South Pacific Design Automation Conference (Sydney)* 1C-3 (IEEE) 79-84
[4] Tang H, Prasad K, Sanjines R, Schmid P E and Levy F 1994 *J. Appl. Phys.* **75** 2042-7
[5] Ielmini D, Nardi F and Cagli C 2011 *Nanotechnology* **22** 254022
[6] Kim D C *et al.* 2006 *Appl. Phys. Lett.* **88** 202102
[7] Kunitsyn A S, Andreeva N V and Petrov A A 2018 *IEEE Conference of Russian Young Researchers in Electrical and Electronic Engineering (EIConRus) (Moscow)* 426-30

IOP Conf. Series: Materials Science and Engineering **443** (2018) 012004 doi:10.1088/1757-899X/443/1/012004

Formation of memristor nanostructures for RRAM memory by local anodic oxidation

V I Avilov, V A Smirnov, A A Fedotov, R V Tominov, N A Sharapov and N A Polupanov

Southern Federal University, Institute of Nanotechnologies, Electronics and Equipment Engineering, 344006 Taganrog, Russian Federation

AvilovVI@sfedu.ru

Abstract. A layout of resistive random access memory was developed and manufactured based on TiO_2 nanostructures obtained by the local anodic oxidation. It is shown that TiO_2 nanostructures exhibit a stable memristor effect without carrying out an additional electroforming operation at 20 measurement cycles. The resistance ratio of the memristor nanostructure in the high-resistance state and low-resistance state was 16.

1. Introduction

The development of modern electronics is associated with the use of new methods of nanotechnology, allowing to produce high-speed components with low energy consumption [1-4]. These components include memristors, which can find its application in the resistive random access memory (RRAM) and synaptronic elements [5-12]. Using memristors in RRAM is due to the fact that these elements have a high operation speed that exceeds the operation speed of modern dynamic random access memory (DRAM) and are nonvolatile; therefore, application of RRAM based on memristors in portable devices will reduce their power consumption and increase the operating time without additional recharging.

The analysis of publications has shown that most part of methods of forming memristor structures for RRAM contains a complex stage of electroforming oxide nanostructures (ONS) to give them memristor properties. However, the application of the local anodic oxidation (LAO) method using atomic force microscopy (AFM) techniques makes it possible to form the titanium ONS exhibiting a memristor effect without additional electroforming of structures, which will simplify the technology of manufacturing RRAM modules [12-16]. Therefore, the goal of the work is the development of manufacturing technology and the formation of RRAM elements based on memristor structures by LAO method.

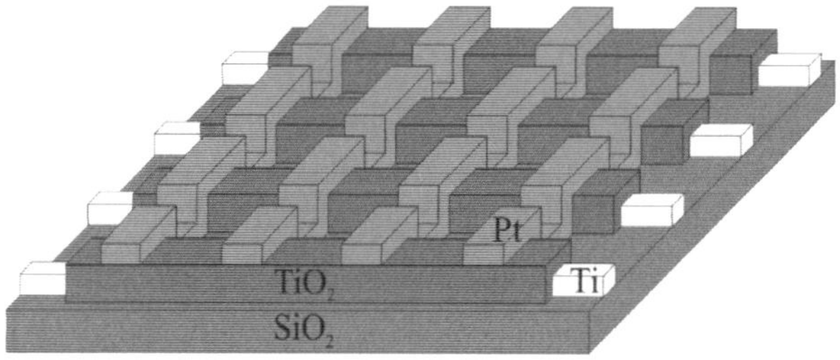

Figure 1. The layout of RRAM elements based on memristor structures.

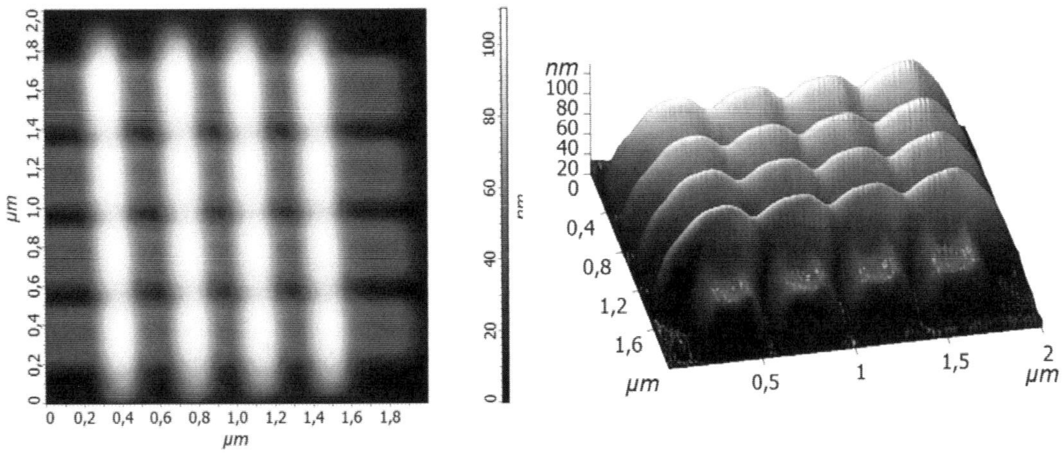

(a) (b)

Figure 2. Topography (a) and 3D-image (b) of the RRAM element layout.

2. Experiment

In this work, a structure of RRAM elements was proposed, representing a cross-bar array of memristor structures containing an insulating SiO_2 substrate, structures of the lower titanium electrodes, titanium ONS obtained by the LAO method, and the structure of the upper electrodes (Fig. 1).

A technological process for manufacturing RRAM elements was developed. For this purpose, a semi-insulating silicon substrate was used, on which a thin titanium film with a thickness of about 20 nm was formed by pulsed laser deposition using an Auto 500 (BOC Edwards, UK) multipurpose installation. Then, by the method of focused ion beams, titanium film lithography by scanning electron microscope Nova NanoLab 600 (FEI, Netherlands) was carried out; as a result, the lower titanium contacts structures were formed. After this, LAO of the lower electrodes was carried out using a scanning probe microscope Solver P47 Pro (JSC "NT-MDT" Zelenograd) and NSG11 silicon cantilevers with a conducting coating, resulting to formation of memristive ONS covering the lower electrodes. At the final stage, the platinum upper contacts structure was formed by ion-stimulated deposition by scanning electron microscope Nova NanoLab 600, resulting to a cross-bar array of 16 memristors (Fig. 2).

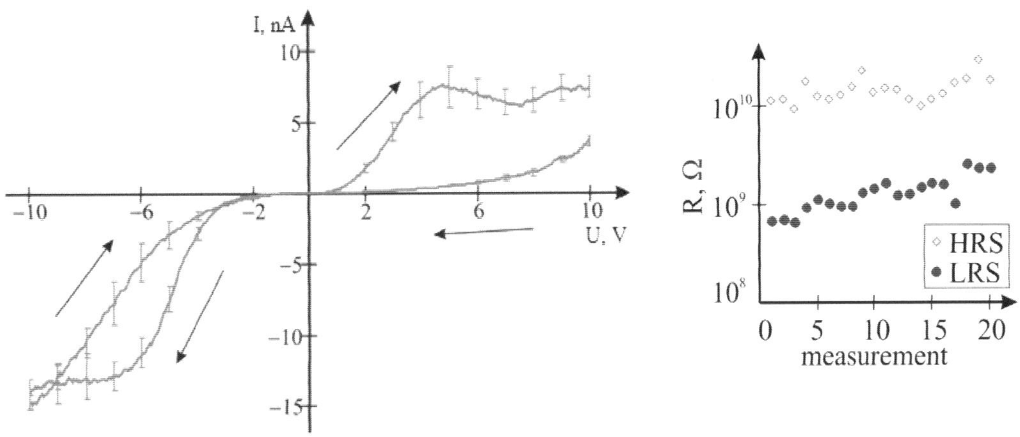

(a) (b)

Figure 3. Current-voltage characteristic (a) and resistance values (b) of memristor nanostructures.

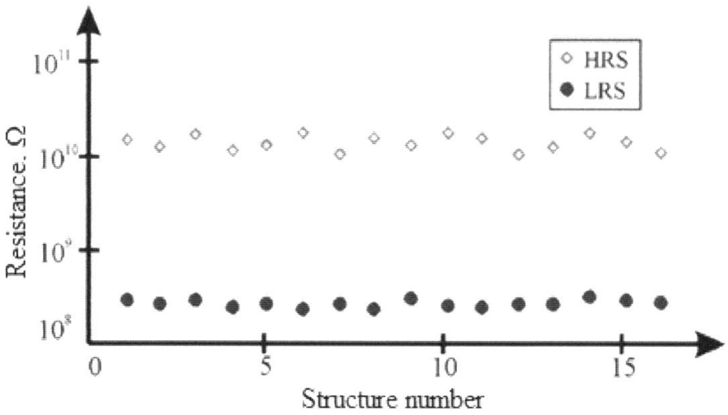

Figure 4. Resistance of titanium oxide nanostructures.

3. Results and discussion

Investigation of the electric characteristics of the array showed that the obtained titanium ONS exhibited a memristor effect and switched between a high $(1.39\pm0.27)\times10^{10}\Omega$ and low $(1.17\pm0.30)\times10^{9}$ Ω resistance states (Fig. 3a). In this case, 20 measurement cycles of a single memristor showed that the produced structures had stable memristive properties (Fig. 3b).

There was an insignificant increase in the resistance for both states, related to the further LAO process of the titanium film, when current-voltage characteristics were measured in air.

Then, the memristive effect reproducibility of the resulting 16 memristors array was investigated (Fig. 4). It is shown that the obtained structures have a uniform memristive effect. The obtained resistance ratio of the structure in the high-resistance state and low-resistance state is $R_{HRS}/R_{LRS} \approx 16$.

4. Conclusion

A layout of RRAM elements was developed and manufactured, consisting of 16 memristor structures based on titanium oxide nanostructures obtained by the local anodic oxidation method. It is shown that titanium oxide nanostructures exhibit a stable memristor effect without carrying out an additional electroforming operation. The obtained resistance ratio of the structure in the high-resistance state and low-resistance state is $R_{HRS}/R_{LRS} \approx 16$.

The obtained results can be used in the development of technological processes for manufacturing the elemental base RRAM on the basis of titanium oxide nanostructures using probe nanotechnology.

Acknowledgements

This work was supported by RFBR according to the research project № 18-37-00299, by Grant of the President of the Russian Federation No. MK-2721.2018.8, and the Southern Federal University (grant No. VnGr-07/2017-26).

References

[1] Ha S D and Ramanathan S 2011 *J. Appl. Phys.* **110** 071101
[2] Ageev O A, Blinov Y F, Ilina M V, Ilin O I and Smirnov V A 2016 *J. Phys. Conf. Ser.* **741** 012168
[3] Chang T C, Chang K C , Tsai T M, Chu T J and Simon M S 2016 *Mater. Today* **19** 254-64
[4] Jityaev L, Ageev O A, Svetlichnyi A M, Kolomiytsev A S and Spiridonov O B 2016 *J. Phys. Conf. Ser.* **741** 012011
[5] Ye C, Wu J, He G, Zhang J, Deng T, He P and Wang H 2016 *J. Mat. Sci. & Tech.* **32** 1-11
[6] Khakhulin D A, Vakulov Z E, Smirnov V A, Tominov R V, Yoon J-G and Ageev O A 2017 *J. Phys.: Conf. Ser.* **917** 092008
[7] Chen A 2016 *Solid-State Electron.* **125** 25-38

[8] Smirnov V A, Tominov R V, Alyab'eva N I, Il'ina M V, Polyakova V V, Bykov Al V and Ageev O A 2018 *Tech. Phys.* **63** 1236-1241
[9] Bousoulas P, Giannopoulos I, Asenov P, Karageorgiou I and Tsoukalas D 2017 *J. Appl. Phys.* **121** 094501
[10] Avilov V I, Ageev O A, Jityaev I L, Kolomiytsev A S and Smirnov V A 2016 *Int. Conf. on Micro- and Nano-Electronics* **10224** 102240T
[11] Zhu L, Zhou J, Guo Z and Sun Z 2015 *Nanotechnology* **22** 285-95
[12] Szot R, Rogala M, Speier W, Klusek Z, Besmehn A and Waser R 2011 *Nanotechnology* **22** 254001
[13] Siles P F, Archanjo B S, Baptista D L, Pimentel V L, Yang J J, Neves B R A and Medeiros-Ribeiro G 2011 *J. Appl. Phys.* **110** 024511
[14] Avilov V I, Polupanov N V, Tominov R V, Smirnov V A and Ageev O A 2017 *IOP Conf. Ser.: Mat. Sci.&Eng.* **256** 012001
[15] Avilov V I, Ageev O A, Blinov Y F, Konoplev B G, Polyakov V V, Smirnov V A and Tsukanova O G 2015 *Tech. Phys.* **60** 717-23
[16] Avilov V I, Ageev O A, Konoplev B G, Smirnov V A, Solodovnik M S and Tsukanova O G 2016 *Semiconductors* **50** 601-6

IOP Conf. Series: Materials Science and Engineering **443** (2018) 012005 doi:10.1088/1757-899X/443/1/012005

Application of acoustic microscopy technique for the bulk visualization and elasticity measurement of nanocomposites

S Bellucci[1], V Levin[2], E Morokov[2], Y Petronyuk[2,3] and A Shubin[4]

[1] INFN-Laboratori Nazionali di Frascati, 00044, Frascati, Italy
[2] N.M. Emanuel Institute of Biochemical Physics RAS, Moscow, 119334, Russia
[3] Scientific and Technological Centre of Unique Instrumentations RAS, Moscow, 117342, Russia
[4] OSTEC Enterprise Ltd, Moscow, 121467, Russia

jps7@mail.ru

Abstract. The impulse acoustic microscopy (AM) has been applied to study 3D microstructure and local mechanical properties of carbon nanocomposites. The particles of exfoliated graphite, graphite nanoplatelets, carbon nanoflakes, and carbon nanotubes have been used as a nanofiller. The AM technique shows the tendency of carbon nanoparticles to clusterization and formation of fractal micron-sized structures. Nanoparticle aggregates are visible on acoustic images for all types of carbon nanocomposites at arbitrary depth of their bulk. The elastic measurement demonstrates a weak dependence on type and content of carbon nanofiller. The impulse AM is the technique that provides direct data on bulk structure of a composite matter; it is characterized by combination of high efficiency and availability.

1. Introduction

The idea of nanocomposites implies that substantial changes in composite properties are realized at minimal content of nanofiller [1,2]. The changes stem from formation of extensive bulk networks of contacting nanoparticles and great value of contact interface of nanoparticles with polymer matrix. Highly dense distribution of contact interfaces results in restructuring polymer matrix and altering properties of the composite material. Therefore, nanoforms with big aspect ratios (nanotubes, nanoplatelets, and nanoflakes) are used as a filler to provide ramified interparticle contacts. Topology of nanofiller distribution is of great importance for nanocomposite properties. Uniform distribution of nanoparticles is needed to provide optimum conditions for forming the continuous clusters of nanoparticles and for maximal influence of nanofiller on composite properties.

The problem of the bulk structure of the filler in nanocomposites is the key for the purposeful creation of nanocomposite materials with desired properties. Progress in the field of nanocomposites is largely associated with the ability to characterize and control their integrated microstructure. In addition, there are new fundamental problems: identification of the variety of fractal structures formed by nanoobjects, the properties influence of these nanoobjects on the nanocomposite properties depending on the morphology of the fractal structure, etc.

Nowadays, the need to control microstructure of nanocomposites is underestimated. Publications on nanocomposites include structural studies using electron and probe microscopies to visualize only the sample surface or the small areas within a region of several microns that is associated with individual cluster of nanoparticles. Nevertheless, the presence of bulk microstructure in nanocomposites was revealed by a small-angle scattering X-ray method. In Ref. [3], heterogeneous and multilevel distribution of nanoparticles in the form of fractal structures and conglomerates was demonstrated. This X-ray method allows measuring the fractal dimension of the structures that gives indirect characterization of the nanocomposite microstructure. The direct observation is possible by X-ray and acoustic microscopy (AM) only. Both techniques provide the same resolution; the acoustic imaging is simpler, safer and allows local measuring the elastic properties.

In the paper, the bulk mesostructure and local mechanical properties of epoxy nanocomposites with diverse carbon nanoforms (particles of exfoliated graphite, graphite nanoplatelets, carbon nanoflakes, and carbon nanotubes) have been studied by the impulse AM methods.

2. Experimental

The impulse AM technique seems to be a powerful tool for visualizing cluster structure inside the nanocomposite bulk [4]. Moderate frequencies of 50-200 MHz provide spatial resolution of 15-60 μm that is substantially bigger than characteristic sizes of carbon nanoparticles, so, the individual particles could not be resolved. But this frequency range is optimal to observe particles distribution, large-scale details of the nanofiller fractal structure, including zones of fractal agglomerations and discrete conglomerates. Visualization of nanofiller conglomerates is performed in the ultramicroscopical regime [5] that is an acoustical analogue of the dark-field light microscopy. Conglomerates generate scattered radiation similar to point sources. Focusing receiver collects the scattered radiation just at the position above the scatterer. Estimations show that fractal agglomerates of carbon nanoparticles could be observed in the images, if their sizes are not smaller than 0.5 μm for the indicated frequency range. So, focused ultrasound is a sensitive instrument for evaluating distribution of carbon nanofiller over the composite bulk, revealing formation of aerogel conglomerates, and studying their fractal nature.

In the paper, epoxy-carbon nanocomposites were under investigation. The epoxy matrix was prepared from Epicote Resin 828 (Momentive Specialty Chemicals Inc.) with own curing agent (modified TEPA) [6,7]. Diverse kind of carbon particles was used for the nanocomposite fabrication. One type of them was so-called carbon nanoflakes (CNF) and graphite nanoplatelets (GNP) in the form of graphite stacks (30-40 atomic layers) manufactured by a fine milling of the exfoliated graphite accompanied by thermal treating. Thickness of CNF and GNP particles was in the range of 10-12 and 7-10 nm; their lateral size was in the range of 10-15 and 0.5-5 mkm, respectively. The other kind of filler was carbon nanotubes (CNT). Nanotubes were 20-40 nm in diameter with length of 0.5-20 μm. Filler content was 0.5-1.5 weight % for CNF and GNP epoxy nanocomposites and 0.05-0.1 weight % for the CNT nanocomposite. The procedure included vacuum degassing of liquid matrix, preparation of graphite nanofiller suspension, mixing the liquid matrix and graphite particle suspension in alcohol, its evaporating and degassing the new mixture, adding curing agent and mixing it with the graphite-epoxy suspension, and curing the suspension at normal conditions and at 80°C. Plane-parallel specimens of 1.0-1.6 mm thick have been employed in experiments.

3. Results and discussion

3.1. Imaging of nanocomposites bulk structure with AM technique

The experimental results are represented as microstructural images at the specimen depth (C-scans) and the cross-sections (B-scans). In Figure 1 in-depth acoustic images are shown. The bright spots inside the material bulk are seen in the Figures 1a, 1b for CNF and GNP fillers. Brightness of the spots is caused by high efficiency of ultrasonic scattering at the obstacles, which are caused by nanoparticles agglomeration to the micron-sized fractal conglomerates. The effectiveness is caused by the fact that conglomerates include air and constitute aerogel particles. According to our estimation, such micron-sized conglomerates can be visible. Figure 1c represents the bulk microstructure for specimen with CNT filler. The microstructure demonstrates non-uniform distribution of CNT local density, as it has been shown in our previous studying by scanning electron microscopy [8]. Visible spots in acoustic image are the compact nanotube packages separated from each other and linked by single filaments.

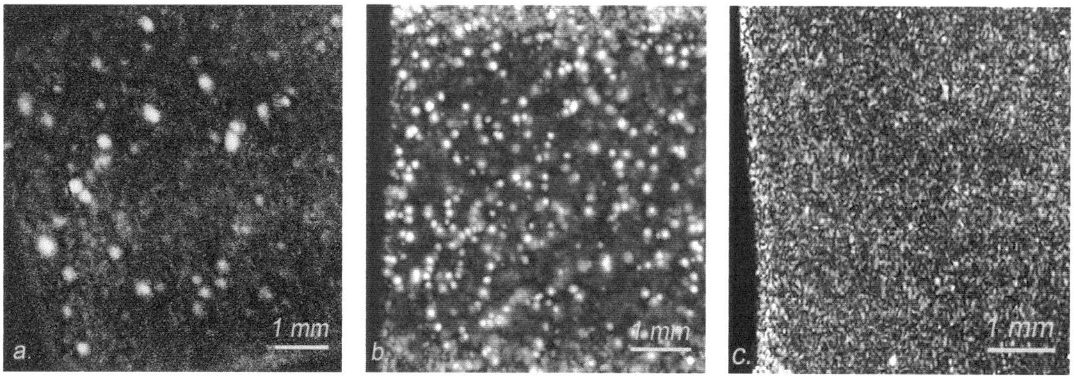

Figure 1. Acoustic imaging (C-scans) of carbon-epoxy nanocomposite with different contents and types of nanofiller at the depth of 0.5 mm: (a) epoxy+1.5 wt% CNF; (b) epoxy+0.75 wt% GNP; (c) epoxy+0.1 wt%.

The impulse AM technique was applied to evaluate homogeneity of elastic properties of the nanocomposites. For the goal, a set of B-scans was made (Fig. 2). The images show high-level elastic homogeneity for all nanocomposite samples. Small inclination of the bottom line in the B-scans is caused by geometry of the samples. The elastic homogeneity did not depend on filler content, at least, within the concentration range that has been employed in the current experiments.

3.2. Measuring the elastic properties of nanocomposites with AM technique

One of the essential goals of nanotechnology is developing nanomaterials with improved elastic and strength properties. Ultrasound is not suitable instrument for studying strength properties, but impulse AM is an efficient technique to study local elastic properties and their distribution over the material bulk by means of sonic velocity measurement. Data on the delay time between echoes reflected from the specimen face and bottom and the thickness measuring make it possible to find value of the sonic velocity inside the focal area of a probe beam. For epoxy-carbon nanocomposites, the longitudinal elastic wave velocity c_L was found (Table 1). Data on the material density ρ have been obtained by hydrostatic (Archimedes) weighing. These data on sonic velocity and density have been used to find longitudinal elastic modulus $C = \rho\, c_L^2$.

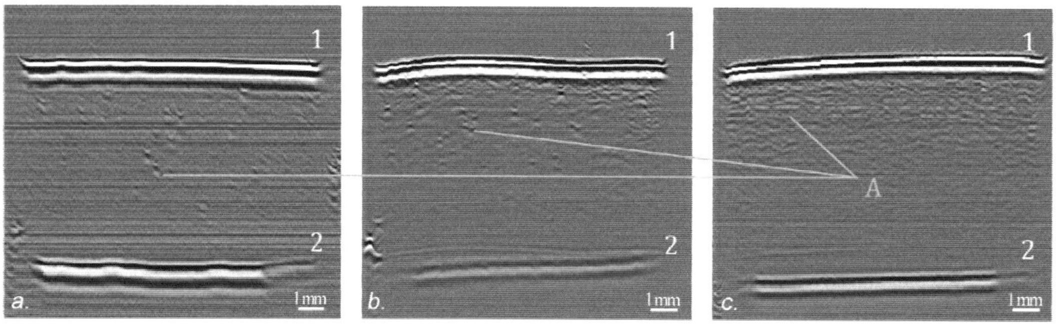

Figure 2. Acoustic imaging transverse cross-sections (B-scans) of carbon-epoxy nanocomposite with different contents and types of nanofiller: (a) epoxy + 1.5 wt % CNF (thickness of 1.42 mm); (b) epoxy + 0.75 % wt GNP (1.33 mm thick); (c) epoxy + 0.1 wt % CNT (1.39 mm thick). 1 – echo-line from the specimen top; 2 – echo-trace from the bottom; A – echoes from the structural elements inside the specimen bulk. Length of the scanning area is 12 mm.

Table 1. Measured longitudinal sonic velocity c_L, material density ρ, and longitudinal elastic modulus C in carbon-epoxy nanocomposites with various types and concentrations of nanofiller.

Material		c_L, km/s	ρ, g/cm^3	C, GPa
pure epoxy		2.90	1.170	9,8
TG	0.25 %	2.97	1.180	11
	0.5 %	3.00	1.186	11
	1.0 %	3.05	1.180	11
	1.5 %	2.97	1.189	11
GNP	0,5 %	2.87	1.179	9,9
	0.75 %	2.88	1.186	10
	1.0 %	2.86	1.186	10
CNT	0.05 %	2.73	1.156	8,4
	0.1 %	2.80	1.180	9,3

Microacoustical measurements have revealed a weak influence of the nanocarbon filler on elastic properties of the composites. In spite of the observed bulk mesostructure including rather large-scaled particles or nanoparticle conglomerates, contribution of the carbon nanofiller in elastic properties is so small that its spatial non-uniform distribution does not induce any inhomogeneity of the material elasticity. The measured sonic velocity values are closed to the measurements for pure epoxy.

In literature, there are evidences of the elastic modulus increment in a polymer material as a result of its transformation into the nanocomposite state [9-12]. Our results are in a contradiction with them, which is caused by diversity in measuring technique and treating the elastic properties concept.

4. Conclusion

The experimental results demonstrate that the pulse acoustic microscopy is a powerful technique for bulk microstructure visualization and local elastic measuring of nanocomposite material. The occurrence of complicated fractal microstructure in the bulk of nanocomposites has been shown. Efficient agglomeration of carbon nanofiller has been demonstrated. Local elastic measurements demonstrate elastic homogeneity of the studied nanocomposites despite the bulk microstructure presents. Ultrasonic elastic measurements demonstrate minimal influence of filler on nanocomposite elasticity.

References

[1] Stankovich S, Dikin D A, Dommett G H, Kohlhaas K M, Zimney E J, Stach E A, Piner R D, Nguyen S T and Ruoff R S 2006 *Nature* **442** 282-6
[2] Potts J R, Dreyer D R, Bielawski Ch W and Ruoff R S 2011 *Polymer* **52** 5-25
[3] Schaefer D W and Justice R S 2007 *Macromolecules* **40** 8501-17
[4] Bellucci S, Micciulla F, Levin V M, Petronyuk Yu S, Chernozatonskii L A, Kuzhir P P, Paddubskaya A G, Macutkevic J, Pletnev M A, Fierro V and Celzard A 2015 *AIP Advances* **5** 067137
[5] Levin V M and Petronyuk Y S 2015 *Physics Procedia* **70** 631-5
[6] Kuzhir P, Paddubskaya A, Plyushch A, Volynets N, Maksimenko S, Macutkevic J, Kranauskaite I, Banys J, Ivanov E, Kotsilkova R, Celzard A, Fierro V, Zicans J, Ivanova T, Merijs Meri R, Bochkov I, Cataldo A, Micciulla F, Bellucci S and Lambin Ph 2013 *J. Appl. Phys.* **114** 164304

IOP Conf. Series: Materials Science and Engineering **443** (2018) 012005 doi:10.1088/1757-899X/443/1/012005

[7] Dabrowska A, Bellucci S, Cataldo A, Micciulla F and Huczko A 2014 *Phys.Status Solidi B* **251** 2599-602
[8] Levin V, Petronyuk Y, Morokov E, Chernozatonskii L, Kuzhir P, Fierro V, Celzard A, Mastrucci M, Tabacchioni I, Bistarelli S and Bellucci S 2016 *Phys. Status Solidi B* **253** 1952-9
[9] Kim H, Abdala A A and Macosko Ch W 2010 *Macromolecules* **43** 6515-30
[10] Mittal G, Dhand V, Rhee K Yo, Park S-J and Lee W R 2015 *J. Industr. Eng. Chem.* **21** 11-25
[11] Coleman J N, Khan U, Blau W J and Gun'ko Y K 2006 *Carbon* **44** 1624-52
[12] Young R J, Kinloch I A, Gong L and Novoselov K S 2012 *Compos. Sci. Technol.* **72** 1459-76

IOP Conf. Series: Materials Science and Engineering **443** (2018) 012006 doi:10.1088/1757-899X/443/1/012006

Measurement of the bending of thin inclined nanowires as a method for determining elastic modulus

M S Dunaevskiy[1], P A Alekseev[1], P Geydt[2], E Lahderanta[2], T Haggren[3] and H Lipsanen[3]

[1] Ioffe Institute,194021, Saint-Petersburg, Russia
[2] Lappeenranta University of Technology, Lappeenranta FI-53851, Finland
[3] Aalto University, Espoo FI-00076, Finland

Mike.Dunaeffsky@mail.ioffe.ru

Abstract. The method of measuring nanowires elastic modulus by scanning probe microscopy is presented. This method uses the measurement of nanowire bending profiles in the precise force control mode. The possibilities of the method are demonstrated by measuring the Young's modulus of thin tapered InP nanowires.

1. Introduction

Recently, the interest of various scientific groups has been attracted to studies of semiconductor nanowires (NW). On the basis of NW, it is planned to create advanced sensoric, nanoelectronic, and photovoltaic devices [1,2]. For producing the NW-based device structures it is important to know not only their electrophysical properties, but also their mechanical properties. Therefore, it is important to have information about the Young's modulus of NW. In a number of works it is shown that the Young's modulus of thin NW can differ from the tabulated values of the bulk materials [3]. The physical properties of a number of $A^{III}B^{V}$ materials (GaAs, InP, InAs, etc.) with a zinc blende structure have been already thoroughly investigated. However, the properties of the same $A^{III}B^{V}$ materials with the wurtzite structure have not been yet studied experimentally. For example, Young's moduli of wurtzite InP, InAs, and a number of other $A^{III}B^{V}$ materials are still unknown. It is worth noting that in sufficiently thin NW the crystalline wurtzite structure may appear. In this work, we present scanning probe microscopy (SPM) method for determining the Young's modulus of sufficiently thin NW. The capabilities of the method are demonstrated by measuring the Young's modulus of thin tapered InP NW [4].

2. Experimental

Thin NW, due to the high length-to-radius ratio may have extremely small flexural stiffness coefficient. An experimental study of sufficiently thin NWs is a serious challenge. When studying thin NW by the SPM methods large bends of NW and even their breakage may occur. In order to prevent this, we used the method of precise force control – Peak Force Tapping (Bruker) [5]. This mode was specially designed to minimize the force of the probe's impact on the surface. Peak Force Tapping mode makes it possible to investigate "very soft media", including NWs with small flexural stiffness. We had the opportunity to reduce the probe's force acting on the surface down to 0.05 nN and maintain it during scanning with a high degree of accuracy. Scanning with ultra-small force allowed us to obtain noiseless images of inclined InP NWs without their deflection (Fig. 1). This was a very important step necessary to further accurately measure the NWs bending.

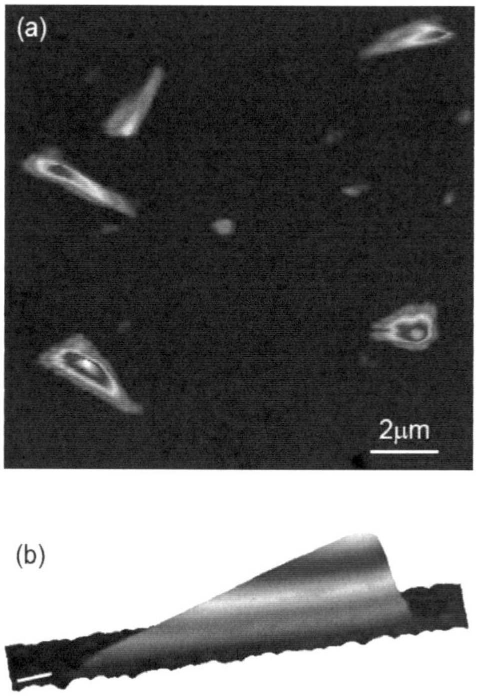

Figure 1. (a) SPM image of an array of inclined InP NW; (b) 3D SPM image of an individual InP NW (the scale bar corresponds to 200 nm).

In this paper, we investigated samples with tapered InP NWs differing both in geometric parameters and in the crystal structure. The length of the studied NWs was from 2 to 3 microns, the average radius was from 15 to 40 nanometres, and the cone angle was 1-1.5 degrees. The samples with NWs array were grown on the Si-substrate by metalorganic chemical vapour deposition method. The growth temperature was chosen so to realize the desired "vapour-liquid-solid" growth mechanism. The geometric parameters of the NW were determined by SPM and scanning electron microscopy (SEM). We used Multimode-8 (Bruker) microscope for SPM measurements. The type of crystal structure of InP NW was determined from electron diffraction data. An important property of the NW array was that a significant number of NW was inclined to the surface at a small angle of 15-20 degrees. This important circumstance made it possible to investigate the bending of these inclined NW by the SPM methods.

In this paper, we used the SPM method of measuring the bending profiles on inclined NW with different peak forces F_{peak} acting on the NW. It was found that with increasing values of the peak force, the deflection of the NW increased, which led to a change in the scanning topography of the inclined NW. It is reasonable to assume that the deflection of the NW is close to zero when working with near-zero forces. Therefore, if we subtract the profile of non-bent NW from the profile of the NW bended by the force F, then we will obtain the bending profile of the NW (for the given force F). During our measurements, we observed a linear increase in the bending profiles w(x) with increasing values of the peak force F_{peak}. This means that in the range of applied forces there is a linear bending of the NW according to the Hooke's law. After the measurement of set of bending profiles measured along the NW with different forces, these profiles were divided by the values of the corresponding forces and thus the inverse stiffness profile was obtained. The resulting inverse stiffness profile is then analysed in accordance with the theoretical deflection model of the NW. In our case, the NW under study had a tapered shape, so we used the bending model from Ref. [4]. By fitting the experimental inverse stiffness profiles by theoretical bending model one can obtain the Young's modulus value of studied NW. It should also be emphasized that this method can be applied to any thin and flexible 1D object (nanoparticles, thin fibrils, asbestos tubes, nanotubes, etc.) as well as thin and flexible 2D membranes (graphene, transition metal dichalcogenide layers, etc.).

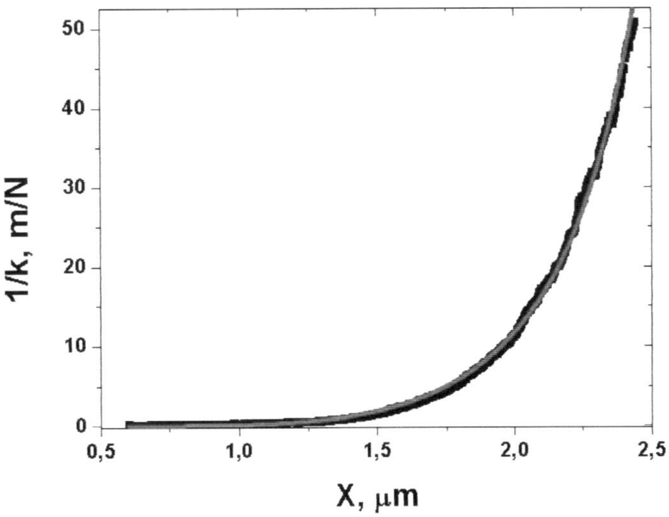

Figure 2. Inverse stiffness profile 1/k(x) measured on the inclined InP NW. Fitting by Eq. (1).

3. Results

Figure 1 shows an image of an array of inclined NW (Fig. 1a), as well as 3D image of one of the NW (Fig. 1b) scanned with a near-zero force (F_{peak}=0.1nN). As it can be seen, the length of this inclined NW is about 2 μm. Figure 2 shows the inverse stiffness profile 1/k(x) measured on this NW and a smooth fitting curve using bending profile formula for a tapered beam [4].

$$\frac{1}{k(x)} = \frac{4}{3\pi}\frac{1}{E}\frac{1}{R_{mid}^{\ 4}}\frac{x^3}{(1+a(x-L/2))(1-aL/2)^3}, \qquad (1)$$

where $a=\alpha/R_{mid}$ (α is the NW cone angle tangent), L – length of the NW. During the fitting one can get the value $E^*R_{mid}^{\ 1}$ (R_{mid} is the average radius of tapered NW and E is the Young's modulus). Knowing this value and measuring R_{mid} by SPM or SEM methods one can obtain the Young's modulus value.

In this work, using the above mentioned method, we calculated the Young's moduli for 17 NW of different radii and with different crystal structures (zinc blende (ZB) and wurtzite (WZ)). It was found that Young's modulus of NW of large radius (R_{mid}> 30 nm) was 65 GPa. With a decrease in the radius of the NW, the measured Young's modulus increased to 130 GPa. Apparently, the increase in the Young's modulus is due to the presence of a wurtzite crystal structure in thin NW. Indeed, this electron diffraction indicates that NW with radii less than 20 nanometres have a wurtzite structure, while larger NW have a mixed ZB/WZ structure. Thus, it is established that wurtzite InP nanoparticles have a Young's modulus of about 130 GPa.

4. Conclusion

Thus, in this work, special SPM procedure was presented for the measurement of Young's modulus of thin and flexible NW. The method consists in measuring the inverse stiffness profiles 1/k(x) in the Peak Force Tapping precision force control mode. To analyse the bending profiles of thin InP NW of tapered shape, the formula was used that related the inverse stiffness 1/k(x) of the tapered NW and the Young's modulus. Using this approach, it was possible to measure the Young's modulus of tapered InP NWs with a wurtzite and mixed ZB/WZ structure. Young's modulus value of mixed ZB/WZ NWs turned out to be close to 65 GPa, while the Young's modulus of wurtzite InP NW was approximately twice as large (130 GPa).

References

[1] Krogstrup P, Jorgensen H I, Heiss M, Demichel O, Holm J V, Aagesen M, Nygard J, Fontcuberta I and Morral A 2013 *Nat. Photonics* **7** 306-10
[2] Kim S K, Day R W, Cahoon J F, Kempa T J, Song K D, Park H G and Lieber C M 2012 *Nano Lett.* **12** 4971-6
[3] Alekseev P A, Dunaevskii M S, Stovpyaga A V, Lepsa M and Titkov A N 2012 *Semiconductors*, **46** 641-6
[4] Dunaevskiy M, Geydt P, Lähderanta E, Alekseev P, Haggrén T, Kakko J and Jiang H 2017 *Nano Lett.* **17** 3441-46
[5] Young T J, Monclus M A, Burnett T L, Broughton W R, Ogin S L and Smith P A 2011 *Meas. Sci. Technol.* **22** 125703

IOP Conf. Series: Materials Science and Engineering **443** (2018) 012007 doi:10.1088/1757-899X/443/1/012007

Mechanical properties of ion irradiated steel EK-181 investigated by dynamic nanoindentation

E V Gladkikh[1,2], K S Kravchuk[1], A S Useinov[1], A A Nikitin[3] and P.A. Fedin[3]

[1] Technological Institute for Superhard and Novel Carbon Materials, Troitsk, Moscow, 108840, Russia
[2] Moscow Institute of Physics and Technology, Dolgoprudny, Moscow Region, 141701, Russia
[3] Institute for Theoretical and Experimental Physics named by A.I. Alikhanov of National Research Centre "Kurchatov Institute", Moscow, 117218, Russia

ekat.gladkih@yandex.ru

Abstract. Heavy ion irradiation is a unique tool that can be used to simulate and study radiation processes in materials. It can reproduce in a small near the surface layer various microstructural features (dislocation loops, precipitates, voids, etc.) observed in neutron-irradiated materials. Because the irradiation damage is restricted by a few micrometres near surface layer, mechanical properties investigation of the samples was carried out by dynamic nanoindentation technique. This method allows obtaining continuously the hardness and elastic modulus depth profile by one indent. Nanoindentation results acquired in this work showed increase in irradiated layer hardness caused by the increase in irradiation temperature and damage dose.

1. Introduction

Material research is an important task for development of energetics and other technical spheres. The most frequent characteristics under review are surface geometry and mechanical properties. Topography could be tested by scanning probe microscopy. Mechanical properties analysis, which is critically important for structural materials, could be carried out via nanoindentation. Current work demonstrates opportunities of nanoindentation to investigate radiation induced processes.

Most materials used at nuclear stations in the reactor pressure vessel constructions are exposed to neutron radiation. Such impact leads to induced radioactivity, heating, and damage of the material lattice structure. It is well established that irradiation with heavy ions can be used to simulate neutron irradiation. Ion irradiation can reproduce in a small near surface layer various microstructural features (dislocation loops, precipitates, voids, etc.) observed in neutron-irradiated materials. As shown in Ref. [1], there is a clear influence of radiation-induced features on material hardening.

SRIM (formerly TRIM) code based on Monte Carlo simulations is used to compute a number of parameters relevant to the ion beam implantation and modification of the materials. It can be used to calculate atomic displacements per atom (dpa), which is a common radiation damage exposure unit. In the work [2], the use of SRIM for calculation of dpa was evaluated and matching with an internationally recognized standard definition of dpa was made.

An austenitic alloy Fe-15Cr-20Ni after ion irradiation and helium implantation was studied in Ref. [3]. Influence on microstructural evolution, hardness, and behaviour of plastic deformation in combined application of ion irradiation was estimated using technique of continuous nanosurface profiling. Austenitic alloy showed a strong (130%) hardening by single-beam irradiation.

In work [4], mechanical nominal hardness of ferritic alloys after irradiation by Fe ions was investigated. Experiments on ion irradiation were conducted at 290°C with 6.4 MeV $Fe3+$ ions. A depth profile of hardness was obtained via the constant stiffness measurement (CSM), being similar to dynamic nanoindentation technique. The results were considered on the basis of the Nix-Gao model and the widened film-substrate hardness model.

2. Material and methods

We investigated properties of promising ferritic-martensitic steel EK-181 after various irradiation modes. Irradiation temperature and damage doses are represented in Table 1. Steel samples were aged for 30 minutes in air at 980°C and tempered for 90 minutes in air at 760°C. Specimens used in irradiation experiments were prepared by the electric spark cutting in the shape of a disk with a diameter of 3 mm. The surface of the samples was subjected to multistage mechanical polishing using SiC abrasive sheets with a consequent reduction in the grain size. To remove mechanical deformations in the near-surface layer, electrochemical polishing was carried out using a solution of 5% $HClO_4$ in ethanol. Heavy ion irradiation with Fe ions was carried out on a linear radio-frequency quadrupole (RFQ) accelerator HIPr-1 at NRC «Kurchatov Institute» - ITEP. Due to the special design of accelerating structure, it is possible to accelerate the ions up to energies of 101 KeV/nucleon. The total energy of Fe ions was 5.6 MeV. The irradiation was carried out under controlled heating of the target to 250, 300, 350, and 400°C. SRIM code was used to calculate damage dose according to recommendations provided in Ref. [2]. Ion implantation and displacement damage profiles are shown in Figure 1.

Nanoindentation is the process of recording loading-unloading curve, while indenter penetrates into the investigated material. One additional mode of the technique is dynamic instrumental indentation. It consists of monotonous indenter penetration at the same time with its harmonic oscillation. Such technique gives an opportunity to obtain depth dependent hardness and elastic modulus in a single experiment. The main disadvantage of this method is the local destruction of the material. For carrying out the tests, it is necessary to have a plastic imprint on the surface of the object under study. However, this feature makes it possible to obtain data on the hardness of the sample.

Table 1. Hardness values of samples under investigation.

Sample number	Temperature (°C)	Doze (ion/cm^2)	Hardness at 500 nm (GPa)	Nix-Gao:		
				$H^{0(irr)}$	$h^{*(irr)}$	$h^{*(unirr)}$
1	initial	-	3.85±0.14	3.70±0.12	62	100
2	250	$3\cdot10^{15}$	4.32±0.21	4.27±0.17	63	236
3	250	$6\cdot10^{15}$	3.98±0.16	3.82±0.09	67	100
4	250	$1\cdot10^{16}$	4.03±0.12	4.06±0.13	41	168
5	300	$3\cdot10^{15}$	4.11±0.22	4.06±0.12	54	190
6	300	$5{,}6\cdot10^{15}$	3.93±0.14	3.93±0.14	52	179
7	300	$1\cdot10^{16}$	4.07±0.19	4.00±0.03	49	162
8	350	$6\cdot10^{15}$	4.05±0.17	3.98±0.14	55	160
9	400	$3\cdot10^{15}$	4.04±0.18	3.93±0.13	63	152
10	400	$5{,}6\cdot10^{15}$	4.12±0.14	4.05±0.12	76	216
11	400	$1\cdot10^{16}$	4.29±0.22	4.26±0.21	84	274

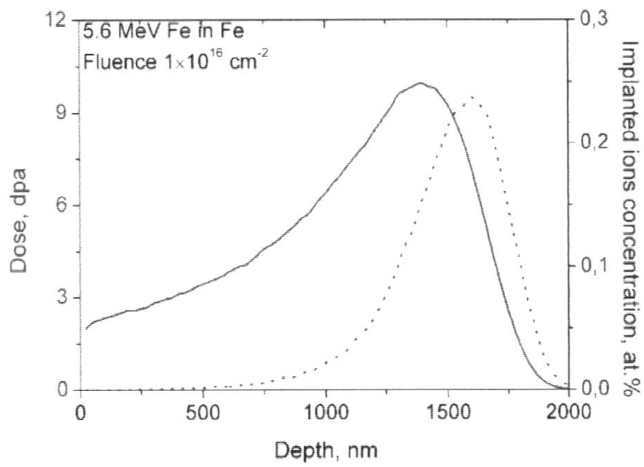

Figure 1. Depth profile of displacement damage (solid line) and implanted ions concentration (dots) in the EK-181 steel irradiated by 5.6 MeV Fe ions.

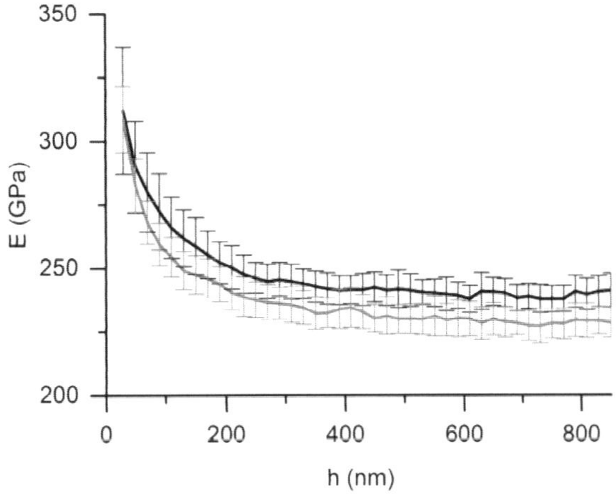

Figure 2. Elastic module of irradiated sample (grey line) and initial sample (black line).

A study of the mechanical properties of the samples was carried out on the NanoScan-4D nanohardness tester [5] (FSBI TISNCM, Russia) with a Berkovich type indentation tip using the dynamic nanoindentation. Calibration of the indentation tip shape and hardness calculation are based on the Oliver-Pharr method. We used the following test parameters: indentation depth 2000 nm, oscillation amplitude 10 nm, frequency 10 Hz, and loading rate 200 nm/min. A series of 30 measurements was carried out for each sample.

3. Results and discussion

Figure 2 demonstrates depth profiles of elastic modulus in the sample before and after irradiation. Black line shows data obtained on the unirradiated sample and grey line refers to the sample, which was exposed to 5.6 MeV Fe^{+} ions with total fluence of $1 \cdot 10^{16}$ ion/cm^2 at 350°C. The observed difference is rather small (less than 5%) and only slightly exceeds the measurement error. Measurement of the rest irradiated samples demonstrated similar values of elastic modulus. The nearness of the elastic properties of the irradiated layer and the substrate allows calculating the true contact area and true hardness [6], irrespective of the effects of pile-up or sink-in.

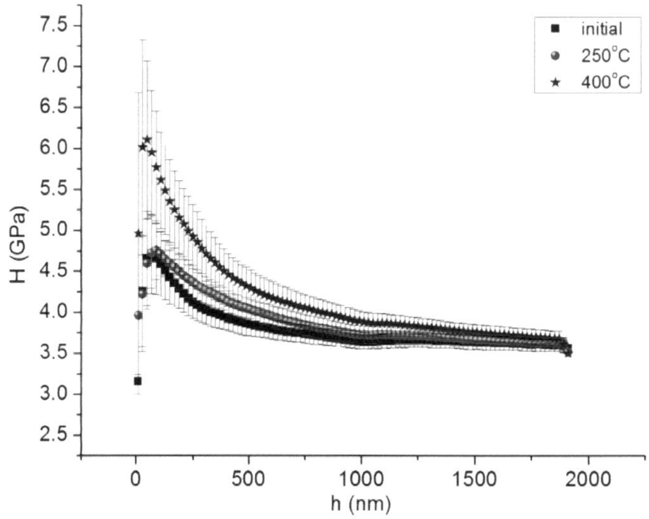

Figure 3. Indentation-depth profiles of the averaged hardness at 10 dpa damage dose and heating to 250°C or 400°C comparing to hardness of initial sample.

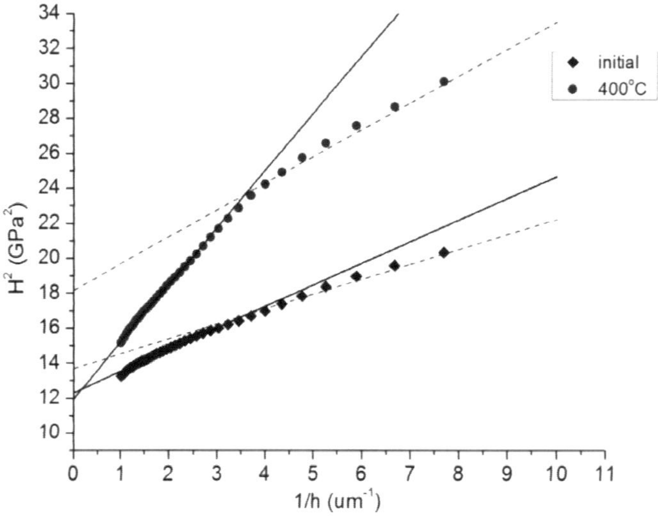

Figure 4. Plot of H^2 versus $1/h$ for initial sample and irradiated at 10 dpa damage doze and temperature 400°C material. The lines show approximation of data groups of near-surface layers (dashed lines) and deep layers (solid lines).

Figure 3 shows indentation-depth profile of hardness of EK-181 steel before and after irradiation at 10 dpa damage dose. Hardness dependence on irradiation temperature was clearly observed. Data up to 100 nm was ignored due to test artefacts inherited from the indenter-surface contact uncertainties.

The measured hardness is maximal in the near-surface layer and decreases uniformly with increasing depth, tending to the hardness of the non-irradiated substrate material. The similar behaviour can be seen in the initial sample, which can be explained by the well-known indentation size effect [7-8]. In order to explain the ISE, Nix and Gao suggested a model based on a concept of geometrically necessary dislocation. The Nix–Gao model states that the hardness depends on the indentation depth as follows:

$$H = H^0\left(1+\frac{h^*}{h}\right)^{0.5} \quad (1)$$

where H^0 is the bulk hardness of the sample, h^* is a specific length, which relies on the material and the shape of indenter tip. To calculate the H^0 and h^*, data was plotted as H^2 versus $1/h$.

Figure 4 illustrates an example of such plot. Table 1 shows H^0 for irradiated layers of the samples and h^* for both regions. Average hardness value of unirradiated layers of all the samples is 3.50±0.07, which is close to hardness of initial sample. A border between irradiated and bulk material layers was estimated at a depth of 300 nm. This choice is due to the softer substrate effect (SSE), which can be seen in the various systems of hard thin film on soft substrate [9]. Besides, the table includes results of hardness at a depth of 500 nm, the values obtained are similar to results of Nix-Gao model.

Obtained data show that with the increase in damage, dose hardness of irradiated layer increases. This may be explained by formation of radiation induced defects. Number density of these features strongly correlates with the level of radiation damage. Maximum increase in hardness (0.76 GPa) was observed for the steel sample irradiated up to 10 dpa at 400°C.

4. Conclusions

Dynamic nanoindentation technique was used to investigate influence of Fe^+ ion irradiation on mechanical properties of f/m steel EK-181. Depth dependences for the hardness and elasticity modulus are constructed. The elasticity modulus of the samples does not change after irradiation. Gradient of hardness can be explained not only by the indentation size effect, but also a softer substrate effect. Strong dependence of irradiated layer hardness on the damage dose and irradiation temperature is observed.

Acknowledgements

This work was supported by the Ministry of Education and Science of the Russian Federation (project ID RFMEFI57717X0274; the agreement No. 14.577.21.0274); the work was done using the Shared research facilities "Research of Nanostructured, Carbon and Superhard Materials" FSBI TISNCM.

References

[1] Hosemann P, Vieh C, Greco R, Kabra S, Valdez J A, Cappiello M J and Maloy S A 2009 *J. Nucl. Mater.* **389** 239-47
[2] Stoller R E, Toloczko M B, Was G S, Certain A G, Dwaraknath S and Garner F A 2013 *Nucl. Instruments Methods Phys. Res. Sect. B Beam Interact. with Mater. Atoms.* **310** 75-80
[3] Katoh Y, Ando M and Kohyama A 2003 *J. Nucl. Mater.* **323** 251–62
[4] Kasada R, Takayama Y, Yabuuchi K and Kimura A 2011 *Fusion Eng. Des.* **86** 2658-61
[5] Maslenikov I, Useinov A, Birykov A and Reshetov V 2017 *IOP Conf. Ser.: Mater. Sci. Eng.* **256** 012003
[6] Saha R and Nix W D 2002 *Acta Mater.* **50** 23-38
[7] Huang Y, Zhang F, Hwang K C, Nix W D, Pharr G M and Feng G 2006 *J. Mech. Phys. Solids* **54** 1668-86
[8] Nix W D and Gao H 1998 *J. Mech. Phys. Solids* **46** 411-25
[9] Manika I and Maniks J 2008 *JoP D: Applied Physics* **41** 074010

IOP Conf. Series: Materials Science and Engineering **443** (2018) 012008 doi:10.1088/1757-899X/443/1/012008

Structure of natural impact glasses by microscopic data

Ye A Golubev, T G Shumilova and S I Isaenko

Institute of Geology of Komi SC of Uralian Branch of RAS, Syktyvkar, Russia

golubev@geo.komisc.ru; shumilova@geo.komisc.ru

Abstract. Newly found impact vein-bodied ultrahigh pressure high temperature (UHPHT) glasses from Kara astrobleme (Russia) and impact drop-shaped glasses from Ries crater suevites (Germany) have been investigated. The nanostructure of the glasses has been analysed using atomic force microscopy (AFM) supported by a complex of optical microscopy, scanning electron microscopy (SEM) combined with EDS analysis, and Raman spectroscopy. The received data demonstrate the different levels of impact melt differentiation through partial crystallization and polymerization; the amorphous substance is presented by similar 60-80 nm globular nanostructural composition.

1. Introduction

Glasses are the most important materials among disordered substances. However, the structure and features of phase composition of glasses remain not quite clear, in spite of actively provided studies in materials science. The behaviour of the disordered systems under strong compression attracts a special interest for fundamental understanding and application needs [1-4]. To the date, numerous studies have been carried out mostly at room temperature in microvolume output within diamond anvil cells under pressure of up to 100 GPa. At the same time, the natural glasses from impact craters can have a specific importance being formed with hyperpressures (35-90 GPa and above) and high temperatures (up to 3000°C). In comparison to experimentally possible high-pressure products, the natural impact glasses form rather essential volumes and can keep structural features for many millions of years, which allows using them not only for the geological tasks and modelling of astroblemes formation, but also for fundamental studies of noncrystalline substances under extreme conditions, to analyse their stability, and also to estimate them as possible materials for hi-tech applications, in particular for high-energy laser technologies and microelectronics [5,6].

The impact glasses have specific structural composition and physical features [3,4]. However, so far the essential attention to natural impact glasses as to a possible new type of high pressure materials is not fully studied. Due to the above specified points, the detailed study of nanostructure of the natural impact glasses, especially for UHPHT varieties, is urgent and very actual both for fundamental problem of disordered substance behaviour at extreme conditions and its subsequent relaxation, and for the applications.

2. Objects and methods

For the studies, we used natural glasses from drop-shaped impact glasses from suevites of the Ries crater (Bavaria, Germany) and UHPHT vein-bodied impact glasses from the Kara astrobleme (Pay-Khoy, Russia). The samples were provided by T.G. Shumilova during expeditions in 2013 and 2015 years, correspondently.

At the present study, the impact glass samples have been analysed using atomic force microscopy (AFM) supported by a complex of optical microscopy, scanning electron microscopy (SEM) combined with microprobe analysis, and Raman spectroscopy.

Initially, the impact glass samples have been observed with optical microscopy in transparent light with and without polarizer for selecting amorphous matter. For this purpose, a microscope POLAM-312 and Nikon Eclipse E400 Pol equipped with a camera DCMOS14000KPA were used.

Raman spectroscopy measurements were provided at room temperature by the spectrometer LabRam HR800 instrument (Horiba, Jobin Yvon) equipped with the He-Ne laser (632.8 nm, 2 mW), the optical microscope Olympus BX41, and the Si-based CCD detector. Spectra were received in the range of 100-4000 cm^{-1} with grating of 600 g/mm, a slit 100 μm, and a confocal hole size 300 μm. Zero position, spectra background correction, and individual bands deconvolution have been provided with a curve-fitting procedure from the LabSpec 5.36 software.

The chemical composition of the glasses has been analysed for cleaved surfaces of the samples using a Tescan Vega LMH scanning electron microscope equipped by an X-ray energy-dispersive spectrometer (EDS) accompanied with AZTEC software (Oxford Instruments) processing.

The nanostructure has been studied here with AFM analysis by surface morphology analysis in a semicontact mode using an Integra Prima (NT-MDT, Russia) by super sharp silicon cantilevers SSS-NCH (Nanosensors) with the resonant frequency 330 kHz, the radius of the tip 2-4 nm, and the stiffness constant about 35 N/m. To avoid the effect of static electricity, the sample surfaces were grounded by a silver paste. The technique of sizes measure was described in detail in Refs. [4,7].

3. Results

The studied impact glasses are presented with preferably amorphous substance that is evidently observed on the optical images in transparent polarized light (Fig. 1). Both the Kara and Ries glasses demonstrate well seen or partially present fluidal texture contrasted with different iron oxide presence illustrated by brown and yellow colours of varied density depending on iron oxidation/reduction level and its concentrations.

According to microprobe measurements, the analysed impact glasses are presented by aluminosilicate type having varying silica content with different concentrations of cation modifiers of the glass network with the biggest variation of Na_2O and Fe_2O_3. The SiO_2 content corresponds to intermediate (andesite) level. In Table 1, the rock-forming oxides contents are presented for the samples studied by AFM. In addition to the listed components, quite large content of H_2O has been detected within the Kara UHPHT impact glasses counting up to several percent.

Table 1. Composition of rock-forming oxides in the silicate matrix of glasses (at.%).

Sample	Impact object	SiO_2	MgO	Al_2O_3	Na_2O	K_2O	CaO	TiO_2	Fe_2O_3
KR-2-15	Kara astrobleme	71.8	4.0	16.1	5.1	1.6	1.1	0.3	0.5
R-13-5/1	Ries crater	67.6	2.6	16.5	2.7	2.1	3.2	0.6	4.0
R-13-8a	Ries crater	58.0	2.6	14.6	1.6	2.9	2.2	0.9	3.7

IOP Conf. Series: Materials Science and Engineering **443** (2018) 012008 doi:10.1088/1757-899X/443/1/012008

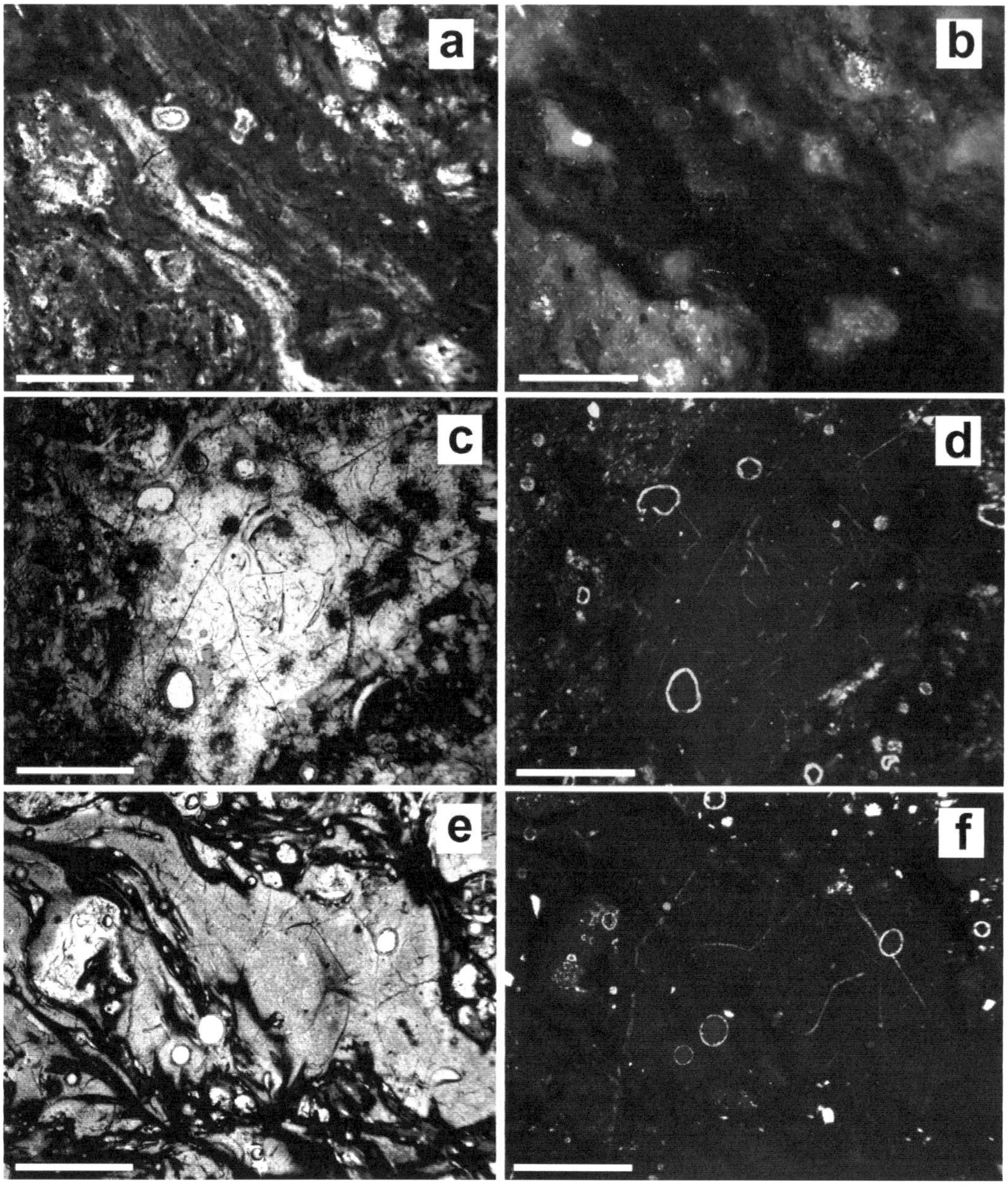

Figure 1. Optical images of the impact glasses in transparent polarized light: (a,b) Kara glass, KR-2-15; (c,d) Ries glass, R-13-5/1; (e,f) Ries glass, R-13-8a; (a,c,e) parallel polarizers; (b,d,f) crossed polarizers. The scale bar 50 μm.

The Raman spectroscopy supports an amorphous state of the studied glasses for the both impact varieties. At the same time, the spectra allow observing the difference in structural ordering of the glasses and partial impact melt crystallization. Thus, we can evidently obtain specific Raman characteristics both for the glasses of the impact craters and for the different samples within the same crater that demonstrate quite wide variety of the structural features.

The low-frequency range 400-700 cm^{-1} is associated with a level of glass polymerization toward the Raman shift increasing. Glasses usually have general spectral characteristics corresponding to their structural features in the range 800-1200 cm^{-1} of Raman spectra [8,9]. The band around 800-850 cm^{-1} is explained by isolated SiO_4 tetrahedra vibrations, where oxygen atoms have non-bridging state (Q^0). The maxima at 900 cm^{-1} and 950-980 cm^{-1} are associated with terminal groups in SiO_4 tetrahedra, with three and two non-bridging oxygen atoms (Q^1 and Q^2, correspondently). Spectral range 1050-1200 cm^{-1} is described by non-bridging bonds in SiO_4 tetrahedra with unbridged oxygen atoms presence (Q^3).

The studied impact glasses have mixed distributions with Q^n (Fig. 2), where n is the number of bridging oxygen per a tetrahedrally coordinated cation. The Ries impact glass has the most three-dimensionally ordered structure with the largest statistical output of the Q^{32} and Q^{33} units (Fig. 2). The Kara UHPHT glass with coesite (KR-2-15) has the dominant structural unit Q^0 corresponding to the lowest order compared with the shortest tetrahedral chains to the other glasses [4]. In spite of high water presence within the impact glasses, according to the provided Raman spectroscopy and other studies [3,4], it rather has free molecular state without chemical bonding with the glass structural networks. The Q^{33}/Q^0 ratio shows that the most ordered (with a high degree of polymerization) glass is present in the sample R-13-8a.

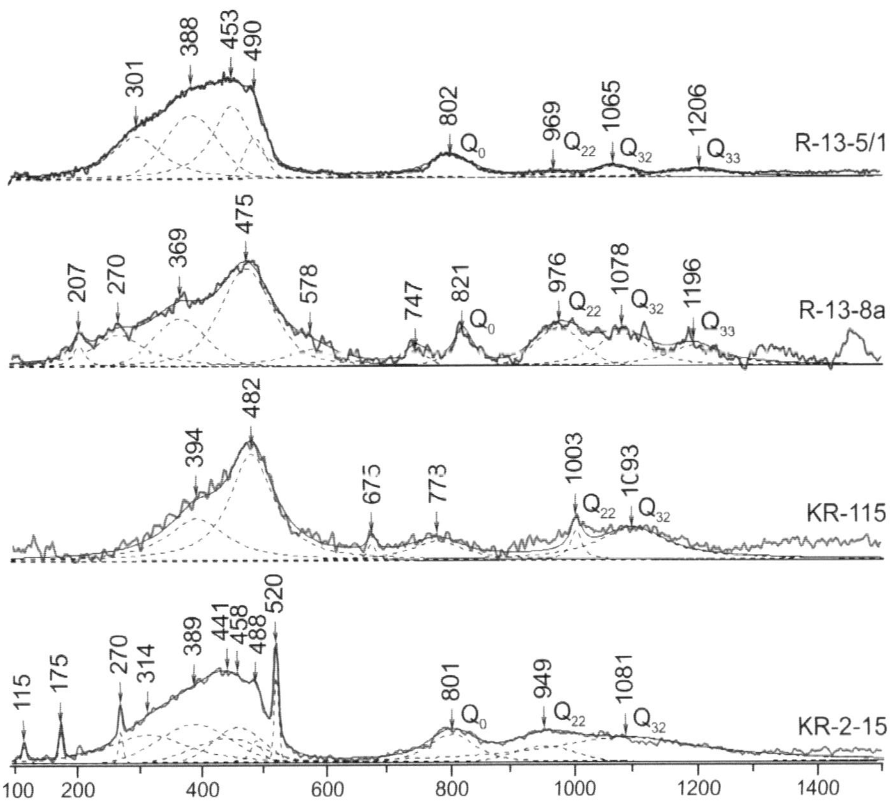

Figure 2. Raman spectra of the studied impact glasses: R13-5/1 and R13-8a – the Ries crater glasses; KR-115 – the Kara astrobleme UHPHT glass, KR-2-15 – the Kara astrobleme UHPHT glass with coesite bands.

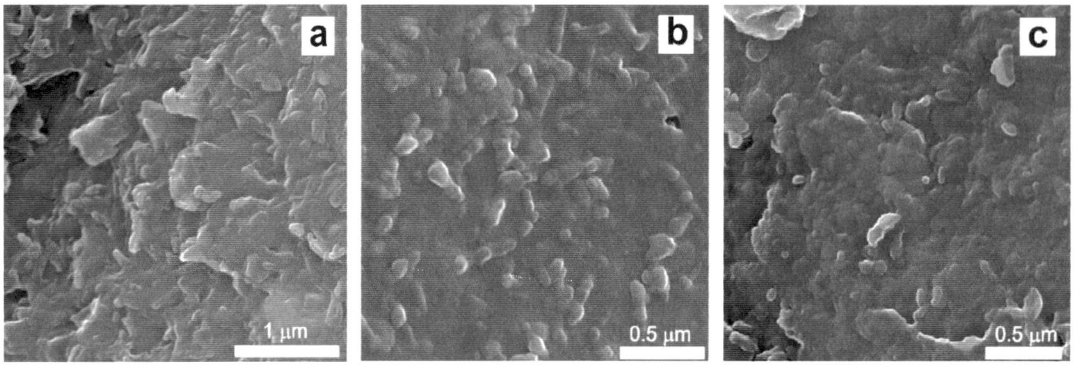

Figure 3. Typical SEM images in secondary electron mode of the glass surfaces: (a) KR-2-15; (b) R-13-8a; (c) R-13-5/1.

Table 2. Silica states of the Kara and Ries impact glasses by the data of Raman spectroscopy.

Sample	Q^0		Q^{22}		Q^{23}		Q^{33}		Q^{33}/Q^0
	Position, cm^{-1}	FWHM, cm^{-1}	Position, cm^{-1}	FWHM, cm^{-1}	Position, cm^{-1}	FWHM, cm^{-1}	Position, cm^{-1}	FWHM, cm^{-1}	
R13-5/1	802	81	969	80	1065	67	1206	100	0.44
R13-8a	821	38	976	90	1078	98	1196	100	1.38
KR-115	–	–	1002	22	1093	137	–	–	–
KR-2-15	801	88	948	113	1081	300	–	–	0

SEM images show a complicated surface relief (Fig. 3). The surface of the sample R-13-8a presents globules with diameter of about 80 nm and rod-shaped morphological elements about 500 nm in length and 60-100 nm in thick. These elements are set in disorder; the elongated morphological elements are reoriented. The surface of the sample R-13-5/1 shows a hilly relief with short flakes about 50-80 nm in thick that are often released from the hillocks. The relief of the surface of the KR-2-15 sample is more flat, having the characteristic surface of a shell-like fracture of a cleavage surface.

Numerous experiments indicate that the glasses are heterogeneous at the scale of a few nanometers [11]. This heterogeneity can be associated with density fluctuations in pure silica glasses or with chemical composition fluctuations in glasses. This heterogeneity is well manifested on the cleavage surface and is detected by AFM, which is actively used to study the nano-heterogeneity of natural and artificial glasses [4,11,12].

Typical AFM images of the samples KR-2-15 and R-13-5/1 show the initial surface roughness as elongated rounded hillocks (Fig. 4), with diameter of 60-70 nm and height of several nanometres. The hillocks are general morphological elements of the glass surface in presence of micro-islands and cracks. Usually, the hillocks are observed being separated from each other with a slight contrast at the images. Moreover, the surface of the sample KR-2-15 is more homogeneous at the microscopic level than the sample R-13-5/1.

The cleavage surface of the sample R-13-8a is more complicated. It consists of islands with diameter about 1 μm and height about 100 nm, which are covered with elongated hillocks about 60-70 nm in diameter. There are numerous rod-like nanocrystals on the surface. Many of them grow directly from the surface, which indicates that they are not debris formed during cleavage. These crystals make it difficult to access the surface to estimate roughness and measure the size of the hillocks. However, in general, the nanoscale surface roughness was also detected for this sample (Fig. 5 – inset).

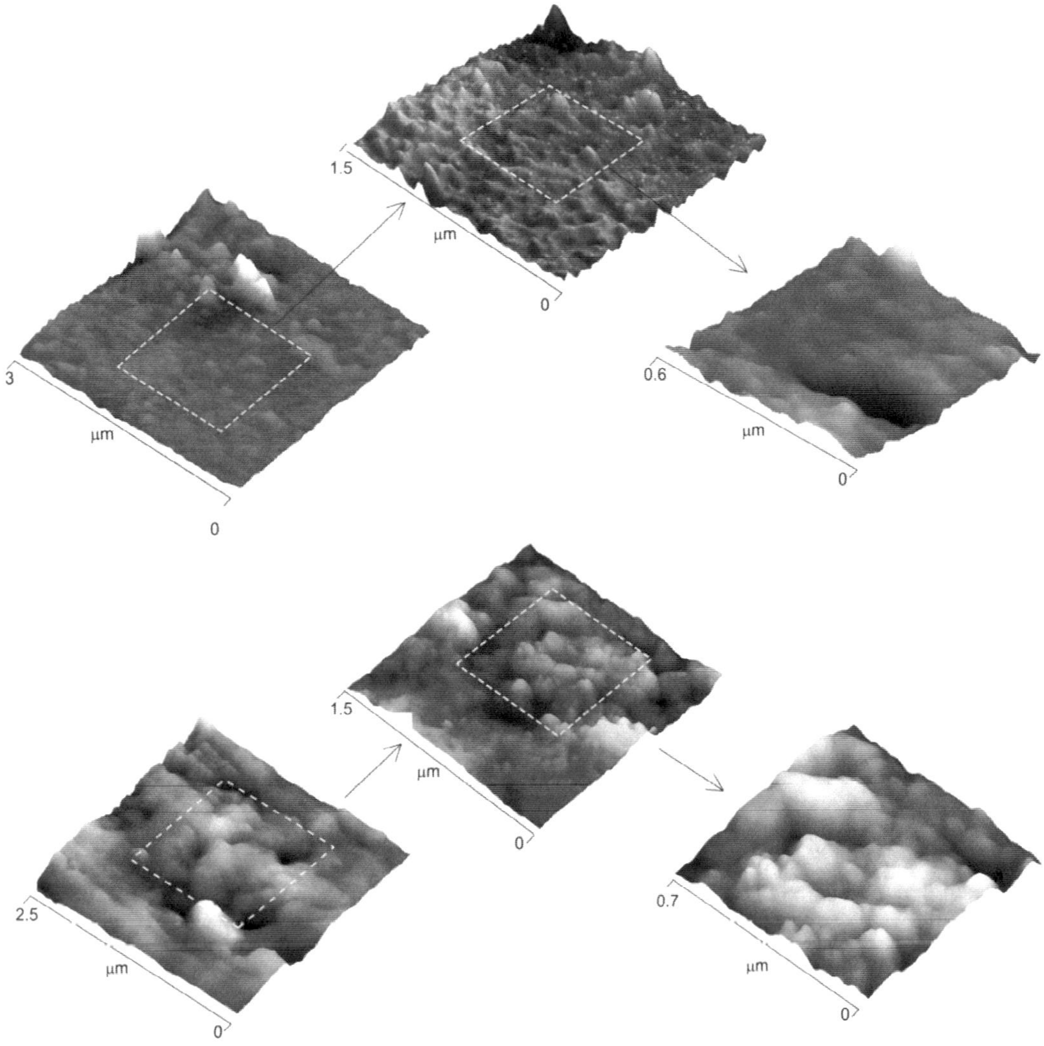

Figure 4. Typical AFM-images of the samples KR-2-15 (top) and R-13-5/1 (bottom)

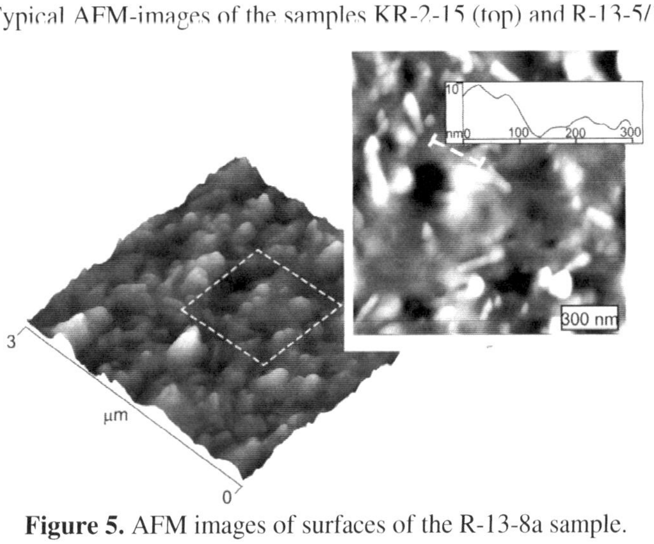

Figure 5. AFM images of surfaces of the R-13-8a sample.

According to the EDS results, the studied samples are multicomponent aluminosilicate glasses with impurities of sodium and magnesium. For similar multicomponent synthetic melts, the manifestation of phase separation as a nanoscale roughness is shown in Ref. [12]. The studied samples looked quite homogeneous on the EDS-images, but their elemental inhomogeneity could occur at a nanoscale level that was beyond the sensitivity of the EDS analysis. In our case, the phase separation manifested itself as the surface nanoscale roughness.

Unlike in "pure" silicate impact glasses (as described in Ref. [4]), nanocrystallites and larger morphological elements of the surface occur in multicomponent impact glasses. Morphological elements uniformly cover the surface and are densely packed. If these elements are associated with the distribution of impurities in the melt, it can be assumed that the distribution of impurities in the solidified melts of the impact glasses studied was fairly uniform.

4. Conclusion

We have studied natural glasses solidified from silica-rich melts formed under UHP shock metamorphism from Kara astrobleme (Russia) and Ries crater suevites (Germany). Rounded and elongated morphological nano-elements (60-80 nm in size) of the impact glass surface are observed by AFM and SEM. They cover the surface uniformly and are densely packed. We attribute the roughness and its changing to nanoscale compositional and structural variations within the glass microstructure. With the help of microscopy, nanocrystallites were found in the sample R-13-8a, which was the most ordered according to Raman spectroscopy.

Acknowledgments

The study was supported by the Russian Science Foundation (project No. 17-17-01080). The authors thank V.A. Radaev for the sample preparation and for assistance in the AFM measurements, Ye.M. Tropnikov and A.A. Utkin for SEM and microprobe observations and measurements, and A.A. Zubov for technical assistance. The equipment of the Centre for Collective Use "Geoscience" of the Institute of Geology of the Komi SC UB RAS was used for our study.

References

[1] El'kin F S, Brazhkin V V, Khvostantsev L G, Tsiok O B and Lyapin A G 2002 *J. Exp. Theo. Phys. Lett.* **75** 342-7
[2] Shumilova T G, Isaenko S I, Makeev B A, Zubov A A, Shanina S N, Tropnikov Ye M and Askhabov A A 2018 *Doklady Earth Sciences* **480** 595-8
[3] Shumilova T G, Lutoev V P, Isaenko S I, Kovalchuk N S, Makeev B A, Lysiuk A Yu, Zubov A A and Ernstson K 2018 *Sci. Rep.* **8** 6923
[4] Golubev Ye A, Shumilova T G, Isaenko S I, Makeev B A, Utkin A A, Suvorova E I and Ernstson K 2018 *J. Non-Cryst. solids* **500** 388-400
[5] Bolmatov D, Brazhkin V V and Trachenko K 2013 *Nature Commun.* **4** 2331
[6] Borisova P A, Blanter M S, Brazhkin V V, Somenkov V A and Filonenko V P 2015 *J. Phys. Chem. Solids* **83** 104-8
[7] Golubev Ye A and Isaenko S I 2017 *IOP Conf. Ser.: Mat. Sci.Eng.* **256** 012019
[8] McMillan P F, Wolf G H and Poe B T 1992 *Chem. Geol.* **96** 351-66
[9] Anfilogov N N, Bykov N N and Osipov V N *Silicate melts*, Nauka, Moscow, 2005 (in Russian)
[10] Greaves G N, Smith W, Giulotto E and Pantos E 1997 *J. Non-Cryst. Solids* **222** 13-24
[11] Hervé A and Daniel A 2000 *Ceramics-Silikáty* **44** 121-8
[12] Wheaton B R and Clare A G 2007 *J. Non-Cryst. Solids* **353** 4767-78

IOP Conf. Series: Materials Science and Engineering **443** (2018) 012009 doi:10.1088/1757-899X/443/1/012009

Adhesive coatings based on aligned arrays of carbon nanostructures

O I Il'in, M V Il'ina, N N Rudyk and A A Fedotov

Southern Federal University, Institute of Nanotechnologies, Electronics and Electronic Equipment Engineering, Taganrog, 347922, Russia

oiilin@sfedu.ru

Abstract. The article presents the results of establishing the influence of the carbon nanotubes growing modes by plasma enhanced chemical vapour deposition method on their geometric dimensions and adhesion properties. The results of experimental research of the activation time, plasma power, temperature, and growth time effects on the parameters of carbon nanotubes are presented. It is shown that control of growth technological modes makes it possible to create carbon nanotubes arrays of various types: bundled, individual, branched, and disoriented. The adhesion of experimental samples was investigated by atomic force microscopy. The greatest value of the adhesion force (11.12 μN) is demonstrated by aligned arrays of carbon nanotubes, which during the growing process were grouped into bundles with the diameter of individual nanotubes 51 nm, height 0.69 μm, and density 69 μm^{-1}.

1. Introduction

The development of nanotechnologies and the possibility of creating spatially oriented nanoscale structures have generated a wide interest in research for creating not only promising devices for nanoelectronics [1-3], but also artificial bio-like structures. Reproduction of parameters of biological objects [4-7] allows solving a number of problems in the field of mechanical contacts of high strength and repairing materials with "dry" adhesion working in vacuum. Among the artificial biomimetic nanostructures, carbon nanostructures based on arrays of vertically aligned carbon nanotubes (VA CNT) possess the best adhesion characteristics [4,8-11], which is associated with an increase in the adhesion force due to an increase in the number of nanostructures contacting the surface [12]. For controlled growth of CNTs with specified characteristics, the most promising method is plasma enhanced chemical vapour deposition (PECVD) [13,14]. However, a large number of parameters of the PECVD method require establishing the relationship between their influence on the growth of CNTs with controlled parameters.

The aim of this work is establishing the influence of the CNTs growing modes by PECVD method on their geometric dimensions and adhesion properties.

2. Experiments and methods

Si/TiN/Ni structure was used for creating experimental samples. VA CNT were grown by PECVD using specialized module cluster complex nanotechnological NANOFAB NTC-9 (NT-MDT, Russia). To provide different CNT parameters, values of the activation time, plasma power, temperature, and growth time were changed during PECVD process. The samples with 4 types of VA CNT were grown: bundles (Fig. 1a), individual (Fig. 1b), branched (Fig. 1c), and disoriented (Fig. 1d).

Figure 1. SEM images of (a) CNTs bundles, (b) individual VA CNTs, (c) branched VA CNTs, (d) disoriented CNTs. (e) Experimental dependence ΔJ_{DFL} for individual VA CNTs. Scale segment 500 nm.

Cultivation of CNTs was carried out in 3 stages: "heating", "activation", and "growth". In the "heating" stage, the formation of Ni catalytic centres from a continuous film of a catalytic material was carried out by annealing the samples at specified temperatures for 20 minutes under argon atmosphere (40 sccm) and ammonia (15 sccm) at a pressure of 4.5 Torr. After that, the "activation" stage was carried out – exposure of the substrates in the ammonia atmosphere (210 sccm). Depending on the type of grown samples, plasma could also be initiated using a high-voltage DC source. At the stage of "growth", acetylene (70 sccm) was fed into the chamber together with ammonia. To create a vertical directional growth of CNTs relative to the substrate, plasma was initiated. CNTs grew by top-mechanism. The values of the activation time (t_{act}), plasma power (W), temperature (T), and growth time (t_{growth}) were varied to create samples with different parameters of CNTs. The estimation of the geometric parameters of the CNT arrays experimental samples was carried out using a scanning electron microscope (SEM) Nova NanoLab 600 (FEI, The Netherlands). Growth parameters of the CNT arrays and their geometric dimensions are presented in Table 1.

Investigation of the adhesion strength of the experimental samples was carried out using a probe nanolaboratory Ntegra (NT-MDT, Russia). A colloid probe of the CPC_SiO_2-20/Au series was used with a radius of 20 μm and a stiffness coefficient k = 0,3 N/m. Measurements of the adhesion value were carried out at 10 points of the array of each experimental sample in the force spectroscopy mode. During the process, the dependence of the cantilever bending value (J_{DFL} signal) on the degree of extension of the z-piezoelectric element of the scanner (signal h) in direct (solid line) and the inverse (dashed line) cantilever courses was measured. The experimental dependence $\Delta J_{DFL}(h)$ for samples with individual VA CNT is shown in Figure 1e.

Using the resulting force spectroscopy dependencies, the array of adhesion force F_a of VA CNT to the surface of the probe was evaluated by the method described in Ref. [15]:

$$F_a = \Delta J_{DFL}(\mathrm{d}h / \mathrm{d}J_{DFL})k, \quad (1)$$

where ΔJ_{DFL} is the difference of J_{DFL} signal at the point of complete detachment of the probe from the CNT array and the one at the point of the maximum negative value of the J_{DFL} signal.

3. Results and discussion

Analysis of SEM images of experimental samples *A1-A3* has shown that, with the parameters presented in Table 1, exposure in ammonia plasma leads to etching of Ni catalyst centres and reduces their diameter from 44 to 36 nm in samples *A1* and *A2*, respectively. However, as the activation time increases, the probability of combining small catalytic centres (CCs) into larger ones due to surface diffusion increases, which leads to an increase in diameter and a decrease in density of CNTs in *A3* sample. Also, the CNT growing modes with the "activation" time of less than 1 min promote to the formation of CNT bundles.

In *B1-B3* samples, the increase in the activation time was also accompanied by combining small CCs into larger ones with increasing diameter and decreasing density. At the same time, an increase in the plasma power up to 40 W contributed to the complete removal of small CCs and a higher electric field strength, which made it possible to obtain arrays of individual VA CNTs.

One of the specificity of the CNT growth by PECVD is that the temperature value at the stage of "heating" is retained at the "activation" and "growth" stages. Thus, with a change in the growth temperature from 645 to 675°C on the SEM images of *C1-C3* samples the integral effect of temperature can be seen associated with the two processes:

1) formation of CCs at the stages of "heating" and "activation";
2) the growth of CNTs on CCs formed at a given temperature.

At the same time, a short activation time promotes an increase in the adhesion of the CCs to the substrate. As a result, there is a growth both top- and base-growth mechanisms with the formation of arrays of branched CNTs. With increasing temperature up to 675°C, there is the simultaneous occurrence of sublimation and surface diffusion processes, which reduces the diameter, height, and density of CNTs. Also, an increase in temperature leads to an acceleration of the desorption of acetylene from the sample surface and the carbon-containing gas is evacuated by the vacuum system without reacting with the CCs.

Table 1. Parameters of CNTs and calculated values of adhesion of experimental samples.

Sample №	Diameter, nm	Height, μm	Density, μm^{-1}	CNT growth parameters: Constant	CNT growth parameters: Variable		Adhesion force, μN
A1	43.8	0.65	82	$T = 660°C$,		20	5.03
A2	35.6	1.21	72	$W = 25$ W,	T_{act}, s	40	8.53
A3	51	0.69	69	$t_{growth} = 12$ min		60	11.12
B1	44.1	0.35	95	$T = 660°C$,		120	2.30
B2	62	0.28	54	$W = 40$ W,	T_{act}, s	180	0.67
B3	63.2	0.65	38	$t_{growth} = 12$ min		240	1.56
C1	42.1	0.42	71	$t_{акт} = 5$ s,		645	3.22
C2	50.1	15.97	44	$W = 50$ W,	T,°C	660	6.42
C3	44.7	12.45	26	$t_{growth} = 12$ min		675	2.24
D1	34.6	2.11	89	$T = 660°C$,		4	1.09
D2	28.7	2.87	52	$t_{act} = 60$ s,	t_{growth}, min	8	7.38
D3	28.8	3.07	45	$W = 0$ W		12	2.70

For the experimental samples *D1-D3*, plasma was not initiated during the growth. The absence of a vector of electric field strength, which determines CNT growth direction, led to the formation of CNTs disoriented array. The increase in the growth time from 4 to 12 min allowed obtaining CNT arrays with a height from 2.11 to 3.07 μm.

Adhesion values shown in Table 1 were calculated based on the formula (1). Analysis of the measurement results has shown that the largest average values of the adhesion strength (8.22 μN) are demonstrated by *A1-A3* experimental samples. This effect can be due to the fact that during the measurement, the longitudinal deformation of the CNT predominated and there was no additional adherence of the nanotubes to each other when interacting with the probe.

Individual VA CNT of B1-B3 experimental samples showed the lowest adhesion value among all series of samples (1.51 μN), which might be due to the prevalence of transverse CNTs deformation and their coalescence during measurement. It should be noted that among B1-B3 samples, VA CNT adhesion strength increases with increasing aspect ratio and their density in the array.

Analysis of the results of adhesion measurements in samples with branched (*C1-C3*) and disoriented (*D1-D3*) CNTs showed that the average adhesion strength was 3.96 and 3.72 μN, respectively. It is seen that the adhesion strength of these samples is lower than that of *A1-A3* samples, but exceeds the adhesion value of *B1-B3* samples (Table 1). This dependence can be due to the fact that during the measurement of *C1-C3* and *D1-D3* samples, the probe interacted not with the top of the nanotubes, but with their side walls, in contrast to the measurements of *A1-A3* samples. However, there was no sticking of the CNTs tops with each other, in contrast to samples *B1-B3*.

It should also be considered that the change in the geometric parameters of the grown CNTs leads to a change in their Young's modulus, which affects the value of the adhesion strength. Thus, it is shown that the growth regimes significantly influence the parameters of CNTs, which allows controlling the value of the CNTs adhesion from 0.67 to 11.12 μN. The maximum value of the adhesion strength of 11.12 μN is achieved under the mode of growing CNT bundles at 60 sec activation time, 660 °C growth temperature, 25 W plasma power, and 12 min growth time.

In this case, the use of a 20-μm colloidal probe allows us to determine the value of the CNT macroadhesion, which is several times larger than the value of the local adhesion of CNT obtained by measuring with a standard silicon probe having radius of 20 nm [15].

4. Conclusion

In this work, an experimental research of CNT arrays growth and measuring of the adhesion strength value from the CNT parameters is carried out. It is shown that control of the activation time, plasma power, temperature, and growth time allows creating CNT arrays of various types: bundled, individual, branched, and disoriented.

It is established that the greatest value of the adhesion force is shown by CNT bundles arrays. Among individual CNTs, to achieve the maximum adhesion strength, it is necessary to grow CNTs with high aspect ratios (more than 10).

The results of the research can be used to create adhesive coatings and mechanical contacts of high strength for the space industry and robotics, as well as for the creation of carbon nanoelectronics elements.

Acknowledgments

The reported study was funded by RFBR according to the research projects № 16-29-14023, № 18-32-00652 and by grant of the Southern Federal University (project No. VnGr-07/2017-26).

References

[1] Ageev O A, Blinov Y F, Ilina M V, Ilin O I and Smirnov V A 2016 *J. Phys. Conf. Ser.* **741** 012168

[2] Avilov V I, Ageev O A, Smirnov V A, Solodovnik M S and Tsukanova O G 2015 *Nanotechnologies in Russia.* **10** 214-9

[3] Avilov V I, Polupanov N V., Tominov R V., Smirnov V A and Ageev O A 2017 *IOP Conf. Ser. Mater. Sci. Eng.* **256** 8-12

[4] Qu L and Dai L 2007 *Adv. Mater.* **19** 3844-9

[5] Röhrig M, Thiel M, Worgull M and Hölscher H 2012 *Small* **8** 3009-15

[6] Jeong H E, Lee S H, Kim P and Suh K Y 2008 *Colloids Surf. A Physicochem. Eng. Asp.* **313-4** 359-64

[7] Majidi C, Groff R E, Maeno Y, Schubert B, Baek S, Bush B, Maboudian R, Gravish N, Wilkinson M, Autumn K and Fearing R S 2006 *Phys. Rev. Lett.* **97** 076103

[8] Ge L, Sethi S, Ci L, Ajayan P M and Dhinojwala A 2007 *Proc. Natl. Acad. Sci. USA* **104** 10792-5
[9] Qu L, Dai L, Stone M, Xia Z and Wang Z L 2008 *Science* **322** 238-42
[10] Yurdumakan B, Raravikar N R, Ajayan P M and Dhinojwala A 2005 *Chem. Commun.* **30** 3799
[11] Ageev O A, Blinov Y F, Il'ina M V., Il'in O I, Smirnov V A and Tsukanova O G 2016 *Phys. Solid State* **58** 309-14
[12] Greiner C, del Campo A and Arzt E 2007 *Langmuir* **23** 3495-502
[13] Il'ina M V., Il'in O I, Blinov Y F, Konshin A A, Konoplev B G and Ageev O A 2018 *Materials* **11** 638
[14] Purohit R, Purohit K, Rana S, Rana R S and Patel V 2014 *Procedia Mater. Sci.* **6** 716-28
[15] Il'ina M V, Konshin A A, Il'in O I, Rudyk N N, Fedotov A A and Ageev O A 2018 *J. Phys. Conf. Ser.* **993** 012025

IOP Conf. Series: Materials Science and Engineering **443** (2018) 012010 doi:10.1088/1757-899X/443/1/012010

Dependence of Young's modulus of vertically aligned carbon nanotubes on their aspect ratio

M V Il'ina, O I Il'in, A A Konshin, A A Fedotov and O A Ageev

Southern Federal University, Institute of Nanotechnologies, Electronics and Electronic Equipment Engineering, Taganrog, 347922, Russia

mailina@sfedu.ru

Abstract. The experimental studies of the dependence of Young's modulus of vertically aligned carbon nanotubes (CNTs) on their aspect ratio were carried out. It was shown that the Young's modulus of the individual CNTs increased from 1.63 to 1.83 TPa with an increase in their aspect ratio from 8 to 15.8. This dependence is probably due to the fact that the relative number of the structure defects in the CNT decreases with increasing its aspect ratio. The Young's modulus of the CNT bundles also increased with the growth of the aspect ratio of CNTs. However, the calculated values of the Young's modulus of the CNT bundles with a large aspect ratio were smaller than ones of the individual CNTs with a smaller aspect ratio. It is due to the fact that CNTs that form the bundles are in strained state initially. A smaller strain is required to form a given deformation in CNT bundle than that of individual nanotube. The obtained results can be used to develop and create nanoelectronics devices based on vertically aligned CNTs, in particular adhesion coatings and nonvolatile memory elements.

1. Introduction

The rapid development of nanotechnology and the new opportunities in the field of nanoscale structures created a wide interest in the research of carbon nanotubes (CNTs) [1]. However, for the application of CNTs as a material for the elemental base of new generation electronics, it is necessary to study in detail their mechanical parameters, in particular, the Young's modulus [2,3]. Theoretical studies of the Young's modulus of CNTs are conducted mainly using models that are valid only for ideal (physically and chemically homogeneous over the entire length) nanotubes [4,5]. Experimental studies of the CNTs Young's modulus are not numerous [6-9], since the use of traditional research methods, such as direct tensile loading, is difficult due to the nanoscale material.

As a consequence, the conclusions about the dependence of the CNTs elastic properties on their geometric parameters are also not numerous and contradictory: some note the "insensitivity" of Young's modulus to the nanotube diameter [10], others demonstrate an increase in the Young's modulus with an increase in diameter for a fixed CNT length [11,12], and, conversely, its decrease [13,14]. Thus, there is a need for further studies of the Young's modulus of CNTs on their geometric parameters, the main of which is the aspect ratio – the ratio of the length to diameter.

The aim of this study is to investigate the dependence of the Young's modulus of vertically aligned CNTs on their aspect ratio using the previously developed measurement technique based on the nanoindentation method.

2. Experimental studies

As the studied samples there were CNT arrays grown by plasma enhanced chemical vapour deposition method on a silicon wafer with a Ni/Ti structure using NANOFAB NTC-9 (NT-MDT, Russia) [15-17]. Six samples of two types were selected for investigation: individual vertically aligned CNTs (Fig. 1a) and vertically aligned CNTs combined in bundles (Fig. 1b). The geometric parameters of the experimental samples were evaluated on the basis of scanning electron microscopy (SEM) images obtained using SEM Nova NanoLab 600 (FEI, the Netherlands).

Experimental studies of the Young's modulus of experimental samples were carried out using the previously developed technique based on the nanoindentation method [18,19]. The measurements were carried out at the Ntegra probe nanolaboratory (NT-MDT, Russia) using a scanning hardness nanotester integrated in it. The indenter was the diamond tree-sided Berkovich pyramid with the apex angle $\theta = 70^o$ between the edge and height.

The nanoindentation process was carried out with the application of loads of 100 μN at 20 different points of the CNT array on the distance about 10 μm from each other. The dependence of the indentation force P_{IND} on the corresponding penetration depth h of the indenter in the material was obtained during the measurement. The dependences $P_{IND}(h)$ for each experimental sample are shown in the Figure 2.

Figure 1. SEM images of experimental samples of CNT arrays: (a) type 1 and (b) type 2. Scale segment 500 nm.

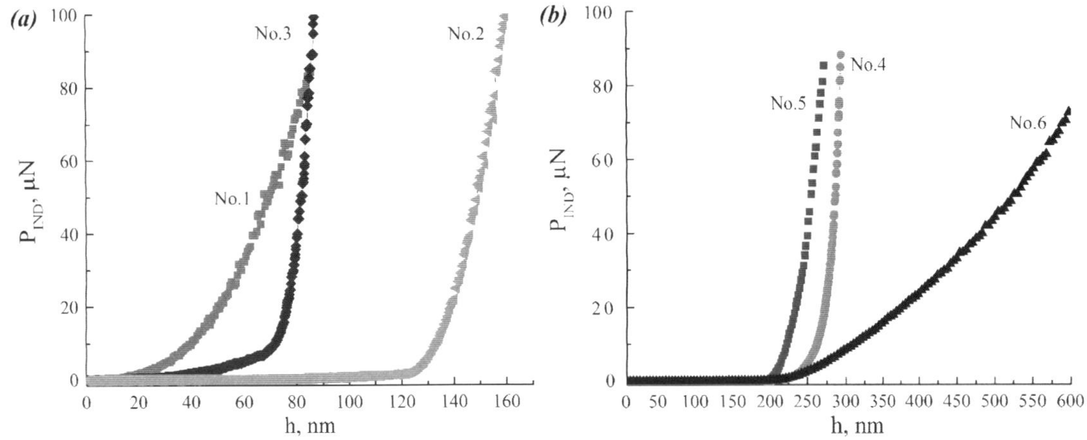

Figure 2. Averaged load curves of vertically aligned CNT for samples of (a) type 1 and (b) type 2.

On the basis of the obtained experimental dependences using the technique from Ref. [18], a Young's modulus of vertically aligned CNTs was calculated by the formula:

$$E = \frac{64(P_{IND} - P_i)\cos\theta}{\pi i D^4 k^2}, \quad (1)$$

where P_{IND} – indentation force, P_i – indentation force corresponding to the depth of the *i*-tube touch, *i* – amount of nanotubes interacting with the indenter at P_{IND}, D – a CNT diameter, $k = (P/(EI)_{eff})^{1/2}$ – the coefficient, $(EI)_{eff}$ – a effective bending stiffness of CNT.

Geometric parameters and calculated values of Young's modulus of vertically aligned CNTs are given in Table 1.

Table 1. The calculated values of the Young's modulus and geometric parameters of experimental samples.

Sample No.	Diameter, nm	Length, μm	Density of CNTs in array, μm^{-2}	Aspect ratio	Young's modulus, TPa
			Type 1		
1	44.1	0.35	95	8	1.63
2	63.2	0.65	38	10.2	1.69
3	44.9	0.71	60	15.8	1.83
			Type 2		
4	43.8	0.65	82	14.8	1.15
5	35.6	1.21	72	34	1.29
6	51	0.69	69	13.5	0.59

It should be noted that the $P_{IND}(h)$ dependence sections corresponding to the elastic interaction area of CNTs with an indenter were used in order to calculate the Young's modulus of the experimental samples.

3. Results

The results of experimental studies have shown that the Young's modulus of individual vertically aligned CNTs increases from 1.63 to 1.83 TPa with an increase in their aspect ratio from 8 to 15.8 (Fig. 3a). This dependence is probably due to the fact that the relative number of the structure defects in the CNT decreases with increasing its aspect ratio. Thus, the defects are concentrated near the base of the CNT during the growth of CNTs and their relative size number with increasing CNT length. In addition, the number of defects can increase with increasing diameter of the multi-walled CNT due to the increase in the number of layers in it.

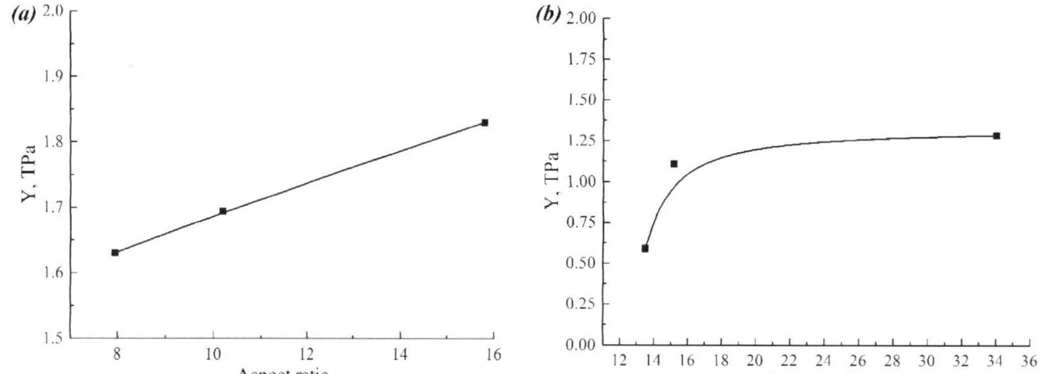

Figure 3. Dependences of Young's modulus of CNTs on their aspect ratio for (a) type 1, (b) type 2.

The Young's modulus of experimental samples of type 2, which are bundles of CNTs, also increases with an increase of the CNT aspect ratio (Fig. 3b). However, the calculated values of the Young's modulus for samples of type 2 with a large aspect ratio are smaller than ones for samples of type 1 with a smaller aspect ratio (Table 1). This may be due to the fact that CNTs forming the bundles are initially in a strained state. As a result, a smaller load is required to form a deformation in bundles than that for individual CNTs in the nanoindentation process. This fact must be taken into account when measuring the mechanical parameters of combined into bundles CNTs using by the nanoindentation method.

4. Conclusion

The dependence of Young's modulus of vertically aligned CNTs on their aspect ratio is established experimentally. It is shown that the Young's modulus of the individual CNTs increases from 1.63 to 1.83 TPa with an increase in their aspect ratio from 8 to 15.8. The Young's modulus of the CNT bundles also increases from 0.59 to 1.29 TPa with the growth of the aspect ratio of CNTs from 13.5 to 34. This is due to the fact that the relative number of the structure defects in the CNT decreases with increasing its aspect ratio. The obtained results are in good agreement with the literature data [6-9,18-20]. The obtained results can be used to develop and create nanoelectronics devices based on vertically aligned CNTs, in particular adhesion coatings and non-volatile memory elements [2,3,21,22].

The results were obtained using the equipment of the Research and Education Centre and Centre of Common Using “Nanotechnologies” of Southern Federal University.

Acknowledgments

The reported study was funded by RFBR according to the research projects No.16-29-14023 ofi_m, No.18-32-00652 and by grant of the Southern Federal University (project No. VnGr-07/2017-26).

References

[1] Ren Z, Lan Y and Wang Y *Aligned Carbon Nanotubes, NanoScience and Technology* (Springer-Verlag Berlin Heidelberg) (2013)
[2] Il'ina M V, Il'in O I, Blinov Y F, Konshin A A, Konoplev B G and Ageev O A 2018 *Materials* **11** 638
[3] Il'ina M V, Il'in O I, Blinov Yu F, Smirnov V A, Kolomiytsev A S, Fedotov A A, Konoplev B G and Ageev O A 2017 *Carbon* **123** 514-24
[4] Zhou X, Zhou J J and Ou-Yang Z C 2000 *Phys. Rev.* **62** 13692-6
[5] Batra R C and Sears A 2007 *Int. J. Solids Struct.* **44** 7577-96
[6] Ding W, Calabri L, Kohlhaas K M, Chen X, Dikin D A and Ruoff R 2007 *Exp. Mech.* **47** 25-36
[7] Lourie O and Wagner H D 1998 *J. Mater. Res.* **13** 2418 22
[8] Wong E W, Wong E W, Sheehan P E and Lieber C M 1997 *Science* **277** 1971-5
[9] Qi H J, Teo K B K, Lau K K S et al. 2003 *J. Mech. Phys. Solids* **51** 2213-37
[10] Wang Q, Vara dan V K and Quek S T 2006 *Phys. Lett.* A **357** 130-5
[11] Li C and Chou T 2003 *Int. J. Solids Struct.* **40** 2487-99
[12] Chang T and Gao H 2003 *J. Mech. Phys. Solids* **51** 1059-74
[13] Lei X, Natsuki T, Shi J and Ni Q Q 2011 *J. Nanomat.* **2011** 805313
[14] Parvaneh V and Shariati M 2011 *Acta Mech.* **216** 281-9
[15] Klimin V S, Il'ina M V, Il'in O I, Rudyk N N and Ageev O A 2017 *IOP J. Phys.: Conf. Ser.* **917** 092023
[16] Il'in O I, Il'ina M V, Rudyk N N, Fedotov A A and Ageev O A 2018 *Nanosystems: Phys. Chem. Math.* **9** 92-4
[17] Il'in O I, Il'ina M V, Rudyk N N, Fedotov A A, Levshov D I and Ageev O A 2018 *IOP J. Phys.: Conf. Ser.* **1038** 012062
[18] Ageev O A, Il'in O I, Kolomiitsev A S, Konoplev B G, Rubashkina M V, Smirnov V A and Fedotov A A 2012 *Nanotechnologies in Russia* **7** 47-53

[19] Ageev O A, Blinov Yu F, Il'ina M V, Konoplev B G and Smirnov V A 2016 *Intelligent Nanomaterials, 2nd Edition.* Eds. A Tiwari, Y K Mishra, H Kobayashi, A P F Turner (Wiley Scrivener Publishing LLC), chapter **11** 361-94
[20] Ageev O A, Ilin O I, Kolomiytsev A S, Rubashkina M V, Smirnov V A and Fedotov A A 2014 *Adv. Mat. Res.* **894** 355-8
[21] Ageev O A, Blinov Yu F, Ilina M V, Ilin O I and Smirnov V A 2016 *IOP J. Phys.: Conf. Ser.* **741** 012168
[22] Ageev O A, Blinov Yu F, Il'ina M V, Il'in O I, Smirnov V A and Tsukanova O G 2016 *Phys. Solid State* **58** 309-14

Influence of AC electric field on the charge generation in albumin solution in a flow-based AFM fishing system

Y D Ivanov[1], A F Kozlov[1], R A Galiullin[1], V Y Tatur[2], N D Ivanova[3] and V S Ziborov[4]

[1] Institute of Biomedical Chemistry, Moscow, Russia.
[2] Foundation of Perspective Technologies and Novations, Moscow, Russia.
[3] Skryabin Moscow State Academy of Veterinary Medicine and Biotechnology, Moscow, Russia.
[4] Joint Institute for High Temperatures RAS, Moscow, Russia.

yurii.ivanov@rambler.ru

Abstract. In the present paper, the influence of AC electric field on charge accumulation in the measuring cell of a flow-based system for atomic force microscope (AFM) based fishing is considered. As a flow-based system, a device for AFM-based fishing of low-abundant proteins has been employed. The charge is generated in a femtomolar albumin solution during its flowing through an injector of this system. We have demonstrated that application of sinusoidal electric field (100 V, 50 Hz frequency) stimulates charge accumulation in the measuring cell upon the input of protein solution at 38°C. Taking this effect into account is of importance for the development of novel highly sensitive flow-based proteomic and diagnostic systems employing AFM-based fishing, for modeling the influence of electric field on hemodynamics, and also for the development of specified theoretical models describing physicochemical properties of water and aqueous solutions.

1. Introduction

The urgency of studying the influence of AC electric fields on the physicochemical properties of water and aqueous solutions is conditioned by the fact that account for this influence can be well used for increasing the sensitivity of biosensor devices. In this way, in our previous paper [1] the stimulating influence of AC electromagnetic field on the efficiency of protein capturing in atomic force microscope (AFM) based fishing system at femtomolar (10^{-15} M) and subfemtomolar concentrations was demonstrated. In the same paper [1], with the example of the detection of albumin, we demonstrated that AC electric fields applied to AFM chips can be well used for the sensitivity enhancement of a proteomic AFM-based fishing system. This allowed us to increase the detection sensitivity, reaching the detection limits of 10^{-18} to 10^{-17} M. As noted therein [1], the efficiency of fishing albumin protein correlated with the generation of charge occurring upon the input of the analysed solution into the measuring cell. The correlation between increasing the efficiency of protein fishing and efficiency of charge generation upon the input of analytic solution into a measuring cell through the injector of an AFM-based fishing system was also discussed. In Refs. [2-4], the efficiency of charge generation in an injector of an AFM-based fishing system in water and protein solutions upon their input with a peristaltic pump with preliminary neutralization of the charge after the pump has been considered. Earlier, the effect of spontaneous generation of charge in water due to a triboelectric effect in the absence of electric field was discussed in the paper by Choi et al. [5].

In our present paper, an injection part of a fishing system (which employs an AFM as a molecular detector) has been used to study the influence of external electromagnetic field on charge generation in flowing water and femtomolar albumin solution at 38°C upon their pumping through an injector (a pipette tip) without preliminary neutralization of charge of solution into a measuring cell containing an AFM chip. The protein molecules captured onto the chip surface in the course of fishing are to be detected on the chip surface by AFM. This scheme of solution input through an injector was employed by us earlier in AFM-based fishing biosensor measurements; it is also used in other analytical biosensor systems for their testing. As an electromagnetic field, sinusoidal AC electric field (of 50 Hz frequency, which occurs during the use of industrial electric devices) was used. As an object to be studied herein, bovine serum albumin has been selected, as it is a well-studied protein highly homologous to human serum albumin. Since analytical systems for studying biochemical processes in near-physiological conditions are operating in the temperature range near 38°C, our experiments have been performed at this temperature.

Herein, a stimulating influence of external electromagnetic field on charge generation in deionized water and albumin solution flowing through an injection part of a system, employing an AFM as a detector, has been observed.

Taking into account that charge state of protein solutions, which are injected through pipes or capillaries into the sensitive area of an AFM-based detector of fishing diagnostic systems (and also other highly sensitive systems, such as nanowire sensors), can have a significant influence on their sensitivity, the results obtained can be well used in analysis of operation efficiency of these systems. Moreover, the data obtained in the present study can be useful in the development of models describing physicochemical properties of protein solutions and hemodynamics in human body in normal and pathological states.

2. Methods

Deionized water (18 MΩ×cm resistivity) was obtained with a Millipore Simplicity UV system (Millipore, France). Fatty acid-free BSA (Sigma, cat. No. A6003) was dissolved in deionized water at 10^{-4}M concentration. 10^{-15}M BSA solution was prepared by sequential dilutions. Electric charge measurements were performed with an electrometer combined with the flow-based sample delivery system for AFM-based fishing [6].

Principal units of the sample delivery system included a peristaltic pump, a pipe with a tip for liquid delivery, and a measuring cell. Upon the measurements, water or protein solution (10^{-15}M) was continuously pumped with an Ismatec ISM 597D peristaltic pump (IDEX Corp., Lake Forest, IL, USA) from a 50-mL polypropylene tube with a grounding stem inside, into the measuring cell. A sterile silicone pipe (40 cm length, 2.0 mm i.d.) with a standard disposable pipette tip (Eppendorf type, 1 to 10 μL nominal capacity, 0.4 mm i.d., Eppendorf AG, Hamburg, Germany) was used for the delivery of either water or protein solution. Flow rate (~15 μL/s) was set so that drops (~15 μL) were formed on the tip nozzle.

Stainless steel inner cylinder represents a measuring cell in the charge measurement system, connected to the electrometer. Charge registration accuracy was 0.1 nC.

The effect of a pulsed electric field on the liquids was produced with a sinusoidal voltage generator developed in IBMC. For this purpose, the tip of the liquid-supplying pipe was placed between two parallel metal plates; an AC sinusoidal voltage (100 V amplitude, 50 Hz frequency) from the generator was applied to the plates. The temperature of the analysed liquid (water or protein solution) was maintained using a Thermomixer Comfort thermoshaker (Eppendorf AG, Hamburg, Germany), into which the liquid-containing tube was placed. The experiments were carried out at 38°C.

3. Results

Figure 1 displays typical time dependencies of the value of charge accumulated in the measuring cell ($\Delta q(t)$ dependencies) upon the input of water in the presence and in the absence of external sinusoidal electric field at 38°C.

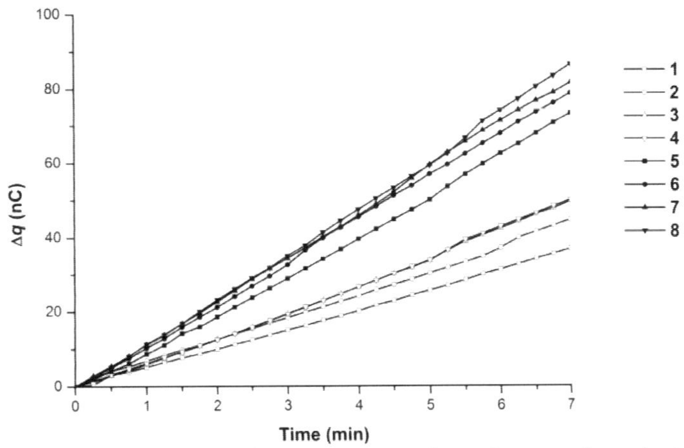

Figure 1. $\Delta q(t)$ dependencies obtained upon pumping of water through the tip in the flow-based system in the absence (curves 1-4) and in the presence (curves 5-8) of sinusoidal electric field at 38°C and 15 μL/s flow rate.

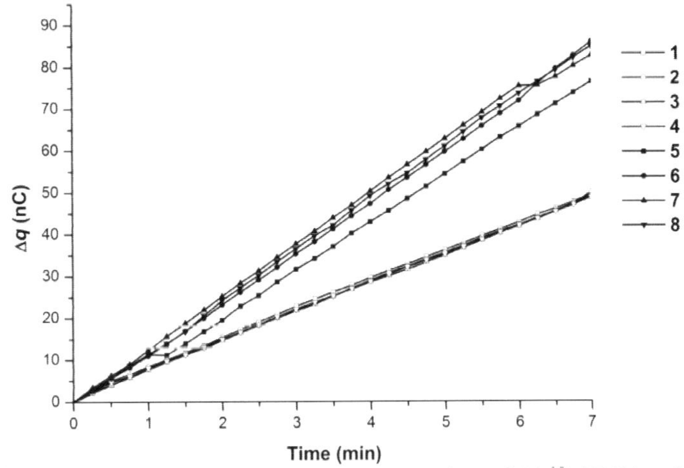

Figure 2. $\Delta q(t)$ dependencies obtained upon pumping of 10^{-15}M BSA solution into the measuring cell in the absence (curves 1-4) and in the presence (curves 5-8) of sinusoidal electric field at 38°C and 15 μL/s flow rate.

As seen from Figure 1, for water in the absence of sinusoidal electric field (curves *1-4*), a generation of charge and its accumulation in the measuring cell with a rate $\Delta q/\Delta t$ = 45±6 nC/7 min is observed, and the $\Delta q(t)$ dependence is linear.

In the presence of the electric field (curves *5-8*), an increase in the level of charge accumulation rate is observed, $\Delta q/\Delta t$ = 80±6 nC/7 min. Thus, a stimulating influence of sinusoidal electric field on charge accumulation in the measuring cell by 78% at 38°C is observed.

Figure 2 displays typical $\Delta q(t)$ dependencies upon pumping of BSA solution in the presence and in the absence of external sinusoidal electric field at 38°C.

As seen from Figure 2, at 38°C, in the absence of external sinusoidal electric field (curves *1-4*), a generation of charge in the protein solution and accumulation of this charge in the measuring cell with a rate $\Delta q/\Delta t$ = 49±1 nC/7 min is observed.

In the presence of the electric field (curves *5-8*), an increase in the level of charge arrived into the measuring cell with the protein solution and accumulated in the cell is observed, $\Delta q/\Delta t$ = 82±4 nC/7 min. Thus, a stimulating influence of sinusoidal electric field on charge accumulation in the measuring cell by 67% at 38°C is observed.

4. Discussion

As noted above, a charge generation occurs upon flow of a liquid through polymer pipes of the AFM-based fishing system [5]. In the same study, a linear character of time dependence of accumulation of charge induced in water was demonstrated.

The data obtained in our present study describe the effect of a 50 Hz sinusoidal AC electric field on charge generation in the course of flowing water and protein solution.

Let us consider summarized data obtained in our experiments. In this way, for water at 38°C, an increase in the value of charge accumulated in the measuring cell in 7 min from 45 nC (in the absence of electric field) to 80 nC (in the presence of electric field) – i.e., by 35 nC (by ~78%) has been demonstrated. For protein solution at 38°C, a stimulation of charge accumulation in the cell from 49 nC (in the absence of electric field) to 82 nC (in the presence of electric field) – i.e. by 33 nC (by ~67%) in 7 min has been observed. That is, the effect of electric field in the cases of water and protein solution is approximately equal.

Physical nature of the stimulation of accumulation of induced charge in water, flowing through a polymer pipe, can be described in the following way. As was discussed by Choi et al. [5], upon pipetting water through a pipette tip, a positive charge is generated at the expense of a triboelectric effect. Pershin et al. [7] reported about a phenomenon of excitation of a conversion of ortho-para isomers in the complex structure of water, representing a mixture of ice-like and liquid phases (para- and ortho-states of water); these states were characterized by different physicochemical properties (surface tension, viscosity etc.). Upon electrical excitation in AC electric field, similar increased conversion of the para-state of water to the ortho-state can occur. The possibility of such transitions in electromagnetic fields of several Hz frequency was reported by Pershin [7].

Thus, we have observed an increase in charge generation rate in water and in protein solution in the presence of a sinusoidal AC electric field. The change in the fraction of ice-like water structures in the presence of AC electric field can be one of the causes of the observed stimulation of charge generation.

It is known that 50 Hz sinusoidal electric fields have an effect on human organism. The effect observed in our study can determine the impact of these fields on hemodynamics in the body, as different phases of water have different physicochemical properties.

It is to be noted that the observed phenomenon can be useful for the development of models describing physicochemical properties of water and aqueous solutions.

5. Conclusion

An effect of a stimulating influence of sinusoidal AC electric field on charge generation in water and protein solution flowing through an injector of a fishing system for the detection of low-abundant proteins (which employs AFM as a molecular detector) without preliminary neutralization of the charge of the liquid after its pumping with a peristaltic pump has been measured.

Account for the discovered effect is important for the development of models describing physicochemical properties of water and aqueous solutions, as well as those describing the influence of sinusoidal electric field on the processes connected with circulatory system (including hemodynamics). It also will be useful in the development of novel highly sensitive diagnostic systems operating in the range of femtomolar concentrations.

Acknowledgements
This study was performed in the framework of the Program for Basic Research of State Academies of Sciences for 2013-2020.

References

[1] Pleshakova T, Malsagova K, Kaysheva A, Kopylov A, Tatur V, Ziborov V, Kanashenko S, Galiullin R and Ivanov Yu 2017 *FEBS Open Bio* **7** 1186-95
[2] Ivanov Y D, Kozlov A F, Galiullin R A, Kolesanova E F and Pleshakova T O 2017 *Biosensors* **7** 60
[3] Ivanov Y D, Kozlov A F, Galiullin R A, Kanashenko S L, Usanov S A, Ivanova N D, Ziborov V S and Pleshakova T O 2018 *J. Electrostat.* **91** 16-20
[4] Ivanov Yu D, Kozlov A F, Galiullin R A, Ivanova N D, Tatur V Yu, Ziborov V S, Yushkov E S and Pleshakova T O 2018 *Pathological Physiology and Experimental Therapy* **62** 109-16 (in Russian)
[5] Choi D, Lee H, Im D J, Kang I S, Lim G, Kim D S and Kang K H 2013 *Sci. Rep.* **3** 1-7
[6] Pleshakova T O, Malsagova K A, Kozlov A F, Kanashenko S L, Ivanova N D, Sadovskaya T A, Archakov A I and Ivanov Yu D 2016 *Patogenez* **3** 23-30
[7] Pershin S M 2009 *Phys. Wave Phenom.* **17** 241-50

IOP Conf. Series: Materials Science and Engineering **443** (2018) 012012 doi:10.1088/1757-899X/443/1/012012

Influence of the FIB parameters on the etching of planar nanosized multigraphene/SiC field emitters

I L Jityaev, A M Svetlichnyi, V I Avilov, I N Kots, A S Kolomiytsev, O A Ageev

Southern Federal University, Institute of Nanotechnologies, Electronics and Equipment Engineering, Taganrog, 347922, Russia

jityaev.igor@gmail.com

Abstract. The possibility of fabrication of planar field emission nanostructures based on multigrafene films on semiinsulating silicon carbide using focused ion beam (FIB) technology is considered in this paper. The effect of FIB-parameters on the etching of the structure was determined using a scanning probe microscope. Conductive probes were used for local studies of electrical properties of nanoscale structures. It was found that etching of planar field emission nanostructures at a current of 1 pA did not lead to a change in the depth of the treated area. An increase in current up to 10 pA was sufficient to initiate the etching process of a multigraphene film on the surface of silicon carbide.

1. Introduction

Electron sources based on field emission from nanostructures are of great interest for the development of modern devices. Devices based on field emission unlike those based on photo- and thermionic emission do not require additional heat loss, they allow obtaining a high emission current density, are stable in a wide range of temperatures, have high speed and low noise level. Studies conducted in the 1970s have shown carbon materials to be promising for field emission application [1].

In this paper, we consider multigraphene films obtained by the thermal decomposition of silicon carbide in vacuum [2]. At present, this method is considered as a high-tech, as well as the most promising to produce graphene and multigraphene films with high homogeneity and high quality for electronic applications. In this case, the multigraphene films are characterized by high adhesion to silicon carbide, low work function, and the possibility of depositing on the entire surface of the substrate. The substrate was semiinsulating silicon carbide 6H-SiC. The choice of silicon carbide is due to the high thermal conductivity, the ability to operate the device at high current densities and high-power levels. In addition, an important and determining factor in the choice of silicon carbide as a substrate was its resistance to external corrosive influences. Silicon carbide is considered as one of the most promising materials for high-temperature radiation-resistant electronics. Emission structures based on graphene on silicon carbide are characterized by low threshold voltages [3-6].

Multigraphene films are an excellent material for planar emission nanostructures due to its two-dimensional structure. The nanometer film thickness of the multigraphene contributes to the high form-factor [7]. The planar design of the field emission structure ensures the emission of electrons from the end of the multigraphene film, when a potential difference is applied. Planar field emission structures are considered promising due to the high amplification of the electric field at the sharp ends [8-10]. In order to reduce the threshold voltages, it is necessary to reduce the interelectrode gap. Providing a nano-sized gap for emission cells with vertical-type emitters is technologically difficult. The use of focused ion beam (FIB) allows nanoscale profiling of substrates. Moreover, the advantages of this method are the ability to process a wide range of materials, the high speed of forming a topological pattern without the use of resists and masks, as well as locality and selectivity of the etching process, achieved by varying the process parameters in a wide range.

Thus, the work is devoted to the investigation of the possibility of manufacturing planar field emission nanostructures based on multigraphene films on silicon carbide using FIB. The etching regimes with the minimum ion beam current (Ii), the minimum time of ion beam exposure at the point (dt), and the number of passes (ne) were determined based on the scanning probe microscopy (SPM).

2. Experiment and results

The electron-ion complex Nova NanoLab 600 (FEI Company, The Netherlands) with electron-beam and ionic columns and a gallium liquid metal ion source (resolution 7 nm) was used to realize the etching of experimental samples by an ion beam and subsequent analysis by the raster electron microscopy method [11]. The possibility of FIB-applying for the etching of multigraphene/SiC experimental samples and subsequent fabricating of planar emission nanostructures was evaluated using a Solver P47 Pro SPM. This is a universal instrument for complex studies with high resolution, both in air and in a controlled gas environment, at temperatures up to 150°C [12-14]. It can be used to control the technological modes of obtaining various coatings, films, and other materials with a thickness of up to 20 mm and a diameter of up to 100 mm. An important feature of SPM Solver P47 Pro is the high measurement accuracy and various measurement techniques.

The experimental samples were oriented in such a way that the flow of accelerated ions hit the surface in the normal direction. Minimum currents (1-10 pA) were used to obtain the minimum possible interelectrode gap for this technology. An increase in the ion current promotes an increase in lateral over-etching. The scanning electron microscope (SEM) image of the experimental sample with 300x300 nm^2 sections treated by FIB at the ion beam current of 1 pA is shown in Figure 1. The upper row was treated at a fixed beam impact time at the point dt = 1 μs, the number of passes was ne = 100, 200 ... 900. The lower row was treated at a fixed number of passes ne = 100; the time of the beam impact at the point dt = 1.2 ... 9 μs was varied. The resulting SEM-image allows us to qualitatively evaluate the effect of the focused beam on the surface of the experimental sample.

Then, the SPM was used for a more detailed study of the FIB-treated sections. The cross sections of the FIB-treated areas were plotted in Figure 2. Based on the obtained cross sections, it was found that etching did not occur at an ion beam current of 1 pA. The increase in impact time of the beam at the point and the number of passes did not contribute to the etching process of the experimental sample. There was a swelling of the surface, which was most likely due to the introduction of gallium ions into the near-surface zone of the experimental sample.

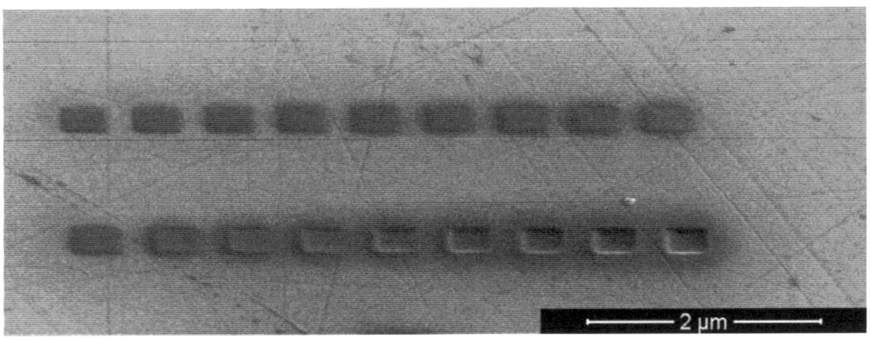

Figure 1. SEM-image of the experimental sample with FIB-treated areas.

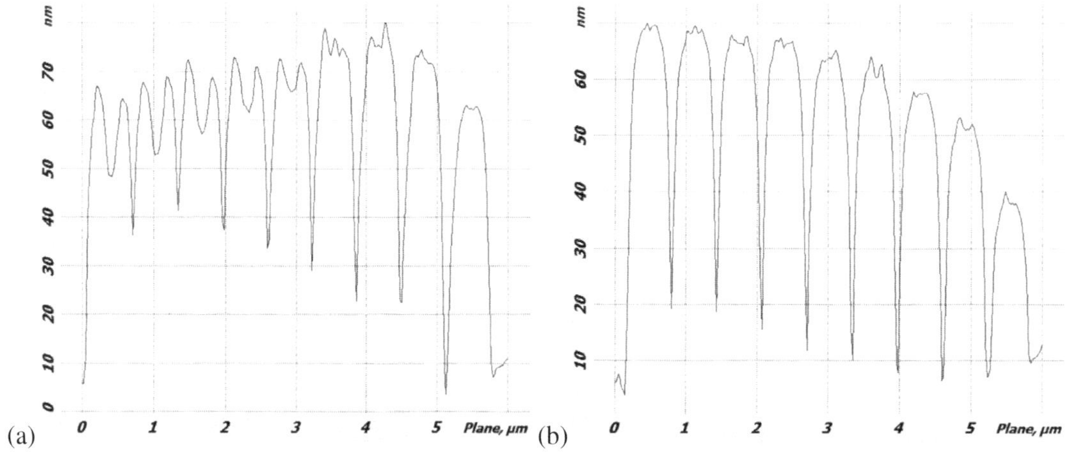

Figure 2. The cross-section of the experimental sample after FIB-treatment at a current of 1 pA: (a) dt = 1, 2 ... 9 µs, ne = 100; (b) dt = 1 µs, ne = 100, 200 ... 900.

Figure 3. SEM-image of planar field emission cell based on multigraphene on SiC.

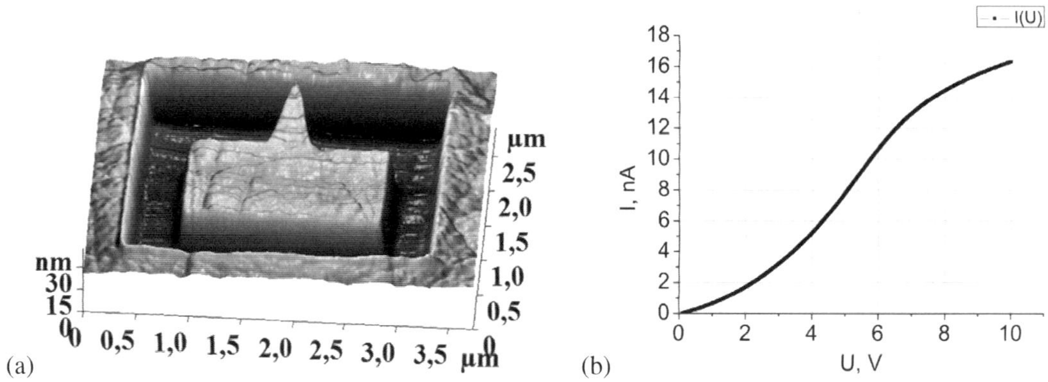

Figure 4. (a) SPM-image of a planar-type field emission cell, (b) current-voltage characteristics of field emission cell.

The subsequent increase of the ion beams current to 10 pA made it possible to initialize the etching process of the experimental sample. An electronic model in the form of a raster template for FIB etching a planar emission cell with a point emitter was developed on the basis of previous theoretical studies of the design parameters of a planar field emission cell. The electronic model of the emission cell took into account the optimal values of the interelectrode distance, the rounding-off radius of the emitter top, the height and the angle of inclination of the lateral surface relative to the central axis. Thus, a planar field emission structure with a tip emitter and an interelectrode distance determined by the diameter of the ion beam was fabricated. The SEM-image of the fabricated emission cell is shown in Figure 3.

It was found that emission cells with an interelectrode distance of ~ 15-30 nm could be produced reproducibly on the equipment used. The rounding-off radius of the emitter top was 30 nm. The FIB application made it possible to fabricate a cathode and an anode in a single production step. The optimal etching time was 2 min 30 sec. Partial etching of the emitter tip was observed with a longer etching time.

Conductive probes were used to study the fabricated nanoscale structures. The SPM-image of a planar-type field emission cell with a tip-shaped emitter is shown in Figure 4a. The etching depth was ~ 25 nm. The pressure contact was located on the surface of the multigraphene film in the anode region. The conductive probe was brought into contact with the investigated region to measure the local electrical characteristics. It was found that etched areas did not conduct electric current. This confirms that the local sections of the multigraphene film treated with FIB are completely etched to semiinsulating silicon carbide. The contact of the conducting probe with the unetched area of the field emission cathode contributed to the appearance of an emission current (Fig. 4b). The emission current was 16.3 V at a potential difference of 10 V. The threshold voltage was ~ 0.2 V.

3. Conclusion

It is shown that the fabrication of the field emission structures with a nanoscale interelectrode distance is possible using the FIB technology. The recommended ion beams current for etching of multigraphene film on silicon carbide is 10 pA. This ensures minimum dimensions of the interelectrode gap and minimizes the processes of over-etching the treated area. The use of SPM made it possible to quantitatively evaluate the FIB effect on the surface of the experimental sample, to check the geometric parameters of the fabricated nanoscale planar field emission structures and local measurements of electrical parameters. The planar design of the tip emitter made it possible to lower the threshold voltages in comparison with emitters of a vertical type based on a multigraphene on silicon carbide [13]. The obtained results make it possible to optimize the technological operations for the fabrication of field emission nanostructures, to reduce the costs of their production, and to improve their emission characteristics. Nanoscale planar field emission structures based on multigraphene films on silicon carbide can form the basis of the element base of energy-efficient nanoelectronics.

Acknowledgments

The equipment of the Research and Educational Centre "Nanotechnologies" of Southern Federal University was used for this study. This work was funded by Internal grant of the Southern Federal University No. VnGr-07/2017-26.

References

[1] Baker F S, Osborn A R and Williams J 1974 *J. Phys. D: Appl. Phys.* **7** 2105-15

[2] Lebedev A A, Kotousova I S, Lavrent'ev A A, Lebedev S P, Makarenko I V, Petrov V N and Titkov A N 2009 *Phys. Solid State* **51** 829-32

[3] Konakova R V, Okhrimenko O B, Svetlichnyi A M, Ageev O A, Volkov E Yu, Kolomiytsev A S, Jityaev I L and Spiridonov O B 2015 *Semiconductors* **49** 1242-5

[4] Konakova R V, Okhrimenko O B, Kolomys A F, Strel'chuk V V, Svetlichnyi A M, Ageev O A, Volkov E Yu, Kolomiitsev A S, Zhityaev I L and Spiridonov O B 2016 *J. Superhard Mater.* **38** 235-40

[5] Eda G, Emrah Unalan H, Rupesinghe N, Amaratunga G A J and Chhowalla M 2008 *Appl. Phys. Lett.* **93** 233502
[6] Lee S W, Lee S S and Yang E-H 2009 *Nanoscale Res. Lett.* **4** 1218-21
[7] Jityaev I L and Svetlichnyi A M 2018 *J. Phys.: Conf. Ser.* **1038** 012055
[8] Gavrilov S A, Il'ichev É A, Poltoratskii É A, Rychkov G S, Dvorkin V V, Dzbanovsky N N and Suetin N V 2004 *Tech. Phys. Lett.* **30** 609-11
[9] Rouhi J, Mahmud S, Hutagalung S D and Naderi N 2012 *Electron. Lett.* **48** 712-4
[10] Aban'shin N P, Avetisyan Yu A, Akchurin G G, Loginov A P, Morev S P, Mosiyash D S and Yakunin A N 2016 *Tech. Phys. Lett.* **42** 509-12
[11] Orloff J, Swanson L W and Utlaut M 1996 *J. Vac. Sci. Tech. B* **14** 3759-63
[12] Avilov V I, Ageev O A, Smirnov V A, Solodovnik M S and Tsukanova O G 2015 *Nanotechnologies in Russia* **10** 214-9
[13] Jityaev I L, Svetlichnyi A M, Kolomiytsev A S, Volkov Y E, Polyakova V V and Ageev O A 2017 *IOP Conf. Ser.: Mat. Sci. Eng.* **256** 012021
[14] Avilov V I, Polupanov N V, Tominov R V, Smirnov V A and Ageev O A 2017 *IOP Conf. Ser.: Mat. Sci.Eng.* **256** 012001

IOP Conf. Series: Materials Science and Engineering **443** (2018) 012013 doi:10.1088/1757-899X/443/1/012013

Modelling of local anodic oxidation of titanium oxide nanostructures formation process

O G Karen'kih, V I Avilov, V A Smirnov, A A Fedotov, N A Sharapov and N A Polupanov

Southern Federal University, Institute of Nanotechnologies, Electronics and Equipment Engineering, Taganrog, 344006, Russian Federation

AvilovVI@sfedu.ru

Abstract. A model for the oxide nanostructures formation by the local anodic oxidation method is presented. Dependences of height, depth, and diameter of titanium oxide nanostructures on the applied voltage pulse duration are obtained, as well as the oxide nanostructures profile for given technological parameters.

1. Introduction

The development of components of modern electronics, in particular memristor structures for resistive random access memory (RRAM) elements, requires application of new methods for the formation of nanostructures that possess high precision and reproducibility [1-8]. One of such methods is local anodic oxidation (LAO), which has high precision accuracy and allows forming structures with thickness less than 5 nm [9-16]. Other advantage of this method is the absence of complex photoresists and chemical etchants. Oxide nanostructures (ONS) obtained by LAO can be used to form profiled nanostructures and masks for surface selective etching [11-12]. Another promising application of titanium ONS is the memristor structures for resistive memory and synaptronics elements.

However, the LAO process has not been sufficiently studied. The analysis of the publications shows that the LAO mechanism consists of the oxygen ions generation and drift to the metal surface, where an oxidation reaction takes place with the oxide formation (Fig. 1). In this case, the application of voltage pulses plays a key role in the LAO process, since the generated electric field leads to the oxygen ions generation due to the electrical dissociation of water molecules in the air and on the substrate surface, and also stimulates the drift of the formed oxygen ions to the reaction region.

The goal of the work is to develop a mathematical model and to calculate an algorithm of titanium ONS formation obtained by LAO.

2. Model

Since the LAO process is associated with the generation and transfer of oxygen ions in the probe-air-oxide-substrate system and subsequent substrate oxidation, it is necessary to calculate the oxygen ion flux at each point of the substrate surface. For this purpose, it is necessary to solve the system of Poisson and continuity equations:

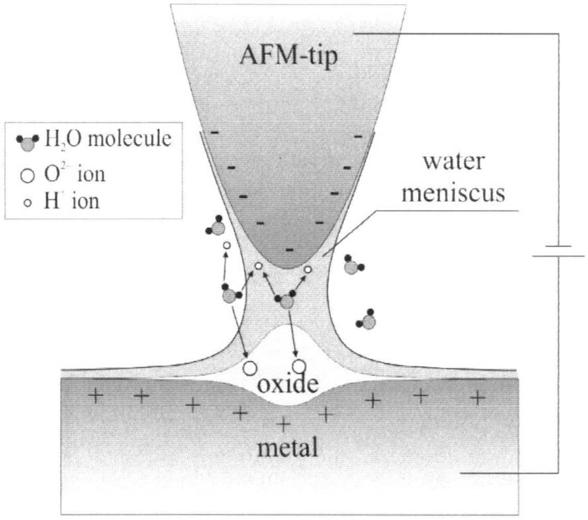

Figure 1. Schematic representation of the oxygen ions generation and drift process in the LAO.

$$\begin{cases} \nabla(\varepsilon\nabla\varphi) = -\dfrac{\rho(N)}{\varepsilon_0}, \\ \nabla(-\mu N\nabla\varphi + D\nabla N) = R(\varphi). \end{cases} \tag{1}$$

where ε is the relative permittivity, φ and N are the distribution of the electric potential and the concentration of oxygen ions in the system, ρ is the volume density of electrical charges, ε_0 is the electric constant, μ and D are the mobility and diffusion coefficient of oxygen ions, and R is the generation rate of oxygen ions in air.

Solution of the equations system (1) will allow calculating the flow of oxidizer ions and the oxide growth rate by formula:

$$v = -\frac{M_o}{\rho_o}(-\mu N\nabla\varphi + D\nabla N), \tag{2}$$

where v is the oxide growth rate, M_o and ρ_o are the molar mass and density of the oxide.

In addition, it was shown in Ref. [17] that the height ratio h_1 and the depth ratio h_2 of the ONS are determined by the expression:

$$h_1 : h_2 = \left(1 - \frac{M_s}{\rho_o}\frac{\rho_s}{M_o}\right) : \frac{M_s}{\rho_o}\frac{\rho_s}{M_o}, \tag{3}$$

where M_s and ρ_s are the molar mass and density of the substrate.

3. Results

Based on the developed model, the titanium film LAO process was numerically simulated using the MATLAB software package. For this purpose, the system geometry containing the AFM-probe, air, oxide, and metal film, for which the equations system (1) solution located was considered. Then, on the basis of equation (2) and the calculated values of the electric potential φ and the oxygen ions concentration N, the oxide growth rate was determined and, based on the relation (3), the oxide height and depth were calculated.

As a result of simulation, it is possible to obtain the titanium ONS height, depth, and diameter dependences from duration of impulses of the applied voltage (Fig. 2), the electric potential and oxygen ions concentration distribution (Fig. 3), and also the profile of the titanium ONS (Fig. 4) at the given values of the applied voltage amplitude and duration, the distance of the probe-substrate, the relative humidity level in the process chamber, and the geometric parameters of the AFM probe.

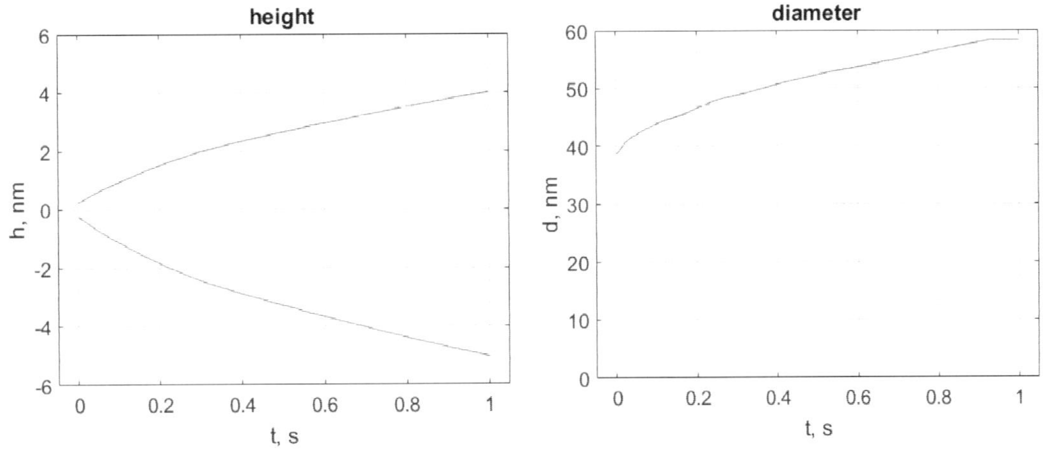

(a) (b)

Figure 2. Dependences of (a) height and (b) diameter at half-height of titanium ONS on the duration of the applied voltage pulses.

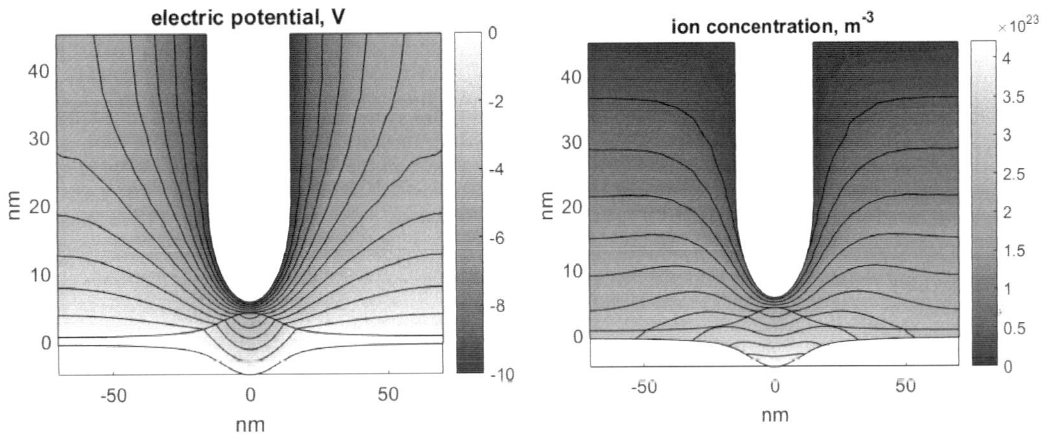

(a) (b)

Figure 3. Spatial distributions of (a) electric potential and (b) oxygen ions concentration.

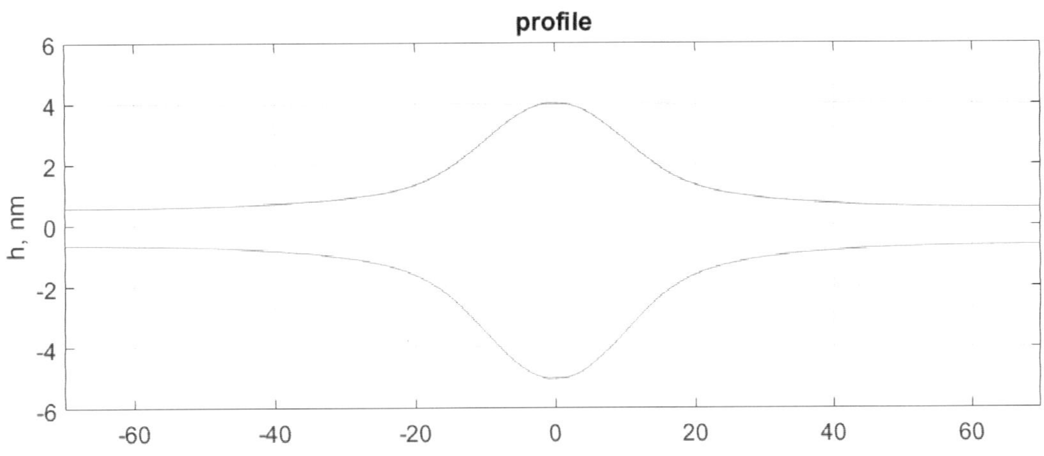

Figure 4. Titanium ONS profile.

4. Conclusion

The paper presents a model for the oxide nanostructures formation by the local anodic oxidation method, based on the electric potential distribution and the oxygen ions distribution calculation in the system. Dependences of height, depth, and diameter of titanium oxide nanostructures from duration of impulses of the applied voltage are obtained, as well as the oxide nanostructures profile for given technological parameters.

The obtained results correlate with the data obtained in Ref. [18] and can be used in the development of technological processes for the fabrication of the RRAM element based on titanium oxide nanostructures.

Acknowledgements

This work was supported by RFBR according to the research project № 18-37-00299, by Grant of the President of the Russian Federation No. MK-2721.2018.8 and the Southern Federal University (grant nos. VnGr-07/2017-26).

References

[1] Ha S D and Ramanathan S 2011 *J. Appl. Phys.* **110** 071101
[2] Pan F, Gao S, Chen C, Song C and Zeng F 2014 *Mat. Sci.&Eng.* R **83** 1-59.
[3] Volkov E, Jityaev I and Kolomiytsev A S 2015 *IOP Conf. Ser.: Mat. Sci.&Eng.* **93** 012031
[4] Avilov V I, Polupanov N V, Tominov R V, Smirnov V A and Ageev O A 2017 *IOP Conf. Ser.: Mat. Sci.&Eng.* **256** 012001
[5] Bersuker G, Gilmer D C, Veksler D, Kirsch P, Vandelli L, Padovani A, Larcher L, McKenna K, Shluger A, Iglesias V, Porti M and Nafria M 2011 *J. Appl. Phys.* **110** 124518
[6] Ageev O A, Blinov Y F, Il'ina M V, Il'in O I, Smirnov V A and Tsukanova O G 2016 *Phys. Solid State* **58** 309-14.
[7] Chang T C, Chang K C , Tsai T M, Chu T J and Simon M S 2016 *Mater. Today* **19** 254-64
[8] Solodovnik M S, Balakirev S V, Eremenko M M, Mikhaylin I A, Avilov V I, Lisitsyn S A and Ageev O A 2017 *IOP J. Phys.: Conf. Ser.* **917** 032037
[9] Popov I, Durišić I and Belić M R 2017 *Nanotechnology* **28** 495706
[10] Walsh M A and Hersam M C 2009 *Ann. Rev. Phys. Chem.* **60** 193-216
[11] Ageev O A, Smirnov V A, Solodovnik M S, Rukomoikin A V and Avilov V I 2012 *Semiconductors* **46** 1616–21
[12] Avilov V I, Ageev O A, Smirnov V A, Solodovnik M S and Tsukanova O G 2015 *Nanotechnologies in Russia* **10** 214-9
[13] Alekseev P A, Borodin B R, Dunaevskii M S, Lebedev S P and Lebedev A A 2018 *Tech. Phys. Lett.* **44** 381-3
[14] Avilov V I, Ageev O A, Konoplev B G, Smirnov V A, Solodovnik M S and Tsukanova O G 2016 *Semiconductors* **50** 601-6
[15] Ageev O A, Klimin V S, Solodovnik M S, Eskov A V and Krasnoborodko S Y 2016 *J. Phys.: Conf. Ser.* **741** 012178
[16] Klimin V S, Tominov R V, Eskov A V, Krasnoborodko S Y and Ageev O A 2017 *J. Phys.: Conf. Ser.* **917** 092005
[17] Avilov V I, Ageev O A, Blinov Y F, Konoplev B G, Polyakov V V, Smirnov V A and Tsukanova O G 2015 *Tech. Phys.* **60** 717-23
[18] Smirnov V A 2015 *Piezoelectrics and Nanomaterials: Fundamentals, Developments and Applications* 85-103

IOP Conf. Series: Materials Science and Engineering **443** (2018) 012014 doi:10.1088/1757-899X/443/1/012014

Microstructure of luminescent $MgAl_2O_4$ nanoceramics

A N Kiryakov[1], A F Zatsepin[1], T V Dyachkova[2], A P Tytunyunnik[2], Yu G Zainulin[2], G Yakovlev[1], V A Pustovarov[1] and D R Bautimirov[1]

[1] Ural Federal University, 620002, Ekaterinburg, Russia
[2] Institute of Solid State Chemistry of the Ural Branch of the Russian Academy of Sciences, 620990, Ekaterinburg, Russia

arseny.kiriakov@urfu.ru

Abstract. Luminescent transparent nanoceramics were obtained by thermobaric treatment (TBT) of magnesium aluminium spinel nanopowder. The morphological features were studied by scanning electron microscopy combined with X-ray powder diffraction. Obtained ceramics do not have any agglomerates and pores larger than 100 nm. Crystallites have a high size uniformity. An increase in the lattice constant of nanoceramics compared to the initial powder is observed. Under the TBT, a decrease in the region of coherent scattering due to elastic deformation of crystallites is found. The absence of cracks, large pores, nanosize grains, and high size uniformity reduce light loss in the material, increasing its transparency. Point defects were characterized by photoluminescence and electron spin resonance (ESR) methods. The glow in the 1.8 eV band is caused by the presence of Ti^{3+} impurity ions. An abnormally wide peak with a maximum at 2.4 eV in the photoluminescence spectrum is recorded. This signal is a superposition of the Mn^{2+} ions emission bands and oxygen vacancies (F and F^+ centres). In the ESR spectrum, signals from impurity ions of iron, titanium, and manganese, as well as an intense signal at g = 2.005 associated with oxygen vacancies in nanoceramics were detected.

1. Introduction

Transparent ceramics with magnesium aluminium spinel (MAS) have promising applications in optoelectronics, photonics, and laser technology [1-3]. A wide light transmission window (0,2-5,5 μm), heat resistance, corrosion and radiation resistance, and high hardness allow such ceramics to compete with aluminium oxide, aluminium nitride, magnesium oxide, and zirconium oxide [4-6].

The transparency of the ceramic material depends on its morphology. First of all, the residual porosity, pore size, and grain size have a strong influence on transparency. According to Refs. [5,6], the main light losses in microceramics are caused by pores formed at the junction of several grain boundaries. The traditional approach of obtaining transparent ceramics including MAS is to reduce the residual porosity. It also implies a significant increase of grains, in order to minimize the reflection of transmitted light at their boundaries. To achieve optimal parameters of porosity and grain size, small amounts of special additives, such as LiF, which increase the mobility of interfaces and grains, are often used. Such microceramics have a crystallite size of 10 μm and higher. However, additives can adversely affect the physicochemical properties of spinel [5,6]. Preparation of the pure transparent ceramics of aluminium magnesium spinel is an important task, since it allows studying the material free from specially introduced impurities.

The pure transparent ceramics are sintered at high pressures and temperatures [7,8]. Interestingly, transparency in this case is achieved by maintaining nanograins with uniform size distribution. The pores in such nanoceramics are replaced due to the fact that the particles of 30-50 nm size are subjected to plastic deformation, filling the entire free volume of the material without agglomerating. On the other hand, the formation of a large number of grain boundaries leads to an increase in the defectiveness of ceramics associated with dangling bonds [9]. The combination of scanning electron microscopy (SEM), photoluminescence (FL), and electron spin resonance (ESR) techniques allows for morphological analysis and determination of the nature of point defects in the material.

The aim of this work is to study the morphological, luminescent, and paramagnetic features of transparent nanoceramics of magnesium aluminium spinel.

2. Materials and methods

Nanopowder of magnesium aluminium spinel is obtained by precipitation from aluminium and magnesium nitrates with the addition of ammonia to control the pH of the system. The suspension was dried overnight at a temperature of 150°C. The gelling was carried out at 1000°C for 1 hour in air.

The resulting nanopowder was pressed into a cylindrical tablet 4×5 mm^2 in size and calcined in a vacuum oven for 3 hours at a temperature of 500°C to remove the sorbed impurities. The tablet was then placed in a standard pre-calibrated high-pressure chamber of the toroid type. The temperature was controlled by the power of the current passing through the graphite heater. To prevent carbon contamination, the inner walls of the graphite heater were insulated with platinum foil. The sample was compressed between the anvils of the press to the required pressure, and then the temperature was raised. After holding at the set temperature for 60 minutes, the sample under pressure was quenched by sudden dropping of temperature and pressure. Transparent nanoceramics of MAS were obtained at P = 4 GPa and T = 600°C.

The purity of synthesized product was checked using X-ray powder diffraction (XRD) patterns collected at room temperature on a STADI-P (Stoe) diffractometer equipped with a linear mini-PSD detector using $CuK\alpha^1$ radiation in the 2θ range 5° to 120° with a step of 0.02°. Polycrystalline silicon (a = 5.43075 (5) Å) was used as external standard. The PDF2 database of powder standards (ICDD, USA, Release 2016) was used to identify possible impurity phases. The unit cell parameters and mass fractions of $MgAl_2O_4$ were refined by Rietveld method using XRD data and the GSAS software [10,11].

The surface structure was studied using SIGMA VP scanning electron microscope (SEM, Carl Zeiss, Germany) and secondary electron detector (in-lens) in high vacuum at a 3 kV accelerating voltage. Registration of point defects in transparent nanoceramics of MAS is performed by photoluminescence measurements. The photoluminescence spectra measured at room temperature on a system equipped with a double prism monochromator type MDR-4, using a 400 W deuterium lamp and a photomultiplier R-6358-10 (Hamamatsu). Excitation spectra were corrected for the wavelength-dependent intensity variation using yellow lumogen with energy-independent quantum yield over the studied spectral range. Emission spectra are presented taking into account the correction for the spectral sensitivity of photomultiplier. An additional method of studying the heterogeneity of the crystal structure is to measure the electron spin resonance (ESR) spectra of the ceramics. The ESR signal was recorded at room temperature on a ELEXSYS 580 spectrometer (Bruker) with resonant frequency of 9.27 GHz. Photoluminescence and electron paramagnetic resonance (EPR) methods are applicable to the study of point defects, because they are sensitive to electrons weakly bound to an atom.

3. Results

3.1. X-ray powder diffraction

According to the results of XRD of the initial powder and transparent nanoceramics of MAS (Fig. 1), the samples have cubic symmetry ($Fd\bar{3}m$ space group). There is a small amount of MgO impurity phase (~3 mass%) both in the initial powder and in nanoceramics. TBT does not change the phase composition of ceramics. The size of the coherent scattering region (CSR) of the initial powder determined for the $MgAl_2O_4$ phase is 38 nm, nanoceramics – 30 nm. Lattice constant of the initial powder a = 8.08344(7) Å, transparent ceramics a = 8.08440(8) Å. The impurity phase of MgO has a spatial group $Fm\bar{3}m$ and a = 4.4124(5) Å for the initial powder and a=4.2126(5) Å for nanoceramics.

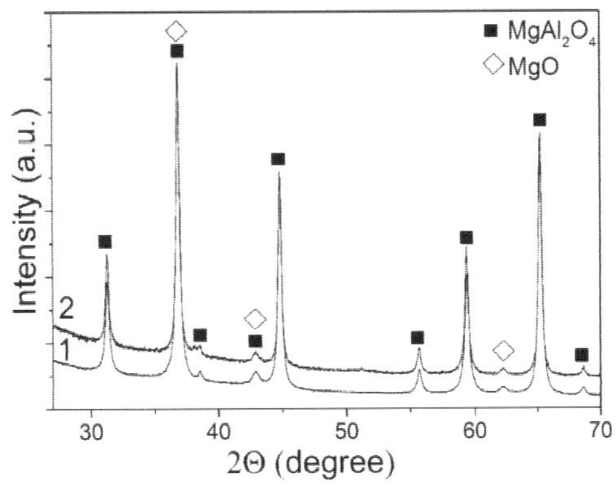

Figure 1. XRD pattern of (1) initial powder and (2) transparent nanoceramics.

3.2. Scanning electron microscopy

SEM images of initial powder (Fig. 2a), as well as nanoceramics (Fig. 2b,c) are presented for different types of magnifications to show the absence of large agglomerates and pores. It is seen from Figure 2a that the initial powder consists of nanograins, agglomerated in large particles. The surface of MAS nanoceramics looks like extended plateaus without large agglomerates. It is important to note that in nanoceramics there are no pores with a diameter of 100 nm or more. Such pores are formed at the boundaries of several grains and are characteristic of microceramics. Another observed feature is the narrow particle size distribution. The distribution of crystallites by size is one of the main parameters that determine the optical and mechanical properties of ceramics. There is also a lack of cracks capable of scattering electromagnetic radiation. Earlier, we showed that they were responsible for transparency [12]. The inset in Figure 2c shows a photograph of the transparent nanoceramics studied.

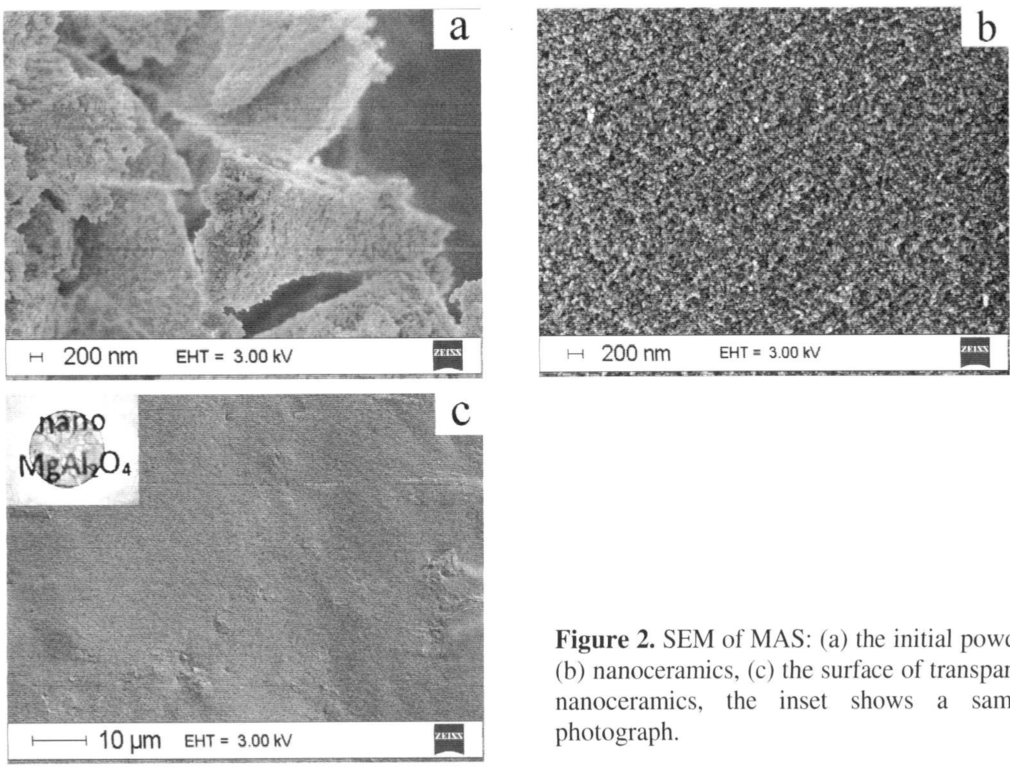

Figure 2. SEM of MAS: (a) the initial powder, (b) nanoceramics, (c) the surface of transparent nanoceramics, the inset shows a sample photograph.

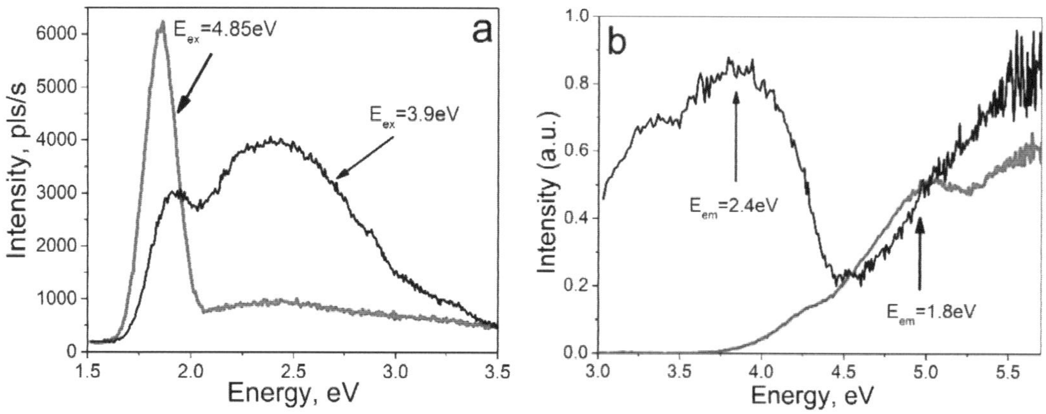

Figure 3. (a) Luminescence and (b) excitation spectra of transparent MAS nanoceramics.

3.3. Photoluminescence

The luminescence and excitation spectra are shown in Figure 3a,b. Two intense bands with maxima at 1.8 and 2.4 eV are observed in the luminescence spectrum. The maximum of excitation band at 1.8 eV is located at 4.85 eV. The 2.4 eV band is most effectively excited at 3.9 eV. It is also worth noting that the luminescent peak at 2.4 eV is greatly broadened. Excitation efficiency of the marked luminescence bands increases when approaching the vacuum UV.

3.4. Electron spin resonance

The ESR spectra are shown in Figure 4a. A wide signal ESR g = 4.39 is observed. In the range of 3000-4000 G there is a series of narrow lines (Fig. 4b). The most intense signals are with g = 2.005 and 1.981. Six narrow, equally spaced lines (A = 90 G) correspond to hyperfine structure.

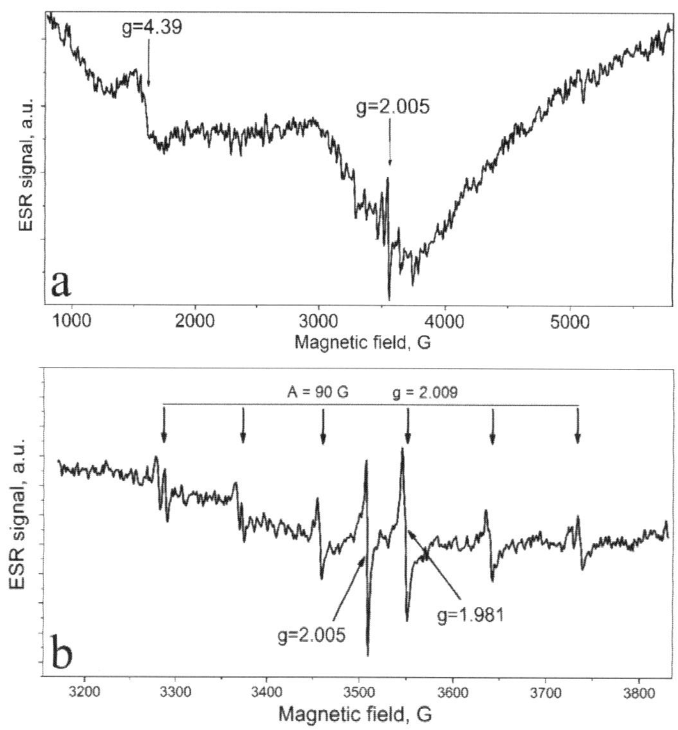

Figure 4. The ESR spectra of nanoceramics MAS.

4. Discussion

Comparing the results of SEM and XRD, we should note the correlation in the recorded particle size (not more than 50 nm). Reduction of CSR in nanoceramics compared to the initial powder is associated with the processes of elastic deformation in the material [7]. The noted growth of the lattice constant for MAS nanoceramics is caused by a large number of substitution defects $[Mg]_{Al}$ and vice versa [13,14].

Thermobaric treatment of the material results in a high degree of compaction, as evidenced by the SEM study. An important characteristic of ceramics is defectiveness. Combining the methods of photoluminescence and EPR allows characterizing the point defects. The luminescence peak at 1.8 eV is due to the presence of trace impurities of octahedral Ti^{3+} in the row materials [15]. The registered excitation band in the UV region is typical for the specified defect. It is worth noting that the maxima positions of the excitation and luminescence bands shift to lower wavelengths region (high energies). The reason for this blue shift is the low grain dimension, which is confirmed by CSR and SEM.

The half-width of 2.4 eV luminescent peak as well as the maximum on the excitation spectrum do not correspond to the literature data. On the one hand, a wide luminescence band with noted maximum can be associated with trace impurities of tetrahedral Mn^{2+} [16]. On the other hand, TBT of $MgAl_2O_4$ could lead to the formation of oxygen vacancies, the luminescence of which in covalent crystals is characterized by significant broadening [17].

ESR spectra make it possible to detect point defects associated with an unpaired electron due to high sensitivity to them. The resonance absorption at g = 4.39 is caused by the trace amount of iron ions in ceramics [18]. According to the characteristic hyperfine structure splitting (hyperfine structure, 6 lines equally spaced from each other), it can be concluded that the nanoceramics contain an uncontrolled impurity of Mn^{2+} ions (g = 2.009). It is known that a narrow signal with g = 2.005 can be associated with oxygen vacancies in MAS [19]. It is worth noting that the 4th line (g = 1.981) of hyperfine structure of manganese ions is much more intense than the others, which may be a consequence of superposition of ESR signals. In the work on the doping of the spinel with titanium ions, the rise of the signal with the specified g-factor is observed [20]. Table 1 contains the main parameters of the registered ESR signals.

Table 1. The main parameters of the ESR signals.

g-factor	4.39	2.009	2.005	1.981
Weight, G	212	8.5	3.8	5.3
Nature	Fe^{3+}	Mn^{2+}	F^+-center	Ti^{3+}

Thus, there is a correlation between photoluminescence and ESR spectra. The luminescence in 1.8 eV is due to impurity titanium ions, the broad luminescence band in 2.4 eV is most likely a combination of the luminescence bands of oxygen vacancies and impurity manganese ions in the tetrahedral Mg^{2+} position. These defects are detected by ESR methods using characteristic lines in the spectrum.

5. Conclusion

The transparent nanoceramics of the aluminium magnesium spinel were obtained by thermobaric treatment in a toroidal chamber. It was established that the main phase of the nanoceramics obtained was $MgAl_2O_4$ with an insignificant fraction of MgO (~3%). The structural features of the resulting transparent nanoceramics are: absence of cracks, narrow grain size distribution, cationic inversion with formation $[Mg]_{Al}$ defects, and vice versa. An important feature of the obtained ceramics is the absence of large pores, which reduce the transparency.

The presence of Ti^{3+} and Mn^{2+} impurity centres was determined by the characteristic bands using luminescence and ESR methods. A blue shift of the emission band of Ti^{3+} ions is associated with the size effect. An abnormally wide 2.4 eV emission band is presumably associated with the luminescence of Mn^{2+} combined with luminescence of oxygen vacancies (F and F^+ centres).

Acknowledgments
The work was done as a part of the government task (№3.1485.2017/4.6) of the Ministry of Education and Science of the Russian Federation and was carried out in accordance with the scientific and research plans and state assignment of the Institute of Solid State Chemistry of the Ural Branch of the Russian Academy of Sciences (AAAA-A16-116122810212-5).

References
[1] Sai Q, Xia C, Rao H, Xu X, Zhou G and Xu P 2011 *J. Luminescence* **131** 2359-64
[2] Aizawa H, Ohishi N, Ogawa S, Watanabe E, Katsumata T, Komuro S, Morikawa T and Toba E 2002 *Rev. Sci. Instr.* **73** 3089-92
[3] Sanghera J, Bayya S, Villalobos G, Kim W, Frantz J, Shaw B, Sadowski B, Miklos R, Baker C, Hunt M, Aggarwal I, Kung F, Reicher D, Peplinski S, Ogloza Al, Langston P, Lamar C, Varmette P, Dubinskiy M and DeSandre L 2011 *Opt. Mat.* **33** 511-8
[4] Clinard Jr F W 1987 *Cer. Int.* **13** 69-75
[5] Rubat du Merac M, Kleebe H J, Müller M M and Reimanis I E 2013 *J. Am. Cer. Soc.* **96** 3341-65
[6] Bergmann C P and Stumpf A 2017 *Topics in mining, metallurgy and materials engineering* Porto Alegre: Springer.
[7] Zou Y, He D, Wei X, Yu R, Lu T, Chang X, Wang Sh and Lei L 2010 *Mat. Chem. Phys.* **123** 529
[8] Lu T C, Chang X H, Qi J Q, Luo X J, Wei Q M, Zhu S, Sun K, Lian J and Wang L M 2006 *Appl. Phys. Lett.* **88** 213120
[9] Vahid B R and Haghighi M 2017 *Energy Convers. Manag.* **134** 290-300
[10] Toby B H 2001 *J. Appl. Cryst.* **34** 210-3
[11] Larson A C and Von Dreele R B 1994 *Report LAUR* 86-748
[12] Kiryakov A N, Zatsepin A F, Slesarev A I, Dyachkova T V, Zainulin Yu G, Mashkovtsev M, Yakovlev G and Vagapov A Sh 2018 *AIP:Conf. Proc.* **2015** 020039
[13] Méducin F, Redfern S A, Le Godec Y, Stone H J, Tucker M G, Dove M T and Marshall W G 2004 *Am. Mineralogist* **89** 981-6
[14] Ball J A, Pirzada M, Grimes R W, Zacate M O, Price D W and Uberuaga B P 2005 *J. Phys.: Cond. Matt.* **17** 7621-31
[15] Sato T, Shirai M, Tanaka K, Kawabe Y and Hanamura E 2005 *J. Luminescence* **114** 155-61
[16] Singh V, Chakradhar R P S, Rao J L and Kim D K 2007 *J. Solid State Chem.* **180** 2067-74
[17] Sawai S and Uchino T 2012 *J. Appl. Phys.* **112** 103523
[18] Cannas C, Gatteschi D, Musinu A, Piccaluga G and Sangregorio C 1998 *J. Phys. Chem. B* **102** 7721-6
[19] Ibarra A, Bravo D, Lopez F J and Garner F A 2005 *J. Nucl. Mater.* **336** 156-62
[20] Lombard P, Boizot B, Ollier N, Jouini A and Yoshikawa A 2009 *J. Cryst. Growth* **311** 899-903

IOP Conf. Series: Materials Science and Engineering **443** (2018) 012015 doi:10.1088/1757-899X/443/1/012015

Fabrication of probes for scanning near-field optical microscopy using focused ion beam

A S Kolomiytsev, N A Shandyba, I V Panchenko and S A Lisitsyn

Southern Federal University, Institute of Nanotechnologies, Electronics, and Equipment Engineering, Taganrog, 347922, Russia

askolomiytsev@sfedu.ru

Abstract. The results of an experimental study of the fabrication of probe tips for scanning near-field optical microscopy (SNOM) using the focused ion beam (FIB) milling and ion beam induced deposition are presented. Methods of the FIB local milling and FIB-induced deposition of tungsten and carbon onto the tip of an atomic force microscope (AFM) probe are studied. In this work, a technique for the formation of AFM probes based on the use of a combination of ion-beam etching and ion-induced carbon deposition was developed. Based on the developed technique, several aperture probes for SNOM with an aperture diameter of 50, 150, and 200 nm, an apex height of about 20 nm, and a cone angle of about 35° were fabricated. The paper presents that the formation of SNOM probes by FIB-induced deposition allows creating a high efficiency tool for nanodiagnostics. The obtained results can be used to develop technological processes for the fabrication of specialized AFM probes and procedures for the express monitoring of technological process parameters in the manufacturing of the elements for micro- and nanoelectronics.

1. Introduction

Modern methods of producing structures and elements of nanoelectronics, sensorics, and nanomechanics are impossible without the high-resolution microscopy methods used to control the production process at all the stages. In nanotechnology, microscopy often appears not only as a method for assessing the topography of a surface, but also as a method for determining local surface properties. The resolving power of traditional methods of optical microscopy is limited by the diffraction limit of the light used, so a near field microscopy method is used to obtain images of nanoscale structures [1,2]. The main feature of the method is that the light source and the sample under study are smaller than the wavelength of light.

At the present stage of nanotechnology development, one of the most promising methods of surface diagnostics is scanning probe microscopy (SPM) [3-5]. The use of SPM methods makes it possible to study the local geometric, electrical, and mechanical properties of the substrate surface, as well as to form nanoscale structures on the surface of solids [1]. The perspective direction of microscopy at nanoscale is scanning near-field optical microscopy (SNOM), which significantly allows increasing the optical microscopy resolution and providing reliable control of the distance between the tip and the sample. SNOM is a microscopy technique for nanostructure investigation that breaks the far field resolution limit by exploiting the properties of evanescent waves. In SNOM (Fig. 1a), the excitation laser light is focused through an aperture with a diameter smaller than the excitation wavelength, resulting in an evanescent field (or near-field) on the far side of the aperture [2]. When the sample is scanned at a small distance below the aperture, the optical resolution of transmitted or reflected light is limited only by the diameter of the aperture. One of the main problems in SNOM technology is the creation of probes with a controlled aperture size. The use of traditional technological processes does not allow varying the aperture diameter and creating probes for SNOM of the required shape.

The application of new methods of local nanostructuring makes it possible to form probes of various geometric shapes for specialized methods of SPM and SNOM [3]. Using local methods of nanoscale structuring, it is possible to create SNOM aperture probes with parameters that differ significantly from standard ones, which will allow minimizing artifacts, reducing costs, and increasing the accuracy and resolution of microscopy. One of the promising methods for forming the SNOM probe tips with high accuracy and resolving power is the use of focused ion beams (FIB) [6-14]. The FIB method allows performing technological operations of local ion beam etching and ion-induced deposition of materials from the gas phase without the needs of resists, masks, and chemical etchants. The essence of the FIB method lies in the local ion-beam sputtering of the material by a flow of gallium ions focused to a diameter of about 7 nm. Combining the FIB and highly selective gas chemistry, allows realizing operations of local ion-induced deposition of materials (carbon, tungsten, platinum, etc.). The combination of precision ion beam milling and ion-induced deposition of materials makes it possible to form the SPM probe tips with parameters that are unattainable when using traditional manufacturing processes.

A wide range of materials deposited by this method allows using FIB in formation of nanoscale structures probes for nanodiagnostics. The aim of this work is to create advanced apertured probes for SNOM using the FIB method and their experimental study [7].

2. Experimental

Traditionally, FIB technology is used to etch the tip of the probe through, thus forming an aperture probe. However, this approach has several disadvantages: the small diameter of the aperture inlet, the complexity of aligning the aperture with the axis of the probe tip, and the negative effect of the redeposited material. To eliminate these drawbacks, it is proposed to use the ion-induced deposition method to form a hollow tip of a probe with a controlled aperture diameter.

Experimental studies were carried out using a FEI Company DualBeam system Nova NanoLab 600, combining a Ga+ FIB and a field emission scanning electron microscope (SEM) and the Ntegra Vita scanning probe microscope (NT-MDT, Russia). The standard atomic force microscope (AFM) cantilevers NSG-10 (NT-MDT, Russia) with broken probes due to operation were used as a substrate.

The scheme of the process of forming a cantilever with an aperture is shown in Figure 1b. At the initial stage of the study, part of the cantilever beam containing the probe was removed. This operation was carried out by ion-beam etching with current of 20 nA.

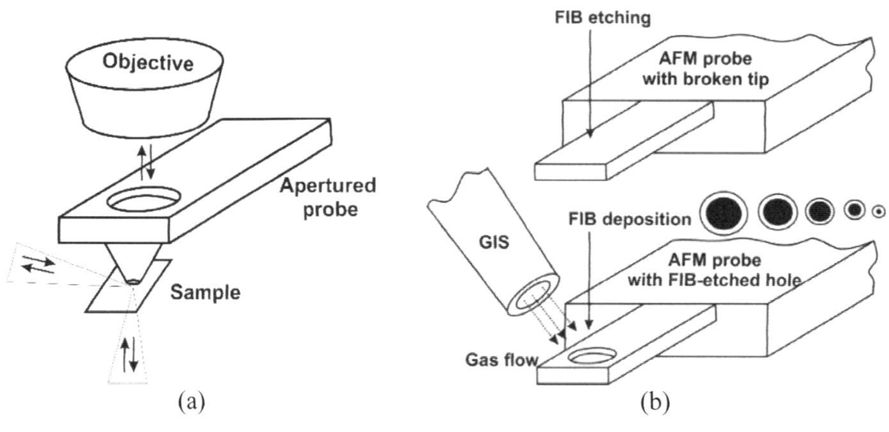

Figure 1. Principle of SNOM with an aperture AFM probe (a) and the process of formation of the SNOM probe tip.

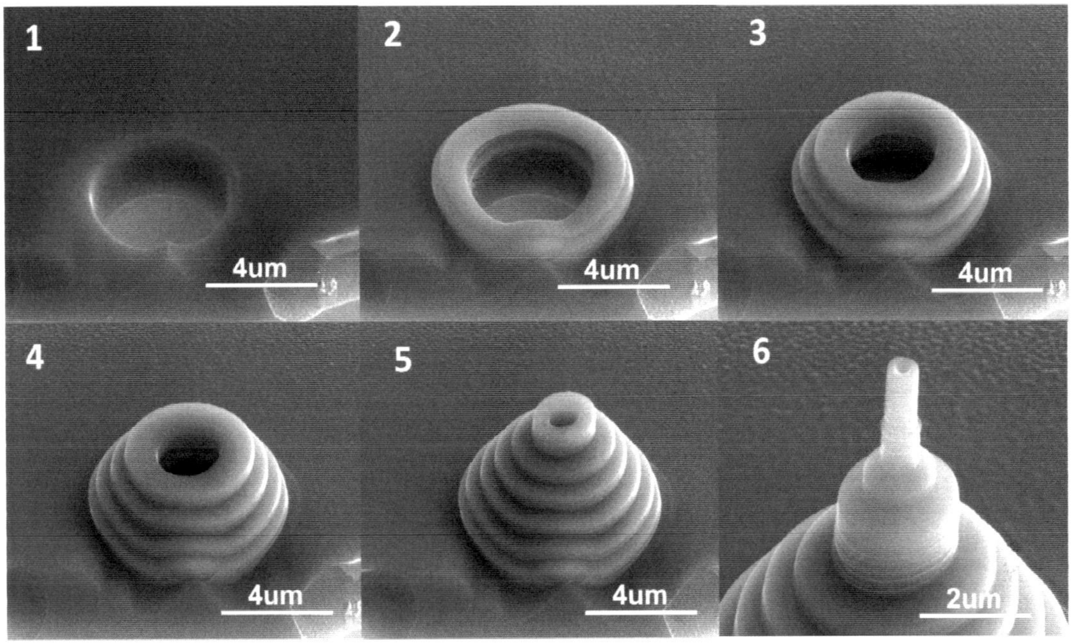

Figure 2. SEM images of 6 stages of forming the tip of an aperture cantilever.

At the next stage of work, a through hole with a diameter of 5 to 10 μm was formed in the remaining part of the beam using FIB etching at maximum current. The resulting hole of 5-10 μm in diameter is the input for optical radiation in the SNOM method. Figure 2 shows SEM images of 6 stages of forming the tip of an aperture cantilever.

After forming the hole through the beam, gas C_6H_{14} was supplied to the area of the formed hole through the gas injector, which, when interacting with the gallium ion beam with an energy of 30 keV, decomposed into volatile components (which were immediately removed by the vacuum system of the microscope) and solid carbon, which was deposited on the surface. The trajectory of the ion beam was determined by digital patterns, so that the deposition of carbon occurred in an orderly manner from the base of the cone (about 5.5 μm in diameter) to its tip. The result was a conical tip formed by the edge height of about 5-6 μm (Fig. 3a).

At the final stage of the work, at the top of the cone, a hole of the required diameter is formed. For this purpose, a focused ion beam is used at ion current values from 1 to 25 pA at an accelerating voltage of 30 keV.

3. Results and discussion

The main advantage of the proposed technology is the ability to form probe tips of different shapes and sizes. The probe shown in Figure 3a has a tip radius of about 200 nm, which also allows using it to obtain AFM images.

The presented technique allows restoring the functionality of worn-out and broken AFM probes and improving their properties and parameters. Restoring the functionality of broken probes by FIB is cost-effective, because the standard probes are not maintainable.

The developed technique for forming probes can be quite simply automated. For example, in the software of the Nova Nanolab 600 system, you can repeat the sequence of actions specified by the operator with the help of simple programs written in a high-level language. Process control is also performed automatically using SEM. Thus, the time taken to produce 1 aperture probe is about 15 minutes.

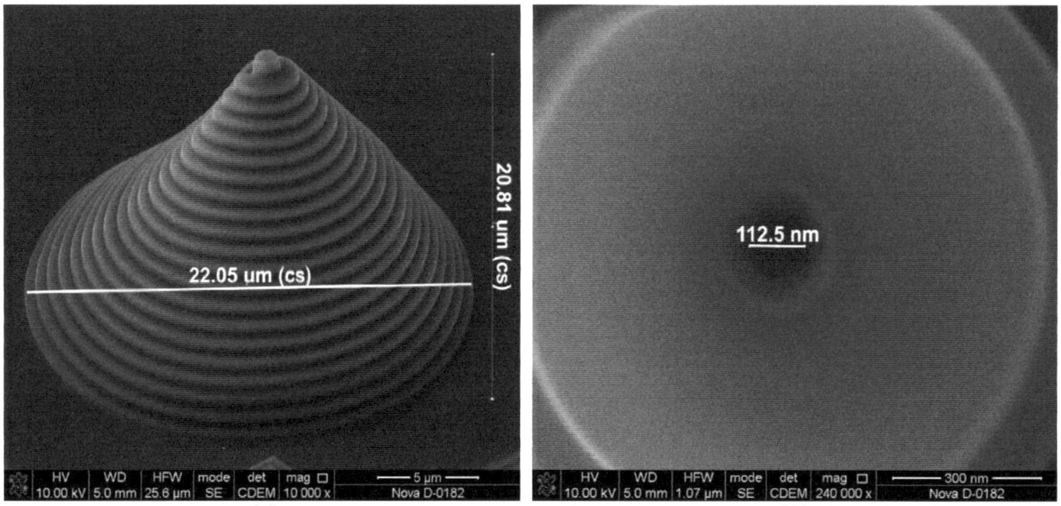

(a) (b)

Figure 3. (a) SEM images of the conical probe of aperture cantilever fabricated be FIB-induced deposition. (b) The diameter of the aperture is 112 nm.

The fabricated probes have been tested and found to have a longer service life as an AFM measurement tool. In addition, the ability to vary the diameter of the aperture makes it possible to form optimized probes for the desired wavelengths. The ability to form aperture SNOM probes based on the broken cantilevers makes it possible to significantly improve the economic efficiency of the proposed technology.

4. Conclusion

The experimental study of methods used to modify probes for AFM by the deposition of materials onto the tip of a worn probe using FIB has been carried out. Application of the technology of ion-induced deposition makes it possible to form probe tips in a wide range of shapes and sizes for different AFM applications. We have illustrated the procedure, based on FIB milling and deposition, to create probes with specific tip shape, starting from commercial ones. It is shown that the FIB method provides the formation of the aperture probes for SNOM method with the aperture diameter of about 112 nm and tip height 20 μm. Unlike many commercial probes for SNOM, manufactured probes can also be used to obtain images of the surface of solids in AFM mode. It is shown that the use of combination of FIB induced etching and deposition allows forming probes with an aperture diameter less than 50 nm.

The results obtained in the study can be used in the development of technological processes for the fabrication and modification of special aperture probes for SNOM, and in the development of procedures for the express monitoring of parameters of the technological process for manufacturing elements for micro- and nanoelectronics and micro- and nanosystems engineering.

Acknowledgments

This work was supported by the Russian Science Foundation Grant No. 18-79-00175. The results were obtained using the equipment of the Scientific-Educational Center and the Center for Shared Use "Nanotechnologies" of the Southern Federal University (Taganrog).

References

[1] Bhushan B *2010 Springer Handbook of Nanotechnology (3rd edn)* (New York: Springer) 1964
[2] Yao N and Wang Z 2005 *Handbook of Microscopy for Nanotechnology* (Kluwer Acad. Publ.: New York) 743

[3] Chang W S, Bauerdick S and Jeong M S 2008 *Ultramicroscopy* **108** 1070-5
[4] Avilov V I, Polupanov N V, Tominov R V, Smirnov V A and Ageev O A 2017 *IOP Conf. Ser.: Mat. Sci. Eng.* **256** 012001
[5] Smirnov V A, Tominov R V, Alyab'eva N I, Il'ina M V, Polyakova V V, Bykov Al V and Ageev O A 2018 *Tech. Phys.* **63** 1236-41
[6] Savenko A, Yildiz I and Petersen D H 2013 *Nanotechnology* **24** 465701
[7] Ageev O A, Alekseev A M, Vnukova A V, Gromov A L, Kolomiytsev A S, Konoplev B G and Lisitsyn S A 2014 *Nanotechnologies in Russia* **9** 26-30
[8] Giannuzzi L A and Stevie F A 2004 *Introduction to Focused Ion Beams: Instrumentation, Theory, Techniques and Practice* (New York: Springer) 357
[9] Lisitsyn S A, Kolomiytsev A S, Ilyin O I, Ilyina M V, Konoplev B G, Bykov A V and Ageev O A 2017 *Russ. Microelectron.* **46** 1-6
[10] Soliman M, Ding Y and Tetard L 2017 *J. Phys.: Condens. Matter* **29** 173001
[11] Wu X-Y, Sun L, Tan Q-F and Wang J 2015 *Chinese Phys. B* **24** 054204
[12] Ageev O A, Konoplev B G and Smirnov V A 2012 *Russ. Microelectron.* **41** 41
[13] Ageev O A, Bykov Al V, Kolomiitsev A S, Konoplev B G, Rubashkina M V, Smirnov V A and Tsukanova O G 2015 *Semiconductors* **49** 1743-8
[14] Ageev O A, Kolomiytsev A S and Bykov A V 2015 *Microelectron. Reliab.* **55** 2131-4

IOP Conf. Series: Materials Science and Engineering **443** (2018) 012016 doi:10.1088/1757-899X/443/1/012016

Change in the surface density of immobilized enzyme molecules induced by photoelectron processes in a silicon substrate

A V Kozlowski, E D Kiseleva, A A Maslennikova and S V Stetsyura

Department of Nano- and Biomedical Technologies, Saratov State University, Saratov, 410012, Russia

kozlowsky@bk.ru

Abstract. We report on a change in the number of adsorbed glucose oxidase molecules depending on both salt concentration and illumination of a silicon substrate during glucose oxidase adsorption on its surface. Glucose oxidase molecules were adsorbed from its aqueous solution onto a silicon substrate either with a polyethyleneimine layer or without it. Based on obtained atomic force microscopy (AFM) data, it was found that addition of salt led to a more prominent effect of photo-assisted adsorption. In addition, illumination led to a monotonic change in adsorbed enzyme coverage (either decrease or increase depending on silicon substrate conductivity type), while variation of salt concentration resulted in the nonmonotonic change. The obtained results can be used for fabrication of biosensors with easily controllable parameters such as sensitivity, lower limit of analytic detection, etc.

1. Introduction

Enzyme adsorption on various surfaces occurs not only in fundamental biological processes, but also in many biotechnological applications, such as biosensors, protein separation, drug delivery, biomaterials, etc. Therefore, enzyme adsorption has attracted the attention of many researchers and has been widely studied by both theoretical [1] and experimental [2] methods to understand, predict, and control it on various surfaces including semiconductors.

The key step for developing biosensor is to maintain the bioactivity of enzyme after its immobilization on a solid surface. When enzyme is adsorbed with the favourable orientation for recognition of the enzymatic reaction by the transducer, the biocatalytic activity of the enzyme is provided. Physical adsorption and chemical attachment are two common methods to immobilize enzymes. Chemical attachment may provide a more stable immobilization of proteins but, frequently, it is accompanied by the loss of their bioactivity. Physical adsorption is a simple method to control adsorbed amount, orientation, and conformation of enzyme molecules. Previous experimental data of glucose oxidase (GOx) adsorption and molecular simulation results of other proteins [1,2] have shown that electrostatic interactions play an important role in controlling the adsorption. Variation of ionic strength and pH of solution is well-known method to adjust the electrostatic interactions between GOx and solid surface. This method is used in Langmuir-Blodgett technique as well as in layer-by-layer adsorption of charged polyelectrolyte molecules onto charged planar or curved surfaces. Atomic force microscopy (AFM) is an effective method for monitoring the surface of hybrid and biosensor structures based on solid substrate and organic components (such as enzyme, DNA, bacteria, etc.).

Semiconductors are widely used as a substrate for enzyme adsorption. Semiconductors are promising materials for biosensor technologies as transducer. Additionally, semiconductors allow remote changing surface charge density (SCD) of a substrate affecting the electrostatic interaction between enzyme molecules and the substrate. Obviously, the remote change of SCD of the substrate can be achieved by either a field effect using external electric field or a super band-gap illumination of the semiconductor substrate. Our previous results demonstrated the effect of photo-assisted adsorption on the processes of immobilization of GOx molecules [3] and branched polyethylenimine (PEI) onto a silicon surface [4].

The mentioned approaches are not state-of-the-art, while the effect of their combinations is not investigated properly. Therefore, in this work, AFM was used to study the GOx adsorption onto differently charged surfaces at different salt concentration and/or SCD through photo-assisted adsorption, with the aim to improve the efficiency of enzyme immobilization method.

2. Experimental

2.1. Materials and methods

Experiments were performed with single-crystal Si wafers of n-type ($\rho \cong 4$ Ωcm) and p-type ($\rho \cong 8$ Ωcm). Initially, the wafers were boiled in a peroxide-ammonia solution and rinsed in deionized water. This treatment led to "reconstruction" of a native oxide layer to SiO_2. It can be noted that isoelectric point (IEP) of the SiO_2 surface is ca. pH=2-3 [5], above this pH value the surface is negatively charged. Afterwards, wafers were cut into substrates of 8×8 mm^2.

GOx molecules from *Aspergillus niger* were used as enzyme molecules. In a wide pH range of the solution, the GOx molecules have an effective negative charge and the size of 6.0×5.2×7.7 nm^3 [6]. A cationic polyelectrolyte polyethylenimine (PEI) with a molecular weight of 25 kDa was used to increase the adsorption of negatively charged GOx onto the silicon substrates. The PEI molecules were adsorbed onto the silicon substrates from the 1 mg/ml aqueous solution during 10 min followed by rinsing in deionized water ($\rho \cong 18.2$ Ω cm) during 10 min and drying in nitrogen flow.

The photo-assisted layer-by-layer adsorption technique suggested in Ref. [7] was used to adsorb GOx from the 0.5 mg/ml aqueous solution onto bare silicon substrates and ones covered with PEI. Additionally, NaCl concentration was varied from 0 to 0.1 M.

2.2. AFM measurements

The topography of the surface was measured using AFM by NTEGRA Spectra (NT-MDT Spectrum Instruments, Russia) in tapping mode with scan frequencies of 0.2-0.5 Hz using HA_NC/W_2C cantilevers of ETALON series with resonance frequency of 140±10% kHz. The Gwyddion software was used for statistical analysis of AFM data. The measurements have not been corrected for the convolution with the tip shape. Therefore, the lateral size of enzyme molecules observed by AFM does not correspond to the real size of the enzyme molecule due to the effect of the instrument on the object resulting in the artefacts appearance ("profile broadening" effect due to the tip-sample convolution). According to Ref. [8], AFM-images of molecules of the GOx with four different tip radii were simulated. As the tip radius increases, the observed lateral dimensions exceed the real size of the molecule significantly and it is impossible to resolve either the structural details or the individual GOx monomers. However, using results from Ref. [8], we can recognize individual molecules of GOx on the scan.

3. Results and discussion

3.1. The effect of salt on the GOx adsorption

Figure 1 shows AFM-images of adsorbed GOx molecules at different NaCl concentrations. According to the Figure 1, adsorption of GOx molecules onto the surface of Si/SiO_2 structure increases as NaCl concentration grows from 0 to 0.1 M. The GOx adsorption efficiency can be estimated using a mean surface coverage calculated as the percentage of area covered with GOx molecules from the total substrate area. Table 1 displays the mean surface coverage (±standard deviation) versus salt concentration. It is obvious that changes of NaCl concentration from 0 to 0.1 M result in corresponding growth of mean surface coverage from 0.37 to 12.63 %. However, according to Ref. [9], at high salt concentrations (e.g., 0.5 M), the number of adsorbed enzyme molecules can decrease.

The results correlate with the double layer theory [10]. Growth of salt concentration in enzyme solution results in screening of electrostatic repulsion between the GOx molecules and the Si surface, which are both negatively charged. Therefore, it enhances GOx adsorption. With a further increase of salt concentration, a decrease in adsorption can occur due to the screening of intermolecular repulsion, which results in protein aggregate formation in solution instead of their adsorption onto a substrate. Since aggregates and the SiO_2 surface are both hydrophilic, the aggregates do not remain on the surface after rinsing, which results in a decrease of the surface coverage.

Additionally, we observed an increase of adsorbed GOx molecules from salt-free solution onto surface of a Si/SiO_2/PEI structure compared to Si/SiO_2 structure.

At low salt concentration in GOx solution, a number of adsorbed molecules reduces sufficiently in comparison with a salt-free solution. It is similar to the result of studying the adsorption of positively charged lysozyme on the negatively charged surface of SiO_2 [11]. High salt concentration in GOx solution leads to aggregates formation, as described above. However, the aggregates do not disappear after rinsing due to strong electrostatic interaction with a positively charged PEI layer (Fig. 1f).

Table 1. The relative change in surface roughness (ΔSa) and average height of irregularities (ΔH), as well as the mean surface coverage (±standard deviation) after GOx adsorption from a salt-free solution and with NaCl.

Concentration of NaCl, M	p-Si/SiO_2/GOx			p-Si/SiO_2/PEI/GOx		
	ΔSa, %	ΔH, %	Surface coverage, %	ΔSa, %	ΔH, %	Surface coverage, %
0	62.13	33.78	0.37±0.03	69.64	62.08	9.32±1.12
0.01	72.75	53.48	4.41±0.45	68.52	77.10	2.17±0.24
0.1	86.88	80.53	12.63±1.38	78.14	69.15	18.86±1.65

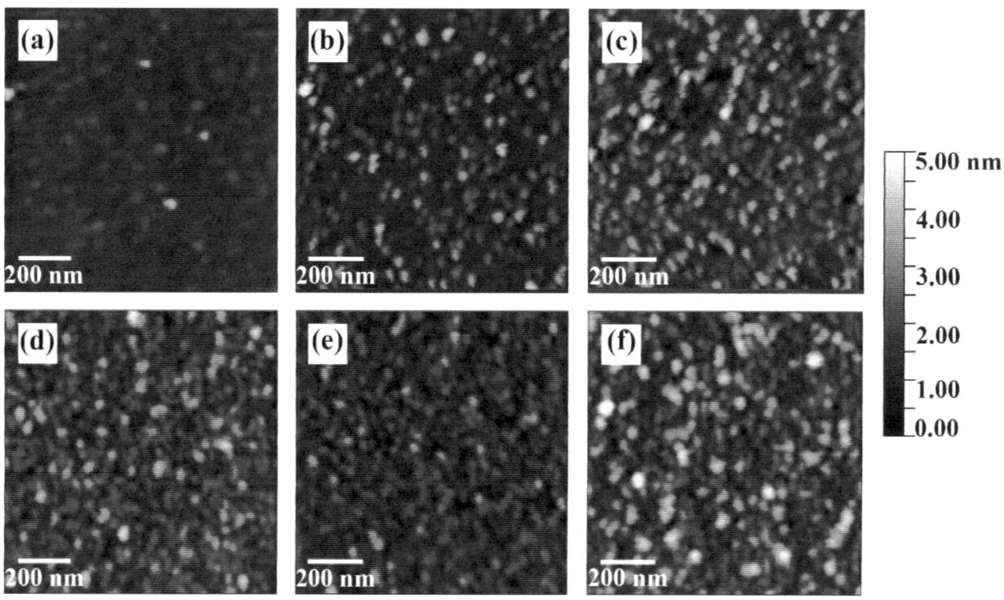

Figure 1. AFM-image: (a-c) Si/SiO_2/GOx and (d-f) Si/SiO_2/PEI/GOx structures. GOx were adsorbed from solution with NaCl concentration: (a, d) 0 M, (b, e) 0.01 M, and (c, f) 0.1 M.

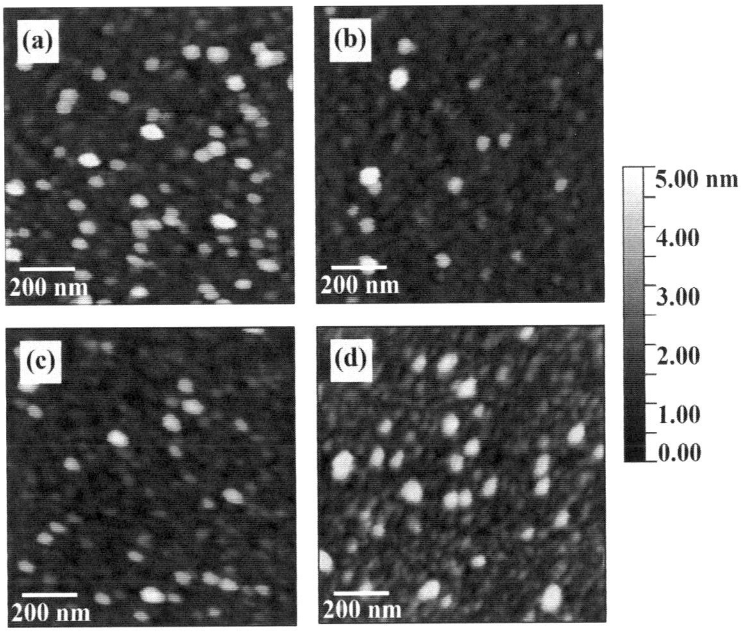

Figure 2. AFM-image: (a-b) p-Si/SiO_2/PEI/GOx and (c-d) n-Si/SiO_2/PEI/GOx structures. GOx were adsorbed from solution with NaCl concentration: (a, c) 0.01 M in the dark and (b, d) under illumination during GOx adsorption.

3.2. Illumination and type conductivity of substrate effect

Analysis of AFM-images of adsorbed GOx molecules on p-Si/PEI and n-Si/PEI substrates has shown that amount of adsorbed molecules is larger on the p-Si than on the n-Si type (Fig. 2a,c). However, illumination of the silicon substrate during the adsorption process alters the adsorption significantly. It decreases the number of adsorbed GOx molecules for p-Si/PEI and increases it for n-Si/PEI.

The influence of illumination on the adsorption of GOx molecules can be explained as follows. At equilibrium conditions (i.e., without illumination and external electric fields), the charge of the space-charge region (SCR) of semiconductor substrate is equal in magnitude and opposite in sign to the charge of surface electronic states (SES). The sign of SES charge usually corresponds to main charge carriers in semiconductor. The deposition of a cationic polyelectrolyte molecules (e.g., PEI) changes the structure surface charge to positive. It increases the band bending of p-Si structures and decreases that of n-Si ones. Thus, the total charge of SES, traps in SiO_2, and PEI in structures based on p-Si is greater than zero, while the one of n-Si is close to zero. When a Si/SiO_2/PEI structure is immersed into an aqueous solution of negatively charged enzyme molecules (e.g., GOx), the enzyme molecules are adsorbed onto the PEI surface predominately due to electrostatic interactions. This factor explains a significant difference of mean surface coverage between GOx adsorption on p-Si and n-Si.

Illumination of Si/SiO_2/PEI structures leads to rectification of their energy bands due to the generation of electron-hole pairs and their separation by SCR field. As a result, the charge of SES at the Si/SiO_2 interface decreases. In n-Si/SiO_2/PEI structures, the SES charge may become neutral or even positive (at high illuminance level). It results in a growth of the effective positive charge near the n-Si/SiO_2/PEI structure surface, an enhancement of the electrostatic interactions with GOx molecules, and, as a result, an increase in the mean surface coverage as compared to the case of adsorption in the dark. SCR field facilitates the drift of photogenerated electrons toward the p-Si/SiO_2 interface decreasing the effective positive charge at the PEI/GOx interface and the mean surface coverage as compared to adsorption in the dark. Table 2 displays the mean surface coverage (±standard deviation) versus salt concentration in case of adsorption in the dark and under illumination.

Table 2. The mean surface coverage (±standard deviation) values and their relative change after photo-assistant GOx adsorption from a salt-free solution and with NaCl.

Concentration of NaCl, M	p-Si/SiO_2/PEI			n-Si/SiO_2/PEI		
	Without illumination	Under illumination	Δ, %	Without illumination	Under illumination	Δ, %
0	11.03±0.99	5.86±0.45	-46,9	2.32±0.30	4.23±0.38	+82.3
0.01	10.59±1.06	2.96±0.34	-72.1	1.99±0.18	4.45±0.42	+123.6

Obviously, addition of salt increases the effect of photo-assisted adsorption. At low salt concentrations, the required surface charge density of the substrate for the adsorption of the enzyme increases [12]. Illumination allows changing the surface charge density of the substrate, thus making possible to control adsorption when the ionic strength changes. Moreover, illumination gives a monotonic change in adsorption (decrease or increase, depending on the type of silicon conductivity), while a change in the salt concentration or pH leads to a nonmonotonic one.

4. Conclusion

In the present study, we have investigated the effects of salt concentration, illumination, and silicon substrate type on the GOx adsorption onto a planar Si substrate. We have shown that both salt concentration and illumination during adsorption affect the enzyme-semiconductor surface interaction. It should be noted that the former leads to a non-monotonous change in protein adsorption. When signs of effective charges of the substrate and molecules of the enzyme are coincident, the changes in salt concentration result in either an increase of protein adsorption (at salt concentration of about 0.1 M) or a reduce due to the formation of massive hydrophilic aggregates (at salt concentration of about 0.5 M). When signs of effective charges of the substrate and molecules of the enzyme are different, the enzyme adsorption decreases at low salt concentration, but the massive aggregates are not removed from the surface after rinsing at high salt concentration.

Depending on the silicon substrate conductivity type, the illumination during adsorption leads either to an increase (for n-Si) or to a decrease (for p-Si) in the number of adsorbed molecules. Moreover, the effect from illumination is enhanced by the addition of salt into the enzyme solution.

As a result of this study, we state that a photo-assisted adsorption on the Si substrate allows controlling over the density of adsorbed enzyme molecules and, as a consequence, different characteristics of biosensor structures (sensitivity, lower limit of analytic detection, etc.). We assume that the effect of photo-assisted adsorption in the dynamics is monotonically decayed. Additionally, we assume that photo-assisted adsorption affects the orientation of the adsorbed enzyme molecule, which also affects the parameters of the biosensor. Further experiments using the scanning probe microscopy methods such as AFM and Kelvin-probe force microscopy might provide additional data to support our observations and hypothesis.

Acknowledgments
The work was supported by the Russian Foundation for Basic Research, projects no. 16-08-00524.

References

[1] Ratner B D, Castner D G, Horbett T A, Lenk T J, Lewis K B and Rapoza R J 1990 *J. Vac. Sci. Technol. A* **8** 2306-17
[2] Xie Y, Li Z and Zhou J 2018 *Phys. Chem. Chem. Phys.* **20** 14587-96
[3] Stetsyura S V and Kozlowski A V 2017 *Tech. Phys. Lett.* **43** 285-8
[4] Malyar I V, Santer S and Stetsyura S V 2013 *Tech. Phys. Lett.* **39** 656-9
[5] Cloarec J-P, Chevalier C, Genest J, Beauvais J, Chamas H, Chevolot Y, Baron T and Souif A 2016 *Nanotechnology* **27** 295602

[6] Hecht H J, Kalisz H M, Hendle J, Schmid R D and Schomburg D 1993 *J. Mol. Biol.* **229** 153-72
[7] Decher G 1997 *Science* **277** 1232
[8] Makky A et al. 2013 *J. Mol. Recognit.* **26** 521-31
[9] Tsapikouni T S and Missirlis Y F 2007 *Colloids Surf. B Biointerf.* **57** 89-96
[10] Bockris J O'M, Devanathan M A V and Muller K 1963 *Proc. R. Soc. Lond. A* **274** 55-79
[11] Su T J, Lu J R, Thomas R K, Cui Z F and Penfold J 1998 *J. Colloid Interf. Sci.* **203** 419-29
[12] Carvalho S J, Metzler R and Cherstvy A G 2016 *New J. Phys.* **18** 083037

IOP Conf. Series: Materials Science and Engineering **443** (2018) 012017 doi:10.1088/1757-899X/443/1/012017

Influence of the third element additives on the surface morphology of the wear-resistant ZrN coatings

T A Kuznetsova[1], V A Lapitskaya[1], S A Chizhik[1], B Warcholinski[2], A Gilewicz[2], A S Kuprin[3]

[1]A V Luikov Heat and Mass Transfer Institute of NAS Belarus, Minsk, 220072, Belarus

[2] Koszalin University of Technology, Faculty of Technology and Education, Koszalin, 75-453, Poland

[3] NSC Kharkov Institute Physics & Technology NASU, Kharkov, 61108, Ukraine

kuzn06@mail.ru

Abstract. ZrN, Zr-O-N, and Zr-Si-N wear resistant coatings were formed using vacuum-arc plasma fluxes deposition at the substrate bias voltage 100 V on HS6-5-2 steel substrates. For comparison, a sample of ZrN coating at the same substrate bias voltage on the same substrate was obtained using magnetron sputtering. The structural and mechanical properties were characterized using X-ray diffraction, atomic force microscopy, and nanoindentation. The influence of O and Si additives on the surface morphology of coating were investigated.

1. Introduction

One of the ways to improve the mechanical and tribological properties of machine parts is surface modification by a hard coating formation [1-4]. The transition metal nitrides coatings are widely applied for cutting tools and tribological surfaces [5,6]. The properties of two-component coatings can be improved by adding the third element or even more of them to the two-element system. ZrN coatings occupy a special place among the transition metal nitrides [7,8] and have a favorable combination of mechanical properties under conditions requiring high thermal stability.

The work is aimed at the study of the effect of Si and O addition on the surface morphology and such structure characteristics of ZrN coatings like lattice parameter, grain size on the surface, and roughness in microscale, at the same substrate bias voltage.

2. Experimental details

ZrN, Zr-O-N, and Zr-Si-N coatings were synthesized using cathodic arc evaporation (CAE) on HS6-5-2 steel substrates in a "Bulat-3T" system. The coatings were deposited at nitrogen pressure of 2 Pa. The deposition was performed at substrate bias voltage of -100 V. Zr-O-N coatings were deposited with relative oxygen concentration $O_2(x) = O_2/(O_2+N_2)$ amounted to 10%. A substrate temperature was 450°C for ZrN and Zr-Si-N and 350°C for Zr-O-N. The deposition time was 30 min and thickness of the coatings was about 3 μm.

X-ray diffraction (DRON-3 M, Cu-Kα) was used to characterize the phase composition of the coatings. Lattice parameter and the average size of the crystallites of the ZrN phase were calculated according to (111) plane of the obtained diffraction patterns.

Table 1. The properties of ZrN, Zr-O-N, and Zr-Si-N coatings, investigated in this work: lattice parameter, crystallite size, Ra, size of grains on the surface, E, and H.

Number of samples	Content of samples	R_a, nm 2x2 μm^2	Lattice parameter, nm	Crystallite size, nm	Size of grains on the surface, nm	E, GPa	H, GPa
1	ZrN (CAE)	6.2±0.3	0.4595*	18±0.2	700	282±14	29±2
2	Zr-Si-N	5.2±0.3	0.4614	6±0.1	583	298±15	22±1
4	Zr-O-N	3.2±0.2	0.4615	15±0.2	300	312±16	31±2
5	ZrN (magnetron)	2.7±0.1	0.4608	11±0.1	100	307±15	29±2

*According to PDF No 35-0753 for ZrN (ASTM), a = 0.4577 nm

Hardness and elastic modulus were determined using nanoindentation. Hysitron Ti 750 Ubi (USA) equipped with the Bercovich diamond indenter with tip radius of 150 nm was used. The standard fused quartz sample was used to control the calibration correctness of the device. The measurements were performed at the load of 10 mN.

The surface morphology of the coatings was investigated using atomic force microscopy (AFM) device HT-206 (produced by MTM Belarus) in the contact mode. AFM is the technique to simultaneously study different surface phenomena, like roughness of initial surfaces [9] or surfaces after mechanical [10,11] or chemical treatment [12,13], friction testing [14], and phase distribution [15]. The surface topography was studied using standard CSC38 silicon probe of beam type with the radius of tip curvature less than 10 nm and stiffness of cantilever of 0.08 N/m produced by "Mikromasch" (Estonia). The roughness of the surface in microscale was investigated in the area of 2×2 μm^2.

The samples under study are listed in Table 1.

3. Results and discussion

According to the X-ray analysis, the addition of alloying elements to ZrN coating in small amounts (6 at.% Si or 10 % O in the working gas mixture) does not lead to formation of silicon containing or oxygen containing phases; the diffraction lines are shifted to lower angles from their positions.

The introduction of small amounts of additives into the ZrN lattice increases the lattice parameter to 0.4614 for Si atoms and to 0.4615 for O atoms. The introduction of additive atoms reduces the crystallite size in Zr-Si-N and Zr-O-N coatings. The average size of the crystallites in ZrN calculated for (111) plane is about 18.3 nm (Table 1). Silicon introduction into ZrN coating deposited at the same substrate bias voltage of − 100 V results in a significant reduction in size of the crystallites to about 6.4 nm. Oxygen introduction into ZrN coating leads to reduction in size of the crystallites from 18 nm to about 15 nm. In the work [16], it was found that oxygen in ZrN coatings obtained by magnetron sputtering led to the crystallite size decrease. Similar values of crystallite sizes in magnetron sputtered coatings ranging from 10 to 12 nm were calculated for argon and nitrogen mixture during deposition. In this work, ZrN coatings deposited by magnetron sputtering at the same bias voltage (UB = −100 V) show a higher lattice parameter than ZrN coatings deposited by CAE equal to 0.4608. The average size of crystallites in the magnetron sputtered ZrN was about 11 nm (Table 1). That value was twice lower than in Ref. [17], where the average size of crystallites was about 22 nm and decreased to 2.5 nm for about 10 at.% of silicon.

The same values of hardness and elastic modulus of obtained coatings show that energetic levels (energy of ion bombardment) of different processes in this work were close to each other (Table 1). It is known that mechanical properties of the coatings deposited using CAE are better compared using magnetron sputtering [8]. Such coatings also show the higher density. But in this research, H and E values in ZrN coatings obtained by CAE and magnetron sputtering are almost the same. The surface microstructure is one of the most important parameters of the wear resistant coatings.

ZrN, Zr-O-N, and Zr-Si-N coatings in the range of areas from 100x100 μm^2 till 10x10 μm^2 show classical surface microstructure characteristic for coatings deposited using PVD methods. The numerous craters and macroparticles are apparent on the coating surface (Fig. 1a). Microparticles in ZrN coating structure have two forms – circular and elongated. The difference in the shape is probably connected with the different phase content of the particles. The circular particles consist of the plates and have obvious boundary between the particle and the surface. Some particles in the form of separate plates are spread over the surface. The elongated microparticles represent a sort of thickened boundary of cells embedded in the surface. During study of nitride and oxinitride coatings microstructure, both structured features («smooth surface» and macroparticles) should be taken into account. Microparticles participate in the formation of the altered secondary structure, which appears in the friction contact under the action of the load and participates in the friction process. Microparticles in Zr-O-N coating deposited at $O_{2(x)}$ = 10 % look like separate plates grouped together around the joint centre or around the elongated feature. There are particles looking like monolithic features. Using lateral forces AFM mode, it is possible to compare the microstructure of the smooth surface and the microplate position in the microparticle. So, according to the investigated microstructure, microparticles appear on the surface as a result of crystallization process of each plate, but not as a result of accidental hitting of microdrops and their solidification.

«Smooth surface» of ZrN coating obtained by AFM is interpreted as a multi-cell structure, in which elevated vertical grain boundaries are visible (Fig. 1b). The average size of cells is 700 nm. The coating microstructure of Zr-Si-N deposited at UB = − 100 V shows also the form of concave cells (Fig. 1c). The addition of Si into coating during deposition reduces the average size of cells from 700 to 583 nm. The roughness of smooth surface Zr-Si-N decreases from 6.2 to 5.2 nm for ZrN coating.

The addition of oxygen instead of Si reverses the microstructure of the surface (Fig. 2a). Cells are not visible. Rounded "large" crystallites with diameter of 300 nm appear. Rounding of faces and grain coarsening take place in comparison with ZrN. The surface looks like the ledges covered with a layer of film and are smooth because of this. Ra is 2.3 nm; there are no "elongated" forms. The lateral forces map reveals a significant difference between the round grains from the rest of the surface (Fig. 2b). Thus, the distribution over the surface of the oxide phases can be detected.

ZrN coating obtained by magnetron sputtering has surface microstructure, which is drastically different from ZrN coating obtained by CAE (Figs. 1b, 2c). It consists from the separate grains with diameter about 100 nm. The surface turns out to be more "friable", which agrees well with the results of work [17].

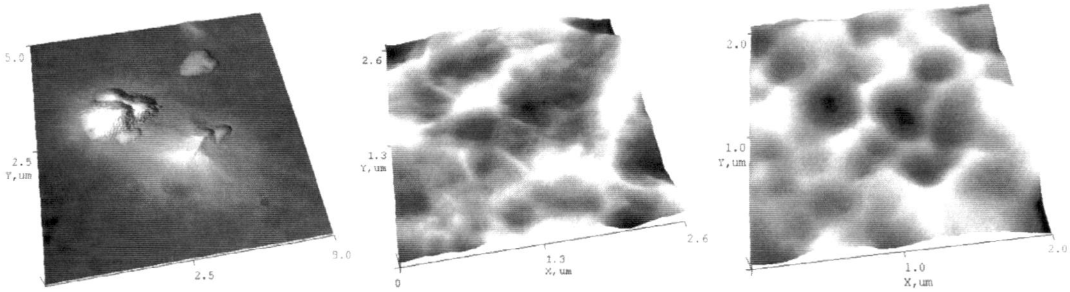

(a) (b) (c)

Figure 1. The topography of ZrN and Zr-Si-N coatings: (a) ZrN, scanned area 4.9x4.9 μm^2; (b) ZrN, scanned area 2.6x2.6 μm^2; (c) Zr-Si-N, scanned area 2x2 μm^2

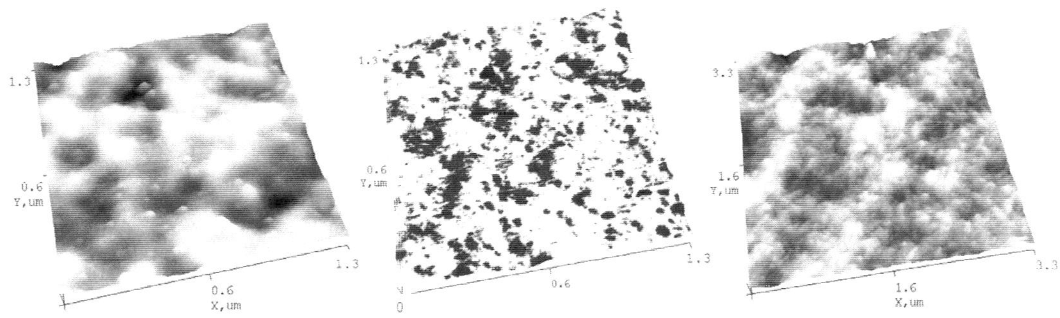

(a) (b) (c)

Figure 2. The surface of Zr-O-N and ZrN coatings, obtained by magnetron sputtering: (a) Zr-O-N, scanned area 1.3x1.3 μm^2; (b) lateral forces map, scanned area 1.3x1.3 μm^2; (c) magnetron ZrN coatings, scanned area 3.3x3.3 μm^2.

4. Conclusion

X-ray did not detect the phases containing silicon and oxygen after doping ZrN coatings with small amount of additives (6 at.% of Si or 10% of $O_{2(x)}$). All coatings had similar values of microhardness (22-31 GPa) and elastic modulus (282-312 GPa). Coatings with small additions of Si and O showed the close parameters of the crystal lattice.

The presence of the additional elements during the formation of ZrN coating changed the morphology of the coating surface. Thus, the initial structure of ZrN coating obtained by the CAE had a honeycomb structure with the average cell size of 700 nm. The addition of 6 at% Si reduced the cell size to 583 nm. The addition of oxygen changed the cellular structure to a smoothed one with circular grains with diameter of 300 nm. The contrast in lateral forces AFM mode allowed us to assume that the oxide phases were in Zr-O-N coating. The method of magnetron sputtering at the same bias voltage on the substrate formed a less dense grain microstructure of the surface with a grain size of 100 nm.

Modification of ZrN coatings with the small amounts of Si and O additives allows forming the surfaces for tribological application with different types of microstructure and, hence, properties of the surface layer in comparison with a pure ZrN coating.

Acknowledgments
This research was supported by the grant of Belarusian Republican Foundation for Fundamental Research BRFFR – SCST – Poland No. T18PLDG-002.

References
[1] Hogmark S, Jacobson S and Larsson M 2000 *Wear* **246** 20-33
[2] Musil J 2000 *Surf. Coat. Technol.* **125** 322-30
[3] Vityaz' P A, Komarov A I, Komarova V I and Kuznetsova T A 2011 *J. Frict. Wear* **32** 313-25
[4] Levashov E A, Shtansky D V, Kiryukhantsev_Korneev Ph V, Petrzhik M I, Tyurina M Ya and Sheveiko A N 2010 *Russian Metallurgy* **10** 917-35
[5] Andreyev M, Anishchik V, Markova L and Kuznetsova T 2005 *Vacuum* **78** 451-54
[6] Andreev M A, Kuznetsova T A, Markov L V and Chekan V A 2001 *Trenie i Iznos* **22** 423-8
[7] Kuznetsova T A, Andreev M A and Markova L V 2005 *Trenie i Iznos* **26** 521-9
[8] Yalamanchili K, Forsén K R, Jiménez-Piqué E, Johansson Jöesaar M P, Roa J J, Ghafoor N and Odén M, 2014 *Surf. Coat. Technol.*, **258**, 1100-7
[9] Anishchik V, Uglov V, Kuleshov A, Filipp A, Rusalsky D, Astashynskaya M, Samtsov M, Kuznetsova T, Thiery F and Pauleau Y 2005 *Thin Solid Films* **482** 248-52
[10] Kuznetsova T A, Chizhik S A and Khudoley A L 2014 *J. Surf. Inv.* **12** 46-56

[11] Ilyuschenko A, Letsko A, Grigorieva T, Vosmerikov S and Tsybulya C 2011 *Proc. Euro Int. Powder Metallurgy Congress and Exhibition, Euro PM 2011 Barselona* **1** 79
[12] Chizhik S A, Kuznetsova T A, Khudolei A L, Komarova V I and Vasilenko M S 2013 *J. Eng. Phys. Thermophys.* **5** 949-59
[13] Ulyanova T M, Titova L V, Medichenko S V, Zonov Yu G, Konstantinova T E, Glazunova V A, Doroshkevich A S and Kuznetsova T A 2006 *Crystallography* **51** 144-9
[14] Chizhik S A, Rymuza Z, Chikunov V V, Kuznetsova T A and Jarzabek D 2007 *Recent Advances in Mechatronics* Ed.: R. Jabłoński [et al.] (Berlin, Heidelberg: Springer) 541-5
[15] Zhdanok S A, Sviridenok A I, Ignatovskii M I, Chizhik S A and Borisevich K O 2010 *J. Eng. Phys. Thermophys.* **83** 1-5
[16] Sandu C S, Cusnir N, Oezer D, Sanjinés R and Patscheider J 2009 *Surf. Coat. Technol.* **204** 969-72
[17] Rawal S K, Chawla A K, Chawla V, Jayaganthan R and Chandra R 2010 *Mater. Sci. Eng.* B **172** 259-66

IOP Conf. Series: Materials Science and Engineering **443** (2018) 012018 doi:10.1088/1757-899X/443/1/012018

Morphology of multilayer AlN/SiN coatings with layers of different thickness

T A Kuznetsova[1], V A Lapitskaya[1], S A Chizhik[1], V V Uglov[2], V I Shymanski[2] and N T Kvasov[2]

[1]AV Luikov Heat and Mass Transfer Institute of NAS Belarus, Minsk, 220072, Belarus
[2] Belarusian State University, Minsk, 220030, Belarus

kuzn06@mail.ru

Abstract. Monolayer AlN coating and multilayer AlN/SiN coatings with single layer thickness of 10 nm and 5 nm were formed using reactive magnetron sputtering. The surface microstructure of these coatings was investigated using atomic force microscopy. Roughness, adhesion forces, and friction coefficient for different layers of coatings were determined.

1. Introduction

Materials multifunctionality is one of the actual trends of modern technology. Coatings based on aluminium nitrides are optimal for operation in the conditions of high temperature and mechanical stresses. Al-Si-N coatings are remarkable by a wide variety of properties, including optical, mechanical, heat-resistant, and tribotechnical ones [1-4]. The combination of different characteristics in one material is possible due to its controlled nanostructuring. On the one hand, the nanostructured state of the films, leads to a significant improvement of their mechanical and functional properties. On the other hand, excess surface energy at the interphase boundaries leads the nanostructure to a nonequilibrium state, which, in its turn, can lead to mechanical degradation of the films, especially at elevated temperatures under open air atmosphere conditions.

For Al-Si-N coatings, nanostructure can be created in two directions: with a uniform distribution of AlN nanocrystals in the amorphous matrix of Si_3N_4 or with alternation of nanosized AlN and SiN_x layers in a multilayer coating. In order to improve the mechanical properties and slow the cracks propagation, the coatings are formed as multilayer structures with a period of layers from several to a dozen nanometers. Such a composition can be a sequence of AlN/SiN layers.

In many applications, the surface is the working element of coatings. The destruction processes due to mechanical or thermal loading also begin on the surface. The adhesive forces and the specific surface energy of sublayers are essential in the process of products manufacturing, as well as in the preparation of surfaces for further tribological and heat resistant applications of coatings. To evaluate the morphology, tribological and adhesion properties, and the individual phases properties on the surface of coating with the 300 nm thickness and single layers of 5 and 10 nm, it is necessary to use a multifunctional atomic force microscopy (AFM) method with a high resolution both for topography and for force loading [5-8]. The forces interaction between the AFM probe and the coating surface in the mode of measuring the adhesion forces makes it possible to estimate the surface energy of the layers. AFM allows modifying the surface and carrying out tests for wear and friction [9-11]. The probe twist due to the contact with the surface can also be used to compare the properties of thin-coating surfaces [12-15]. The properties of Al-Si-N coatings in the form of nanocrystallites distributed in the amorphous matrix are considered in Refs. [1-4]. The aim of this work was to study the surface of AlN/SiN coatings using the AFM method, to determine the morphology of their surfaces and to compare the friction coefficients, adhesion forces on the surface of AlN film, and AlN/SiN multilayer coatings with the thicknesses of individual layers of 10 and 5 nm.

2. Experimental details

AlN and AlN/SiN coatings were formed on silicon single crystal substrates with the (100) preferred crystallographic orientation by reactive magnetron sputtering with consecutive use of aluminum and silicon targets in Ar + N_2 plasma. The coatings thickness was about 300 nm. Multilayer coatings AlN/SiN were obtained with two thicknesses of single layer: 5 and 10 nm. The upper layer in the multilayer coatings was SiN.

Phase composition of the coatings was investigated using X-ray diffractometer RIGAKU Ultima IV with a Cu-Kα source. Hardness and elastic modulus were determined using nanoindentor Hysitron Ti 750 Ubi (USA) equipped with the Bercovich diamond indenter with tip radius of 150 nm. The standard fused quartz sample was applied to control the calibration correctness of the device. The measurements were performed at loads 2-10 mN.

Cross-sectional TEM specimens have been prepared using FEI Helios Nanolab 650 FIB. The specimens were analyzed using JEOL JEM 2100 LaB6 transmission electron microscope operating at 200 kV.

The morphology investigations of the AlN coatings surface and multilayer AlN/SiN coatings surface were performed using Dimension FastScan AFM (Bruker, USA) in the PeakForce Tapping Quantitative Nanoscale Mechanical Mapping (QNM) mode with the standard silicon probe type NSC-11 (MicroMash, Estonia) with the curvature radius according to the passport less than 10 nm and the stiffness of the cantilever of 3 N/m.

After the probe evaluating according to the data processing program "NanoScope Analysis" using the "Tip Qualification" option, the tip radius was determined as 30 nm. The standard titanium sample RS-12M was used to calibrate the probe tip. For these hard coatings, the QNM mode using a silicon probe allowed estimating the distribution of adhesion forces over the surface, constructing an adhesion map, thus better revealing the different phases and boundaries of cluster formations. The adhesion forces were determined from the minimum of the probe detachment curves. The friction coefficient C_{fr} was determined by a two-pass technique of lateral forces microscopy (LFM) in the contact mode. The studies were carried out in air at 25°C and 40% relative humidity.

3. Results and discussion

The TEM image shows the right rows of atoms in the AlN crystal lattice (Fig. 1). The change in the orientation angle of atomic rows (crystalline planes) occurs every 1-10 nm. So the size of single crystallites in AlN layers of 5 nm thickness is 1-2 nm inside the coating. The atoms of the amorphous SiN layers are aligned in chains, most often curved and twisted, with ring formations (Fig. 1). The rare locations of regular crystalline organization include from 6 to 8 atoms.

The elastic modulus E of AlN coating was 189 GPa, in the case of multilayer coatings it was practically equal to 180-182 GPa. The microhardness H of AlN coating was 25 GPa, and that of multilayered coatings was 21-22 GPa (Table 1). The somewhat lower values of the elastic modulus for AlN coatings can be explained by their columnar structure, and for multilayer films – by the amorphous structure of SiN layers.

Typical images of the surfaces in the Al-Si-N coatings are presented in Figures 2 and 3. The monolayer AlN coating is crystalline according to the XRD and TEM results. The surface of AlN coating consists of rounded clusters (grains) with diameter 40 nm (Fig. 1a,b). The term «cluster» is used here as a larger structural formation, consisting of individual crystallites determined by TEM. In the case of using the term "grain" for these surface formation with uniquely defined boundaries, the term "subgrains" should be used for single nanocrystallites. Perhaps, in reality these clusters are somewhat smaller, since the probe radius can overestimate the size of the formations.

IOP Conf. Series: Materials Science and Engineering **443** (2018) 012018 doi:10.1088/1757-899X/443/1/012018

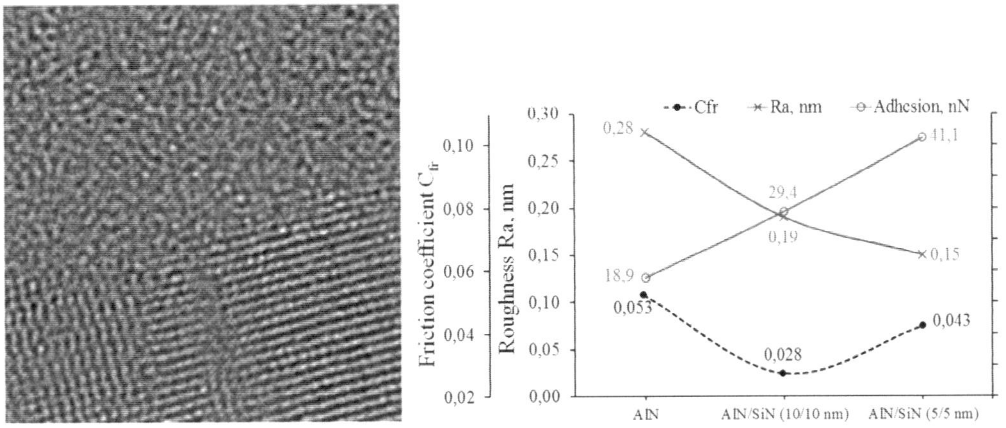

(a) (b)

Figure 1. (a) TEM image of SiN (up) and AlN (down) layers with size of area 10x10 nm^2. (b) Dependences of roughness, adhesion, and friction coefficient from coatings construction.

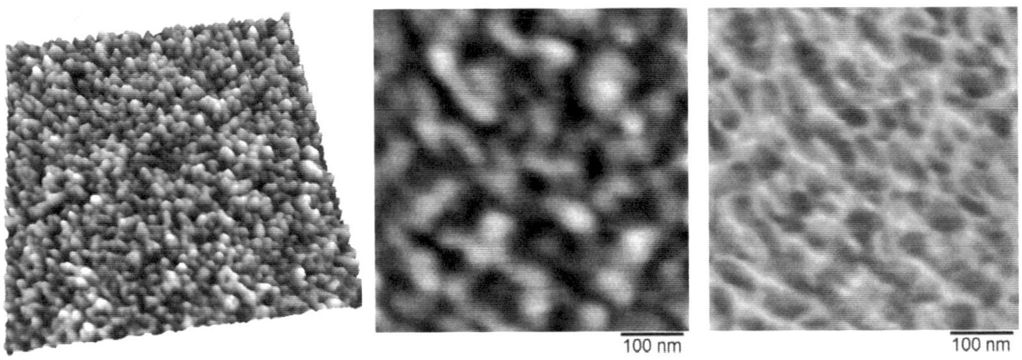

(a) (b) (c)

Figure 2. (a) Three-dimensional image of AlN coating on area 2x2 μm^2. (b) Topography with Z range 1555 pm and (c) adhesion map with Z range 0.213 μN on area 500x500 nm^2.

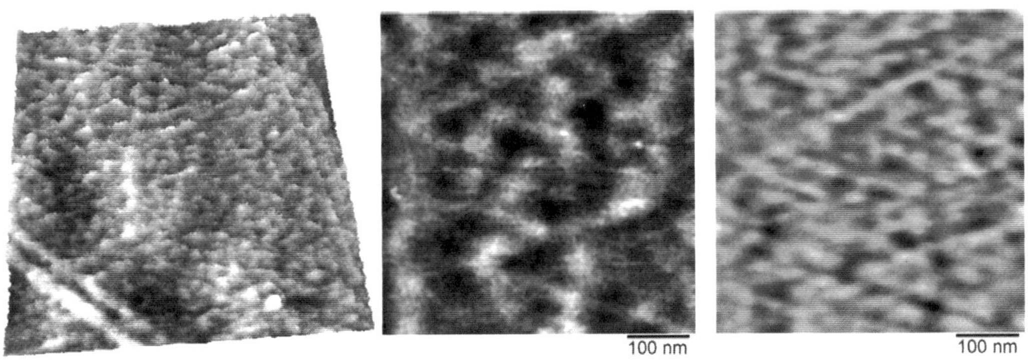

(a) (b) (c)

Figure 3. (a) Three-dimensional image AlN/SiN (5 nm) coating on area 2x2 μm^2. (b) Topography with Z range 561 pm and (c) adhesion map with Z range 0.694 μN on area 500 x500 nm^2.

Table 1. The properties of AlN and AlN/SiN coatings investigated in this work: R_a, F_{ad}, C_{fr}, E, H, and the smallest diameter of structural formations D.

Number of samples	Content of samples	R_a, nm	C_{fr} by LFM	Adhesion forces S_a for F_{ad}, nN	The smallest diameter of structural formations D, nm	E, GPa	H, GPa
1	AlN	0.28	0.053	18.9	40	189±3	25±1
4	AlN/SiN 10/10 nm	0.19	0.028	29.4	50	183±3	22,0±0,7
8	AlN/SiN 5/5 nm	0.15	0.043	41.1	20	180±3	21,3±0,4

During the coatings deposition and AFM investigation, all silicon substrates of the [100] orientations were unrolled in the same way. So, the linear orientation of the cluster chains for all coatings was traced approximately at 45° to the X axis. This direction is typical for the exit of the (100) and (010) planes on the silicon plate surface with orientation [100]. In the crystalline AlN coating, these chains are the shortest, most often composed from two or three clusters, sometimes up to 10 clusters and chains lengths is about 150-300 nm (Fig. 2a,b). The roughness R_a of the AlN surface of 5x5 μm^2 area is 0.28 nm. The map of adhesion forces distribution over the surface shows that the dimensions of the structural formations are much smaller than that revealed on the topography (Fig. 2c). Adhesion forces F_{ad} in this study conditions can be associated with the capillary forces due to the nanometer layer of water between the probe and the coating surface. However, even in this case, the adhesion forces map expands the information about the surface microstructure, since different phases are covered with water atoms in different ways and, as a rule, the water can accumulate at the boundaries. The block structure with the average cluster size of 40 nm is more clearly revealed for AlN coating from the adhesion forces map (Fig. 2c). Z range of adhesion forces is 0.213 μN, and the arithmetic mean Fad over the surface of 2x2 μm^2 was 18.9 nN.

Typical images of the amorphous SiN layers in Al-Si-N coatings are presented in Figure 3. The surface of SiN amorphous layers in AlN/SiN multilayer coatings is susceptible to texturing according to directions (100) and (010) of the silicon substrate in the highest degree. According to the topography image, the surface consists of asperities with fuzzy boundaries. The separate asperities with a diameter of about 100 nm for AlN/SiN (10 nm/10 nm) and about 50 nm for AlN/SiN (5 nm/5 nm) exist on the surface together with extended linear formations from 100 nm to 10 μm and more (Fig. 2). R_a of the surface is 0.19 nm for AlN/SiN (10 nm /10 nm) and R_a = 0.15 nm for AlN/SiN (5 nm/5 nm). The adhesion forces maps make it possible to reveal structural formations on the amorphous films surface and to determine their size more exactly. On the surface of amorphous SiN films, the adhesion forces and, consequently, the specific surface energy are significantly higher than those of crystalline AlN. The Z range of adhesion forces is 0.337 μN for AlN/SiN (10 nm/10 nm) and 0.694 μN for AlN/SiN (5 nm/5 nm), the arithmetic mean values of F_{ad} over the surface of 2x2 μm^2 area are 29.4 and 41.1 nN for these coatings, respectively. The size of the clusters is about 100 nm in length and 40 nm in width in the SiN layer in AlN/SiN (10 nm/10 nm) coating, and it is about 50 nm in length and 20 nm in width in AlN/SiN (5 nm/5 nm) coating.

The values of C_{fr}, determined by LFM for AlN and SiN surfaces differ almost twice being 0.028 and 0.053, respectively. The dependence of C_{fr} on the coating construction is presented in Figure 1b. Often, C_{fr} is bound either to the surface roughness or to the adhesive forces. The results obtained in this work show that none of these characteristics for Al-Si-N coatings have a single-valued relationship. The value of AlN C_{fr} is more likely to be affected by roughness, and the value of SiN C_{fr} in AlN/SiN 5/5 nm coating is more likely to be affected by adhesion forces.

The obtained surface morphology makes it possible to explain the good thermal stability of the studied coatings. The upper layer from amorphous SiN covers the surface of AlN layer with the columnar structure. Adsorption of water and gases in AlN layer occurs along the developed clusters (grain) boundaries, creating conditions for the oxidation reaction propagation into the coating. A layer from amorphous SiN closes the boundaries of AlN. In SiN layer, adsorption goes on the whole surface evenly, without creating nucleus of deepening. The increased thermal stability of multilayer structures is due to the presence of amorphous layers of the SiNx phase, which are a barrier for the diffusion penetration of oxygen into the coating.

4. Conclusions

The surface microstructure of monolayer AlN coating and multilayer AlN/SiN coatings with thicknesses of separate layers of 10 nm and 5 nm was determined using AFM. The adhesion forces maps obtained in QNM mode allowed us to refine the size of structural formations on the surface of both the crystalline AlN film and amorphous SiN films. The dimensions of the cluster formations on the surface were 40 nm for AlN, 50 nm for AlN/SiN (10 nm/10 nm), and 20 nm for AlN/SiN (5 nm/5 nm). The values of the adhesion forces, friction coefficient and surface roughness of the coatings were determined. The increased surface energy for amorphous SiN layers in comparison with crystalline AlN was determined and the uniform distribution of adhesion forces over the surface was shown. The revealed microstructure explains the good thermal stability of multilayer AlN/SiN coatings.

Acknowledgments

This research was supported by the grant of SPSR "Energy Systems, Processes and Technologies"-2.6.

References

[1] Kuznetsova T A, Lapitskaya V A, Chizhik S A, Uglov V V, Kvasov N T and Shymanski V I 2016 *12th Int. Conf. Methodological aspects of SPM: Minsk: Belaruskaya Navuka,* 66-70
[2] Kuznetsova T A, Zubar T I, Lapitskaya V A, Chizhik S A, Shymanski V I and Kvasov N T 2017 *12th Int. Conf. Interaction of Radiation with Solids (IRS-2017) Minsk: BSU*, 256-7.
[3] Kuznetsova T A, Zubar T I, Lapitskaya V A, Shymanski V I, Kvasov N T and Chizhik S A 2017 *Int. Conf. Polymer Composites and Tribology (Polycomtrib-2017),Gomel, MPRI* 29.
[4] Shymanski V I, Kvasov N T, Kuznetsova T A and Kananovich N A 2017 *12th Int. Conf. Interaction of Radiation with Solids (IRS-2017) Minsk: BSU*, 420-2.
[5] Andreyev M, Anishchik V, Markova L and Kuznetsova T 2005 *Vacuum* **78** 451-4
[6] Kuznetsova T A, Chizhik S A and Khudoley A L 2014 *J. Surf. Inv.* **12** 46-56
[7] Chizhik S A, Kuznetsova T A, Khudolei A L, Komarova V I and Vasilenko M S 2013 *J. Eng. Phys. Thermophys.* **5** 949-59
[8] Zhdanok S A, Sviridenok A I, Ignatovskii M I, Chizhik S A and Borisevich K O 2010 *J. Eng. Phys. Thermophys.* **83** 1-5
[9] Kuznetsova T A, Andreev M A, and Markova L V 2005 *Trenie i Iznos* **26** 521-9
[10] Chizhik S A, Rymuza Z, Chikunov V V, Kuznetsova T A and Jarzabek D 2007 *Recent Advances in Mechatronics* Ed.: R. Jabłoński [et al.] (Berlin, Heidelberg: Springer) 541-5
[11] Andreev M A, Kuznetsova T A, Markov L V, and Chekan V A 2001 *Trenie i Iznos* **22** 423-8
[12] Anishchik V, Uglov V, Kuleshov A, Filipp A, Rusalsky D, Astashynskaya M, Samtsov M, Kuznetsova T, Thiery F and Pauleau Y 2005 *Thin Solid Films* **482** 248-52
[13] Ulyanova T M, Titova L V, Medichenko S V, Zonov Yu G, Konstantinova T E, Glazunova V A, Doroshkevich A S and Kuznetsova T A 2006 *Crystallography* **51** 144-9
[14] Ilyuschenko A, Letsko A, Grigorieva T, Vosmerikov S and Tsybulya C 2011 *Proc. Euro Int. Powder Metallurgy Congress and Exhibition, Euro PM 2011 Barselona* **1** 79
[15] Vityaz' P A, Komarov A I, Komarova V I and Kuznetsova T A 2011 *J. Frict. Wear* **32** 313-25

IOP Conf. Series: Materials Science and Engineering **443** (2018) 012019 doi:10.1088/1757-899X/443/1/012019

Lateral force microscopy as a method of surface control after low-temperature plasma treatment

V A Lapitskaya[1], T A Kuznetsova[1], S A Chizhik[1], K A Sudzilouskaya[1], D A Kotov[2], S A Nikitiuk[2], Y V Zaparozhchanka[2]

[1] A V Luikov Heat and Mass Transfer Institute of NAS Belarus, 220072, Minsk, Belarus

[2] Belarusian State University of Informatics and Radioelectronics, 220013, Minsk, Belarus

vasilinka.92@mail.ru

Abstract. The surface properties of titanium, silicon, and a two-layer polymer film PMF-351 are studied after the action of a low-temperature plasma of dielectric barrier discharge (DBD). It is established that such treatment leads to the surface cleaning. This is evidenced by the values of roughness. The friction coefficient C_{fr} is selected as a characteristic that allows the time variation of surface properties after the plasma treatment. C_{fr} is determined using lateral forces regime of atomic force microscopy. The dependences of the friction coefficient C_{fr} on time for titanium, silicon, and a two-layer polymeric film PMF-351 are obtained. The best effect after treatment with low-temperature plasma persists for the first 20-30 minutes for titanium and a polymer film, and 5 minutes for silicon.

1. Introduction

Among the various methods of the surface modification without influence on the bulk properties of the material, the most promising one is the treatment in a low-temperature plasma of dielectric barrier discharge (DBD) [1,2]. Such plasma is used to improve the contact and adhesion properties of materials connection, in order to clean the contaminated surface. Due to the low temperature, this plasma has found application in medicine for sterilization, disinfection, and healing of living tissues. The ability to change the properties and the surface structure for a certain time is a unique property of this method. After that, the surface of the processed materials returns to the initial state.

There are various ways to control the properties and chemical structure of the materials processed in a low-temperature plasma [1]. Atomic force microscopy (AFM) is the most accurate method for estimating topography and surface properties (elastic modulus, adhesion, conductivity, and thermal conductivity) [3-8]. In addition to studying the surface properties of materials, AFM allows modifying the surface, carrying out tests for wear and friction [9-13], and exploring the time-varying processes. Changes in the surface properties after plasma treatment refer to such processes.

Often, modification by plasma does not lead to a change in the surface topography, determined by the high-resolution AFM method. However, other properties can change, for example, adhesion. An effective way of quantitative surface control can be measuring friction coefficient by AFM over time.

The aim of this work was to investigate the effects of low-temperature plasma treatment of a DBD on the surfaces of metal, polymer, and silicon using the method of the friction coefficient control of a surface over time.

2. Experimental details

To study the effect of a low-temperature plasma of a DBD on the surface properties, the following samples were chosen: silicon Si, metal Ti, and polymer 2-layer film PMF-351.

Samples were processed for 10 min by low-temperature plasma of DBD using a special setup (development of BSUIR) with working gas N_2, plasma source power about 10-30 W, and the distance from the source to the sample 20 mm.

The friction coefficient, adhesion, and microscale roughness of the samples were investigated using AFM device NT-206 (produced by MTM, Belarus) with the standard silicon probe type NSC-11 (MikroMash, Estonia) with the stiffness of the cantilever of 3 N/m. The surface roughness and friction coefficient (C_{fr}) were investigated in scanning areas 20x20 μm^2.

The method of measuring the coatings friction coefficient using AFM is based on the estimation of the console twisting angle of the probe around its axis under the action of frictional forces between the surface and its tip [14]. The interaction of the AFM probe and the surface is also strongly affected by the adhesion forces between the probe and the surface; the larger the probe "sticks" to the surface, the greater the friction coefficient. After the plasma action, the adhesion forces increase and, consequently, the twisting of the console and C_{fr} increases.

3. Results and discussion

The microscale roughness R_a is one of the criteria for the presence of unevenness and contamination on the surface.

According to the obtained results (Table 1), the surface microscale roughness R_a for titanium and the polymer two-layer PMF-351 film decreases twice after treatment with low-temperature plasma DBD. These results are in agreement with the data of works [1,2,15], which have established that plasma treatment cleans the surface. For smooth and pure silicon there are no significant changes in R_a before and after treatment.

Table 1. Surface roughness of test samples before and after plasma treatment.

Sample	Roughness before, nm (scanning area 20x20 μm^2)		Roughness after, nm (scanning area 20x20 μm^2)	
	R_a	R_q	R_a	R_q
Ti	90.2	106.2	46.1	59.5
Si	5.8	7.6	4.8	8.7
Two-layer film PMF-351	33.1	41.6	16.5	20.7

Surface treatment by the low-temperature plasma increases the two-layer film PMF-351 adhesion properties [1,15]. The friction coefficient between the surface and the probe tip is a quantitative parameter, depending on the adhesion forces. After treatment with the low-temperature plasma, the friction coefficient was fixed continuously for 75 min for the polymer film, 85 min for the titanium, and 20 min for the silicon.

Table 2 shows the mean values of the friction coefficient for initial samples without plasma treatment, maximum values after plasma treatment, values during 20 min after plasma treatment, and values during 75 min after plasma treatment.

Table 2. Mean values C_{fr} of tested samples before and after plasma treatment.

Sample	C_{fr} (initial samples)	$C_{fr.max}$ (after treatment)	C_{fr} (20 min after treatment)	C_{fr} (75 min after treatment)
Ti	0.21±0.01	0.23±0.01	0.140±0.007	0.050±0.002
Si	0.100±0.005	0.60±0.03	0.25±0.01	–
Polymer PMF-351	0.170±0.009	0.028±0.001	0.023±0.001	0.015±0.001

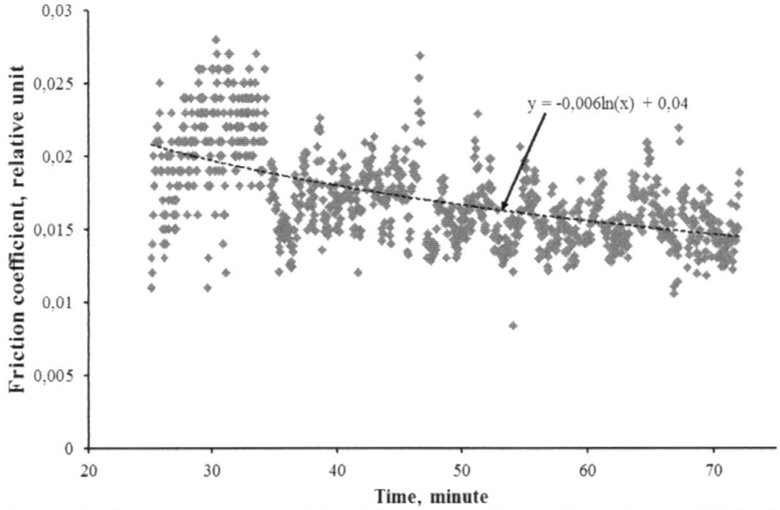

Figure 1. Time dependence of the friction coefficient for polymer PMF-351 after plasma treatment.

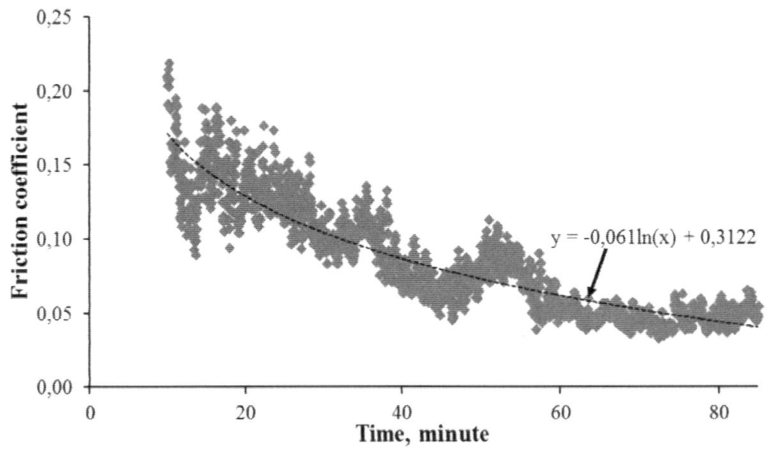

Figure 2. Time dependence of the friction coefficient for Ti after plasma treatment.

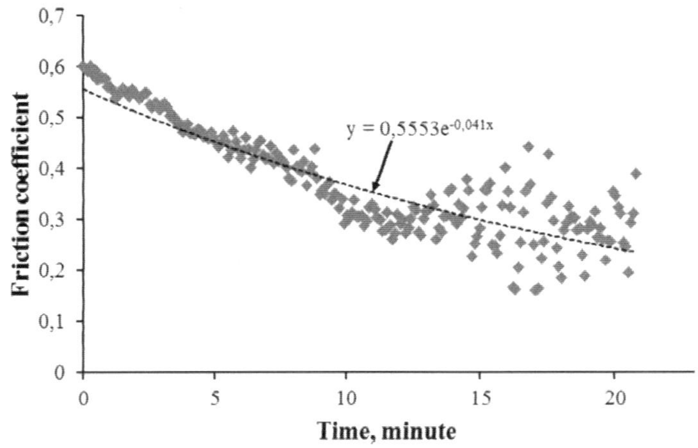

Figure 3. Time dependence of the friction coefficient for Si after plasma treatment.

Table 3. Adhesion force F_{ad} and specific surface energy σ of test samples before and after plasma treatment.

Sample	Time after treatment, [min]	Adhesion forces F_{ad}, [$\cdot 10^{-7}$ N]	Specific surface energy σ, [N/m]
PMF-351	0	0.79±0.04	0.039±0.002
	80	1.57±0.07	0.078±0.004
Ti	0	4.7±0.2	0.23±0.01
	80	4.2±0.2	0.21±0.01

Based on the results of the study, it is established that for all samples the friction coefficient decreases with time. For the two-layer film PMF-351, during the first 35 min after treatment C_{fr} remains in the same range from 0.020 to 0.028 and during the next 35 min it decreases evenly to the value 0.015 (Fig. 1). For titanium (Fig. 2), C_{fr} decreases from 0.23 to 0.05 for the first 60 min, then remains constant at 0.05. For silicon (Fig. 3), C_{fr} decreases from 0.6 to 0.25 through the entire period of time.

Adhesion of the titanium surface (Table 3) decreases with time after plasma treatment. This fact cannot be stated for a polymer film. The adhesion of the surface polymer film increases. The decrease in adhesion in titanium can be due to the cleaning of the surface from the oxide layers. It is assumed that treatment of the surface with low-temperature plasma leads to the change in the chemical structure of the surface layer of the polymer film [15] and leads to an increase in surface adhesion.

4. Conclusion

The properties of the titanium surface, a two-layer polymeric film, and silicon after the treatment of low-temperature plasma of DBD were investigated by atomic force microscopy. It was established that the modified surfaces of various materials lost their active properties in different ways over the time. Processing in the plasma of the DBD cleaned the surface, which was confirmed by the roughness values.

The friction coefficient C_{fr} of titanium immediately after treatment was practically unchanged compared to the initial one. C_{fr} after 80 min has reduced by an order of magnitude.

Modified effect for 80 min better retained for the polymer film. C_{fr} slowly declined over the entire period of time and decreased slightly.

The adhesion forces of the titanium surface after treatment reduced. The reason for the reduction was cleaning of the solid surface Ti from the oxide layers and pollutions, which provided adhesion. The adhesion forces of the polymer film surface after treatment increased due to the change in the chemical structure of the surface layer.

Acknowledgments

This research was supported by the grant of Belarussian Republican Foundation for Fundamental Research BRFFR No. F17-118.

References

[1] Gilman A B, Piskarev M S, Kuznetsov A A and Ozerin A N 2017 *High Energ. Chem.* **51** 136-45

[2] Murokh I and Dugin E 2014 *Electron. Comp.* **9** 1-5 (in Russian)

[3] Ilyuschenko A, Letsko A, Grigorieva T, Talako T, Kuznetsova T, Lyakhov N, Barinova A, Vosmerikov S and Tsybulya C 2011 *Proc. Euro Int. Powder Metallurgy Congr. Exhib., EURO PM* **2**

[4] Zhdanok S A, Sviridenok A I, Ignatovskii M I, Chizhik S A and Borisevich K O 2010 *J. Eng. Phys. Thermophys.* **83** 1-5

[5] Ulyanova T M, Titova L V, Medichenko S V, Zonov Yu G, Konstantinova T E, Glazunova V A, Doroshkevich A S and Kuznetsova T A 2006 *Crystallogr.* **51** 144-9
[6] Anishchik V, Uglov V, Kuleshov A, Filipp A, Rusalsky D, Astashynskaya M, Samtsov M, Kuznetsova T, Thiery F and Pauleau Y 2005 *Thin Solid Films* **482** 248-52
[7] Chizhik S A, Kuznetsova T A, Khudolei A L, Komarova V I and Vasilenko M S 2013 *J. Eng. Phys. Thermophys.* **5** 949-59
[8] Andreyev M, Anishchik V, Markova L and Kuznetsova T 2005 *Vacuum* **78** 451-54
[9] Vityaz' P A, Komarov A I, Komarova V I and Kuznetsova T A 2011 *J. Frict. Wear* **32** 313-25
[10] Kuznetsova T A, Andreev M A and Markova L V 2005 *Trenie i Iznos* **26** 521-9
[11] Kuznetsova T A, Chizhik S A and Khudoley A L 2014 *J. Surf. Invest.* **12** 46-56
[12] Andreev M A, Kuznetsova T A, Markov L V and Chekan V A 2001 *Trenie i Iznos* **22** 423-8
[13] Chizhik S A, Rymuza Z, Chikunov V V, Kuznetsova T A and Jarzabek D 2007 *Recent Advances in Mechatronics* Ed.: R. Jabłoński [et al.] (Berlin, Heidelberg: Springer) 541-5
[14] Kuznetsova T A, Zubar T I, Lapitskaya V A, Kudryavtsev V V and Sukhanova T E 2017 *IOP Conf. Ser.: Mater. Sci. Eng.* **256** 012022
[15] Rybkin V V 2000 *Sorosovsky educational journal. Chemistry* **3** 58-63 (in Russian)

Morphology of multilayer metal-carbon coatings with different arrangements of functional layers

V A Lapitskaya[1], T A Kuznetsova[1], S A Chizhik[1], A A Rogachev[2]

[1] A V Luikov Heat and Mass Transfer Institute of NAS Belarus, 220072, Minsk, Belarus
[2] Francisk Skorina Gomel State University, 246019, Gomel, Belarus

*vasilinka.92@mail.ru

Abstract. The surface morphology of multilayer metal-carbon coatings before and after annealing in the air atmosphere at temperature of 350°C for 60 minutes was investigated by atomic force microscopy. The coatings were obtained by the arc discharge method. The surface morphology, roughness, and adhesion forces largely depend on the type of deposit layers and their sequence (architecture). Annealing in air at temperature of 350°C differently affects the properties of coatings of different designs.

1. Introduction

During the last decades, metal-containing diamond-like coatings (Me/α-C or α-C:Me) deposited by various methods of physical spraying (PVD) or chemical vapour deposition (CVD) have become widespread in various engineering industries due to their high hardness as well as wear and chemical resistance. The wide use of such coatings in pure form is limited by the low adhesion to the substrate.

Various methods are used to increase the adhesion of such coatings by controlling their structure. One of them is the inclusion of metal and nitride layers in the coating. They perform the function of adhesion and damping layers [1]. Metal and nitride layers make it possible to increase the strength of the adhesive bond between the coating and the substrate. It was found that the residual stress in a-C coatings decreased and adhesion increased after depositing a carbon coating on a sublayer of a carbide-forming metal (chromium or titanium) [2].

There are various ways of investigation and control of the thin multilayer coatings properties. Atomic force microscopy (AFM) is one of the multifunctional methods. It allows evaluating the parameters and properties of the surface [3-8]: roughness, average grain size, elastic modulus, adhesion forces, deformation, and others. AFM allows carrying out tests for wear and friction [9-13].

The aim of this work was to investigate the influence of annealing and coating design on morphology, roughness, and adhesion force of the surface of multilayer metal-carbon coatings.

2. Experimental details

Two- and three-layers metal-carbon coatings were produced: Ti-(Ti+α-C) – № 1; Ti-(Ti+α-C)-α-C – № 2; Ti-(TiN+α-C)-α-C – № 3; Ti-(TiN+α-C) – № 4; TiN-(Ti+α-C) – № 5 (Fig. 1).

The carbon component of the coating was deposited by a pulsed arc-cathodic method at the pulse frequency of 15 Hz and the discharge voltage of 350 V. The alloying with the nitrogen was due to the sputtering at the nitrogen partial pressure in the chamber of 5×10^{-2} Pa. The metal component was

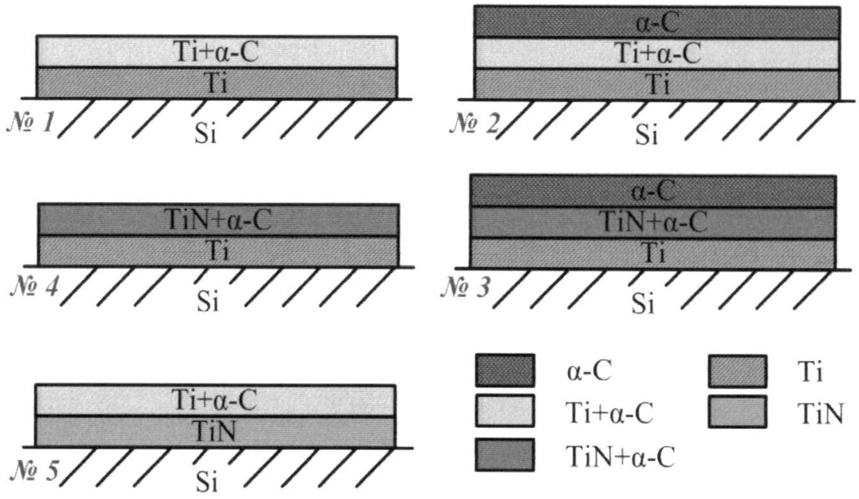

Figure 1. Multilayer metal-carbon coatings. Two-layers coatings: *№1* – Ti-(Ti+α-C); *№4* – Ti-(TiN+α-C); *№5* – TiN-(Ti+α-C); three-layers coatings: *№2* – Ti-(Ti+α-C)-α-C; *№3* – Ti-(TiN+α-C)-α-C

formed using the direct current arc discharge at the arc current of 70 A. The titanium of the VT-100 grade was used as a cathode. The thickness of the coatings depended on its architecture and was in the range from 300 to 500 nm. Two- and three-layers metal-carbon coatings were formed on silicon single crystal substrates with the (100) preferred crystallographic orientation. The samples annealing was carried out in the air atmosphere at the temperature of 350°C for 60 minutes.

The surface morphology, roughness R_a and adhesion forces F_{ad} measurements were performed using Dimension FastScan AFM (Bruker, USA) in the PeakForce Tapping QNM (Quantitative Nanoscale Mechanical Mapping) mode with the standard silicon probe type MPP-12120-10 (Bruker, USA) with the curvature radius 42 nm and the cantilever stiffness of 6.2 N/m.

3. Results and discussion

The surface images of multilayer metal-carbon coatings before and after annealing were obtained by AFM (Fig. 2). The surface is the "smooth coating" with particles evenly distributed over the surface. The "smooth coating" consists of cluster formations of 50-70 nm in size with undefined boundaries. Such morphology is typical for nanocomposite multilayer metal-carbon coatings.

Annealing in air at 350°C affects the metal-carbon coatings differently. The particles quantity on the surface of some coatings decreases: Ti-(Ti+α-C), Ti-(Ti+α-C)-α-C, and Ti-(TiN+α-C). In other coatings Ti-(TiN+α-C)-α-C and TiN-(Ti+ α-C), amount and particles size increase. After annealing, the cluster formations size of the "smooth coating" increases or remains unchanged.

The coatings surface roughness R_a (Table 1) during annealing decreases almost for all samples. The growth of R_a was only on TiN-(Ti+α-C) coatings surface. This is due to the increase in the amount of submicron-particles of 100-230 nm after annealing on the surface of TiN-(Ti+α-C).

The particle size on the surface of Ti-(Ti+α-C)-α-C coating decreases after the annealing from 100-370 nm to 36-97 nm. So, the roughness R_a decreases from 1.04 to 0.15 nm. The same changes proceed in Ti-(TiN+α-C) coating. Annealing practically did not change the properties of Ti-(Ti+α-C) coating.

The adhesion forces of coatings during annealing decrease or increase either. The adhesion forces increase from 1.33 to 4.60 nN and from 3.19 to 6.97 nN, respectively, in the two-layer TiN-(Ti+α-C) coating and the three-layer Ti-(TiN+α-C)-α-C coating. Annealing leads to the decrease in the adhesion forces all the other coatings.

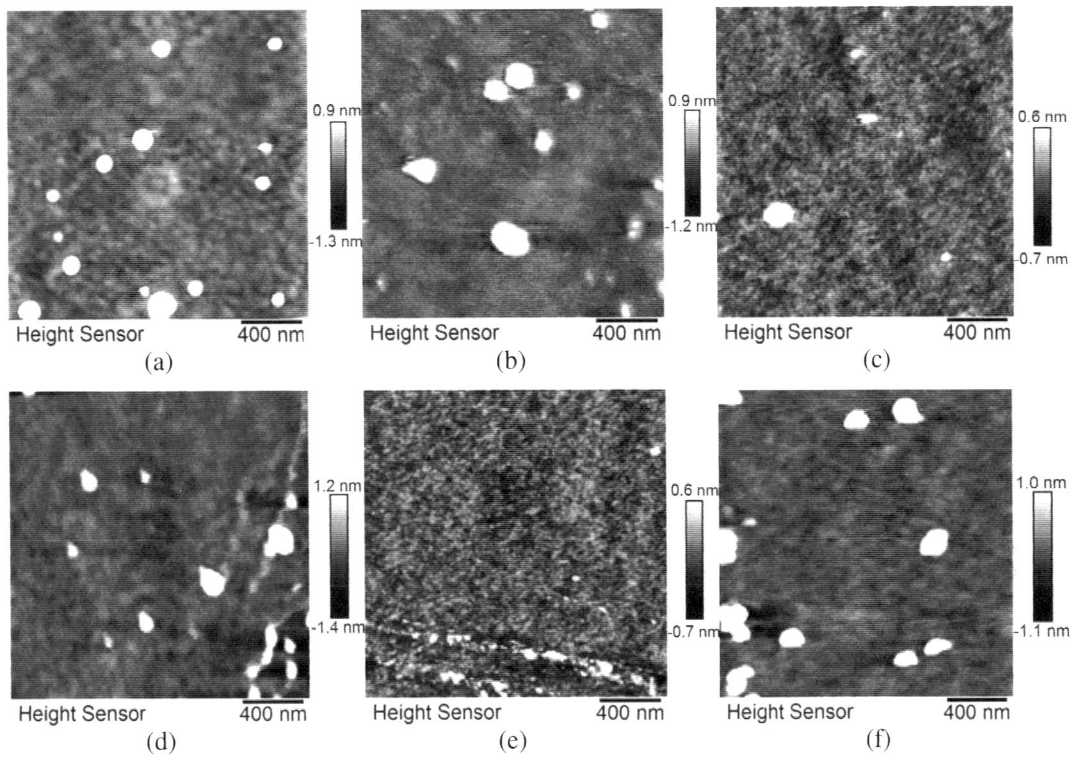

Figure 2. Surface morphology of multilayer coatings before (a,b,c) and after (d,e,f) annealing: (a,d) – №3 (Ti-(TiN+α-C)-α-C); (b,e) – № 4 (Ti-(TiN+α-C)); (c,f) – № 5 (TiN-(Ti+α-C)).

Table 1. Surface roughness R_a and adhesion forces F_{ad} of investigation coatings before and after annealing (scanning area 2x2 μm^2).

№	Multilayer coating	Before annealing		After annealing	
		R_a, nm	F_{ad}, nN	R_a, nm	F_{ad}, nN
1	Ti-(Ti+α-C)	0.11±0.01	1.55±0.08	0.17±0.01	1.03±0.05
2	Ti-(Ti+α-C)-α-C	1.04±0.05	6.39±0.03	0.15±0.01	1.53±0.08
3	Ti-(TiN+α-C)-α-C	0.46±0.02	3.2±0.2	0.40±0.02	7.0±0.4
4	Ti-(TiN+α-C)	0.21±0.01	2.8±0.1	0.14±0.01	0.32±0.02
5	TiN-(Ti+α-C)	0.20±0.01	1.33±0.07	0.63±0.03	4.6±0.2

The increase in the adhesion forces of Ti-(TiN+α-C)-α-C and TiN-(Ti+α-C) coatings during annealing can be due to diffusion between TiN and α-C layers. It is assumed that the annealing increases the adhesion forces of the entire multilayer coating: the adhesion between the layers and the adhesion of the coating to the substrate.

4. Conclusion

The influence of the architecture, the type of deposited layers and the annealing on the morphology of the surface, roughness, and adhesion forces were investigated by atomic force microscopy. It was established that the annealing in air at 350°C differently affected the coatings of various designs. The particles number on the surface and roughness decreased for Ti-(Ti+α-C)-α-C and Ti-(TiN+α-C) coatings. Annealing of Ti-(TiN+α-C)-α-C and TiN-(Ti+α-C) coatings led to the increase of surface adhesion. It was assumed that during annealing process diffusion occurred between TiN and α-C layers. Therefore, it is possible to reduce the surface roughness and increase the adhesion of the multilayer metal-carbon coatings after annealing considering their architecture.

Acknowledgments
This research was supported by the grant of BRFFR – SCST – China No. T17KIG-009.

References

[1] Castillo H A 2011 *Appl. Surf. Sci.* **257** 2665-8
[2] Han X, Zhu J, Han J, Tan M and Gao W 2008 *Appl. Phys. A* **91** 529-33
[3] Anishchik V, Uglov V, Kuleshov A, Filipp A, Rusalsky D, Astashynskaya M, Samtsov M, Kuznetsova T, Thiery F and Pauleau Y 2005 *Thin Solid Films* **482** 248-52
[4] Ulyanova T M, Titova L V, Medichenko S V, Zonov Yu G, Konstantinova T E, Glazunova V A, Doroshkevich A S and Kuznetsova T A 2006 *Crystallogr.* **51** 144-9
[5] Zhdanok S A, Sviridenok A I, Ignatovskii M I, Chizhik S A and Borisevich K O 2010 *J. Eng. Phys. Thermophys.* **83** 1-5
[6] Ilyuschenko A, Letsko A, Grigorieva T, Talako T, Kuznetsova T, Lyakhov N, Barinova A, Vosmerikov S and Tsybulya C 2011 *Proc. Euro Int. Powder Metallurgy Congr. Exhib., EURO PM* **2**
[7] Andreyev M, Anishchik V, Markova L and Kuznetsova T 2005 *Vacuum* **78** 451-4
[8] Chizhik S A, Kuznetsova T A, Khudolei A L, Komarova V I and Vasilenko M S 2013 *J. Eng. Phys. Thermophys.* **5** 949-59
[9] Chizhik S A, Rymuza Z, Chikunov V V, Kuznetsova T A and Jarzabek D 2007 *Recent Advances in Mechatronics* Ed.: R. Jabłoński [et al.] (Berlin, Heidelberg: Springer) 541-5
[10] Andreev M A, Kuznetsova T A, Markov L V and Chekan V A 2001 *Trenie i Iznos* **22** 423-8
[11] Kuznetsova T A, Chizhik S A and Khudoley A L 2014 *J. Surf. Invest.* **12** 46-56
[12] Kuznetsova T A, Andreev M A and Markova L V 2005 *Trenie i Iznos* **26** 521-9
[13] Vityaz' P A, Komarov A I, Komarova V I and Kuznetsova T A 2011 *J. Frict. Wear* **32** 313-25

IOP Conf. Series: Materials Science and Engineering **443** (2018) 012021 doi:10.1088/1757-899X/443/1/012021

Work function tuning of the individual polyaniline/carbon nanotube nanostructures

I A Lobov, N A Davletkildeev and D V Sokolov

Omsk Scientific Center SB RAS, Omsk, 644040, Russia

LI__87@mail.ru

Abstract. Electrostatic force microscopy was used to study the dependence of the electron work function of individual polyaniline/nitrogen doped carbon nanotube (PANI/N-CNT) nanostructures on the duration of synthesis. The work function of nanostructures increases monotonically with the synthesis time increasing. These changes are interpreted as a result of the growth of the polymer layer thickness on CNTs and an increase in the degree of protonation of PANI during the synthesis of PANI/N-CNTs composite.

1. Introduction
Polyaniline (PANI) is the most important representative of the conductive polymers and attracts considerable interest due to its unique physicochemical properties, simplicity of synthesis, and resistance to environmental conditions. Carbon nanotubes (CNTs) are characterized by unique electrical and mechanical properties, which, in combination with their high specific surface area, underlie their use as matrices to create promising nanocomposites. The introduction of nanotubes into the PANI matrix leads to an improvement in the properties of the nanocomposite as a whole: electrical conductivity enhancement, thermal stability and mechanical strength improvement, redox capacity and stability of electrochemical activity increase [1,2]. It has been already shown [3-5] that such composites work well as elements of gas and electrochemical sensors, in energy storage systems, as protective coatings from microwave radiation, as fuel cells, etc.

In some applications of nanocomposites, it is necessary to reduce the size of the elements, up to the use of individual nanoobjects. The prospect of creating electronic components on single nanostructures requires the development of methods for obtaining nanostructures with specified properties. However, as a rule, the properties of massive nanocomposite samples differ from the properties of individual composite nanostructures due to the effect of dimensionality and heterogeneity of component distribution, and also due to the interface interaction of individual nanostructures in the array. Therefore, it is important to develop methods for studying the properties of individual nanostructures using high-resolution methods.

In this paper, we investigated the effect of the synthesis time of the PANI/nitrogen doped CNT (PANI/N-CNT) composite on the change in the electron work function in individual nanostructures using the methods of scanning probe microscopy.

2. Experimental

2.1. Synthesis of PANI/N-CNT nanocomposite
The PANI/N-CNTs composite were synthesized via in-situ chemical oxidative polymerization in the presence of nanotubes. 25 mg of N-CNT obtained by the method described in Ref. [6] were dispersed

in 20 mL of isopropyl alcohol and sonicated for 30 min. 2.5 mL of aniline and 25 mL of 3M HCl were added in this dispersion and sonicated for 30 min. To the N-CNT-aniline dispersed solution, 6 g of ammonium persulphate in 15 mL of 3M HCl were added drop wise at 0^0C. The reaction product was taken from the reaction mixture after 5, 10, 20, 30, 40, and 60 minutes, and then the composite powders were dried at 60°C for 24 h.

To separate individual PANI/N-CNT nanostructures, the bulk PANI/N-CNT composite was dispersed in ethanol by sonication for 30 min. After that, the slurry was precipitated on SiO_2/Si substrate by spin-coating method for further studies.

On SiO_2/Si substrate, individual N-CNT and PANI/N-CNT films (40 nm thick) were planted to determine their work function and compare it with that of the PANI/N-CNT nanostructures.

2.2. Characterization

Morphologies of the PANI/N-CNT composite were characterized by scanning electron microscopy (SEM) on JSM-6610LV microscope, JEOL and atomic force microscopy (AFM) on MFP 3D microscope, Asylum Research.

Electrostatic force microscopy (EFM) was used to evaluate the work function of individual N-CNT, PANI/N-CNT nanostructure, and PANI/N-CNT composite film [6]. Voltage in the range from -7 to +7 V was applied. The HA_FM/Pt (NT-MDT SI) probes with Pt coating were used. Calibration of the probe material work function was carried out at HOPG (W_f = 4.6 eV).

3. Results and discussion

Figure 1a shows the SEM image of a massive PANI/N-CNT composite layer. As can be seen, N-CNTs are completely covered with a layer of polymer. Individual PANI/N-CNT nanostructures are densely packed in a layer and are interconnected by free PANI molecules. Morphology of individual PANI/N-CNT nanostructures was investigated by AFM method (Fig. 1b). Free PANI molecules were removed from PANI/N-CNT nanostructures after sonication. It can be seen that the surface of PANI/N-CNT nanostructure is covered by a polymer layer (Fig. 2a). There are no uncovered sections of N-CNTs.

Figure 2 shows an example of the morphology, the EFM signal, and the section profile along the EFM signal line of the individual PANI/N-CNT nanostructure at a probe potential of 5 V. The bright contrast around the nanostructure in the EFM image indicates its conductive character. The contact potential difference between the probe and the nanostructure was determined and the electron work function was calculated from the profile analysis of the EFM image section of individual nanostructures using the procedure described in Ref. [6].

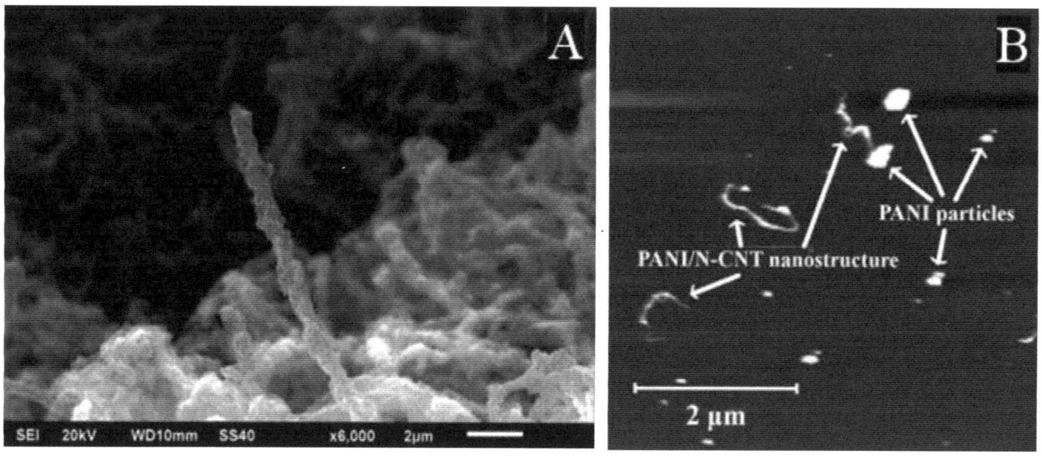

Figure 1. (a) SEM image of PANI/N-CNT bulk composite and (b) AFM image of individual PANI/N-CNT nanostructures on SiO_2/Si substrate.

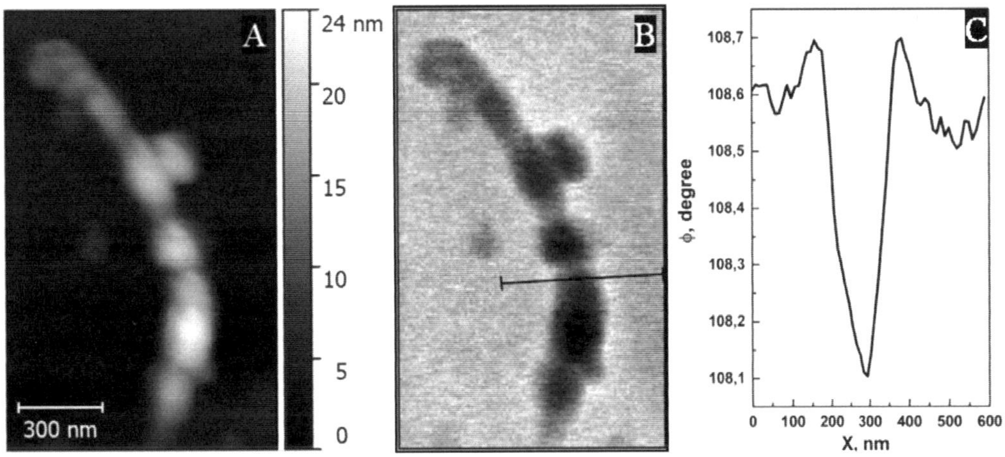

Figure 2. The PANI/N-CNT nanostructure: (a) AFM topography, (b) EFM image, and (c) cross section profile along the line marked at (b).

Figure 3a shows the change in the average value of the electron work function in individual nanostructures with an increase in the synthesis time of the composite. It can be seen that with an increase in the duration of synthesis, the work function in the nanostructures increases monotonically. The mean values and scatter of the electron work function in individual N-CNTs and in continuous thin layer of PANI/N-CNT composite can be also observed in Figure 3. The change in the work function in nanostructures occurs within the limits of these values and reaches a value of 0.4 eV.

We believe that increase of the work function with increasing of the synthesis time is due to two reasons. The polymer layer thickness on the CNT grows with increasing the synthesis time. The change in the work function with increasing thickness of the PANI layer can be demonstrated using the energy diagram of the PANI/N-CNT structure (Fig. 3b). At the polymer/nanotube junction, a space-charge region (SCR) is formed in the polymer. PANI band bending down leads to a decrease in the work function. Therefore, with decreasing thickness of the polymer layer on CNT, the work function of PANI/N CNT nanostructure decreases. The work function of the bulk layer of PANI/N CNT composite considerably exceeds the work function of nanostructure. This is due to the fact that the main contribution in the work function is provided by thick polymer layers on the CNTs or unrelated polymer particles.

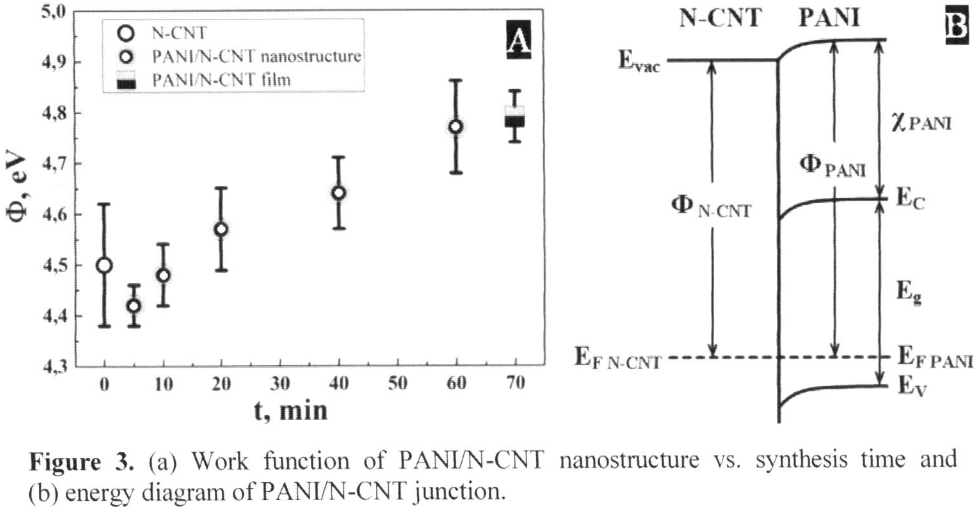

Figure 3. (a) Work function of PANI/N-CNT nanostructure vs. synthesis time and (b) energy diagram of PANI/N-CNT junction.

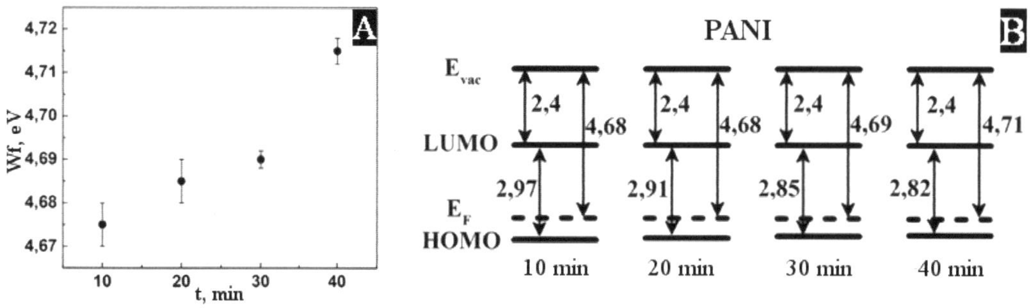

Figure 4. (a) Work function of PANI vs. synthesis time and (b) energy diagram of PANI for different durations of synthesis.

We estimate the thickness of the SCR in PANI using the expression:

$$W = \sqrt{\frac{2\varepsilon\varepsilon_0 \Delta\Phi}{eN}} \tag{1}$$

where ε is the permittivity of the PANI, which we take as 6.6 Fm^{-1} [7]; ε_0 is the electric constant; $\Delta\Phi = \Phi_{PANI}-\Phi_{N\text{-}CNT}$ is the barrier height equal to 0.3 eV; e is the electron charge; N is the concentration of charge carriers, taken as 10^{24} m^{-3}[8].

The estimated value of the SCR width in the PANI is 15 nm. According to this estimate, when the thickness of the polymer layer on an N-CNT is less than this value, the work function of the nanostructure will decrease and tends to a value for N-CNTs with a decrease in the thickness of the PANI. Thus, the variation of the work function in the nanotubes and the thickness of the PANI coating on the CNT makes it possible to adjust the work function in the individual PANI/CNT nanostructures.

Another reason for changing the work function with an increase in the duration of the synthesis may be a change in the electronic properties of the polymer. Earlier, we showed that with an increase in the synthesis time of pure PANI, the work function was increasing [9]. Figure 4a shows the change in the work function in the pure PANI layer depending on the duration of synthesis, as well as change of the band structure of the polymer at different durations of synthesis (Fig. 4b). This change is attributed to the increasing degree of protonation of the polymer, increasing concentration of charge carriers, and the shift of the Fermi level toward the valence band.

4. Conclusion

The electron work function in individual PANI/N-CNT nanostructures was determined by EFM method. It is shown that the work function of nanostructures monotonically increases with the duration of synthesis. This value lies in the range from the work function of N-CNT to that of the bulk PANI/N-CNT composite layer. These changes in the work function of PANI/N-CNT nanostructures are associated with an increase in the thickness of the polymer layer on CNTs and an increase in the degree of protonation of PANI with an increase in the synthesis time of PANI/N-CNTs composite.

Acknowledgments

The work was carried out according to the state task of the OSC SB RAS (project registration number AAAA-A17-117041210227-8) and partially supported by the Russian Foundation for Basic Research (grant 18-48-550009 p_a).

References

[1] Kumar A M and Gasem Z M 2015 *Surf. Coat. Tech.* **276** 416-23
[2] Mariano L C, Salvatierra R V, Cava C E, Koehler M, Zarbin A J G and Roman L S 2014 *J. Phys. Chem. C* **118** 24811-8
[3] Ji T, Tu R, Mu L, Lu X and Zhu J 2018 *Appl. Catal. B: Environ.* **220** 581-8

[4] Liu P, Han J-J, Jiang L-F, Li Z-Y and Cheng J-N 2017 *Appl. Surf. Sci.* **400** 446-52
[5] Ramana G V, Srikanth V V S S, Padya B and Jain P K 2014 *Eur. Polym. J.* **57** 137-42
[6] Davletkildeev N A, Stetsko D V, Bolotov V V, Stenkin Y A, Korusenko P M and Nesov S N 2015 *Mater. Lett.* **161** 534-7
[7] Zakrevskii V A, Kuz'min S V and Sudar' N T 2008 *Polym. Sci. Ser. A* **50** 411-6
[8] Toušek J, Toušková J, Chomutová R, Křivka I, Hajná M and Stejskal J 2017 *Synth. Met.* **234** 161-5
[9] Lobov I A and Shulus I K 2018 *Omsk Scientific Bulletin* **159** 98-101

Influence of silicon dioxide nanoparticles on mechanical properties of erythrocyte and platelet membranes estimated by atomic force microscopy

G B Melnikova[1], E E Konstantinova[1], T N Tolstaya[1], A S Petrovskaya[1], S A Chizhik[1], O N Shyshko[2] and T V Mokhort[2]

[1] A.V. Luikov Heat and Mass Transfer Institute of NAS Belarus, Minsk, Belarus
[2] Belarusian State Medical University, Minsk, Belarus

galachkax@gmail.com

Abstract. In this research, the influence of silicon dioxide nanoparticles on local mechanical properties of erythrocytes and platelets membranes were investigated by atomic force microscopy. It was found that adhesion force and aggregation properties (sedimentation rate of the second stage) of erythrocyte membranes decreased after incubation of cells with SiO_2 nanoparticles that could be justified by changing the charge of cell membranes.

1. Introduction

Silicon dioxide is widely used as excellent adsorbent for production of different drugs and drug carriers, and also for other applications. The research of nanoparticles' interaction with biological cells is important, because not only the drug itself affects a cell membrane, but also excipients do. Application of modern methods for nanoscale objects research allows getting detailed information about changes after different actions.

Erythrocytes are significant part of blood volume (approximately 45%). Viscosity depends on their concentration and rheological properties. There are some methods usually used to determine viscosity, e.g. viscometer in share rates, flow chambers, biomicroscopic data using tweezers and different indirect techniques [1]. One of these methods, which is perfectly and widely developed now, is atomic force microscopy (AFM) for studying local elastic modulus and adhesion force. Rheological properties of erythrocytes are estimated on the base of local mechanical properties of erythrocyte membranes. Changes of structure and mechanical properties of erythrocytes and platelets membranes are related to type of patient pathologies and different external influences. But these changes may not be obvious and cannot be identified using routing methods. S. Baar studied the effect of high temperature on osmotic properties of erythrocytes [2]. J. W. Choi analysed how to change a number of cells, haemoglobin concentration after thermo action [3].

It is known, that nanoparticles have an effect on viscosity and rheological properties of erythrocytes. For example, polymeric nanoparticles reduce the rate of sedimentation of erythrocytes, which is dilute with connecting functional groups of nanoparticle with carboxyl groups of acetyl derivatives of neuraminic acids. The last defines the charge of erythrocytes membranes, which is one of the main factors in the process of aggregation [4]. The second theory is about depleted layer with low concentration, which forms and changes the erythrocyte aggregation.

AFM can be used to investigate the cell properties after incubation with different types of nanoparticles and temperature. AFM methods allow studying mechanical properties and structure at the nanoscale that is very perspective and promising.

2. Experimental details

The research group of patients with average age of 55 (50; 59) and type 2 diabetes mellitus (DM 2) was formed. They did not achieve compensation for HbA1c 9.1 (8.30, 10.4). Men and women were in same age, and they had DM 2 during last 5-10 years. Blood sampling was carried out from the cubital vein not earlier than 12 hours after the last meal.

Erythrocytes were separated from whole blood and stabilized with potassium ethylendiaminetetraacetatic acid dipotassium salt (K_2EDTA) (Sigma-Aldrich) by centrifugation. Platelets were separated from stabilized whole blood by sodium citrate. Obtained cells precipitation was washed with phosphate buffered saline (PBS) 3 times and was added to physiological solution of silicon dioxide nanoparticles (SiO_2 NP, Sigma-Aldrich, d = 10-20 nm) with concentration C = 0.2 and 1 mg/ml. Cells were incubated with NP suspension at room temperature during 40 and 60 min. Then suspensions of cells were washed and fixed with 0.5% buffer solution of glutaraldehyde during 30 min. After that, suspension was washed twice by PBS and water precipitation with cells made a smear on mica plates. Method of preparing cells for AFM researching was described in Ref. [5].

The influence of silicon dioxide nanoparticles on structure and mechanical properties of membranes blood cells by AFM method has been studied. We used AFM device NT-206 (produced by MTM, Belarus) with standard silicon probe NSC11 of V-shaped type (produced by "Mikromasch", Estonia), stiffness 3 N/m, curves radius of probe 50 nm (for spectroscopy in point procedure) and 10 nm (for scanning procedure). Elastic modulus was calculated by Jonson-Kendall-Robertz model [6].

3. Results and discussion

In early research, we showed that gold and polyacrylic acid [5] nanoparticles did not influence the elastic modulus of RBCs membranes; results were based on AFM data. In this research, we found that cell morphology (Fig. 1) and membrane structure did not change after incubation with NP.

The average sizes of erythrocyte is 6.5-7 μm. As a rule, the discocyte form of erythrocytes is prevailing in the whole population of investigated erythrocyte cells. In the case of platelets, size of cells are slightly increased from 2 to 3 μm (Fig. 2). No other changes were observed. The value of roughness of platelets membranes also did not change.

We associate it with NP interaction with membranes of structural element and shaping uniform structure with the cell. It was confirmed that in some cases aggregation stability of RBCs was decreased in presence of NP. It is connected with NP absorbance by cells and as a result changing charge on erythrocytes surface and their aggregation.

The average elasticity modulus of initial erythrocytes membranes is 128.5 ± 10.0% MPa and for platelets it is 151.6 ± 10.0% MPa, adhesion force is 23.0 ± 10.0% and 26.0 ± 10.0% nN, respectively. After cells incubation with silicon dioxide, the elastic modulus changed within experimental error for both cell types. Changes of adhesion force are minor, however insignificant variations were found in the case of incubation of erythrocytes with NP (c= 1.0 mg/ml) during 40 minutes and reached 19.0% of initial values (Fig. 3). In the Figure 3, data of adhesion and elasticity are presented as relation values compared to initial cells membranes. More detailed calculation is presented in Ref. [5].

In our research, we analysed the aggregation properties of erythrocytes in comparison with AFM data for mechanical properties. Erythrocytes aggregation included three stages. The first and the second ones were more indicative, which allowed interpreting interaction between the NP and cell membrane in different moments. We found that at the first stage sedimentation rate changed in different ways and depended only on the biochemical state of patient's blood. At the second stage, after incubation with suspension of SiO_2 NP, the rate increased as a rule. The most significant increase was observed at the concentration 1 mg/ml (Fig. 4), which was a consequence of the increase in charge and repulsion of cells at the first stage.

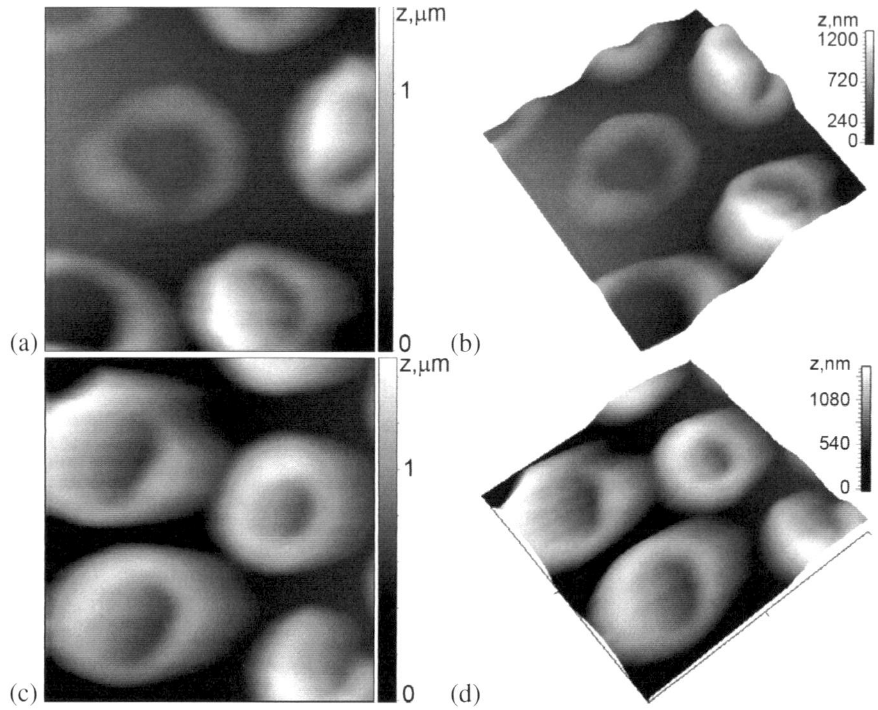

Figure 1. AFM images of (a,b) initial erythrocytes and (c,d) erythrocytes after cells incubation with physiological solution of SiO_2 NP. Scan size 14x14 μm^2.

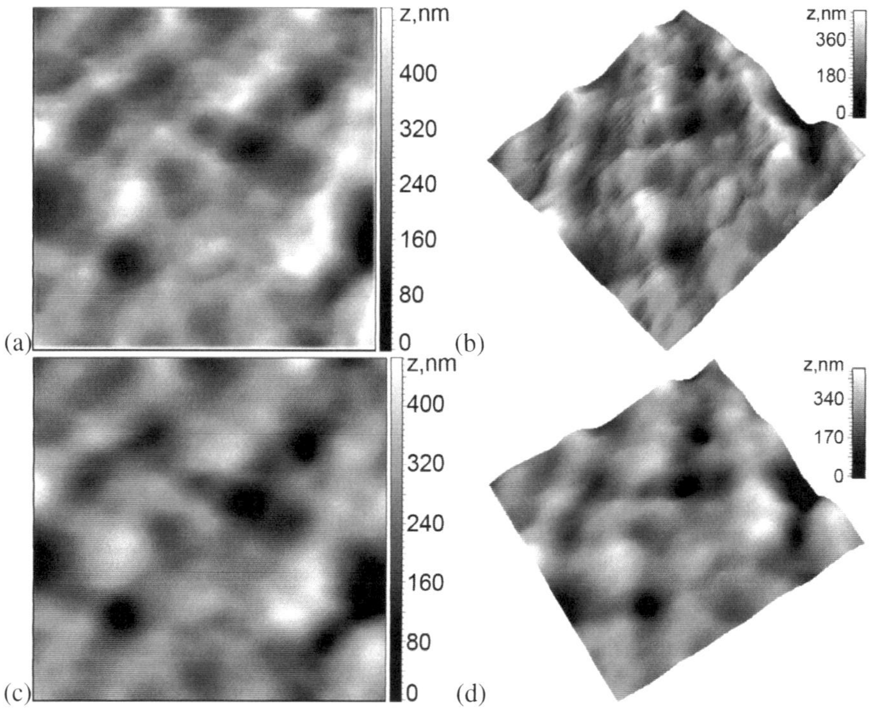

Figure 2. AFM images of (a,b) initial platelets and (c,d) platelets after cells incubation with physiological solution of SiO_2 NP. Scan size 10x10 μm^2.

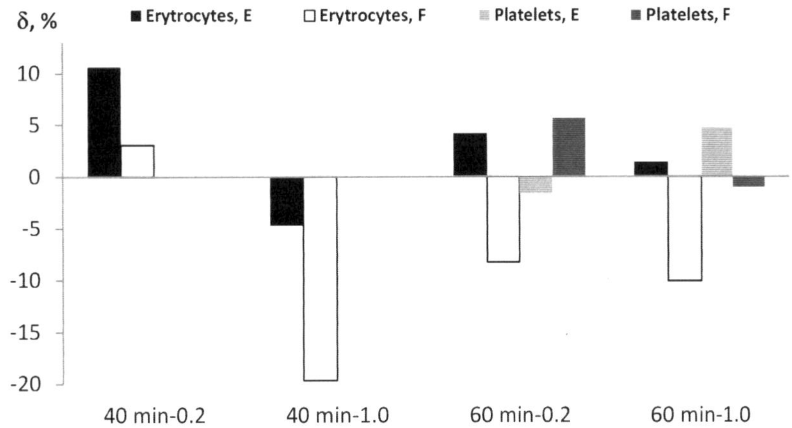

Figure 3. Relative (in percent) changes of elastic modulus (E) and adhesion force (F) of erythrocytes and platelets membranes after cells incubation with physiological solution of SiO_2 NP.

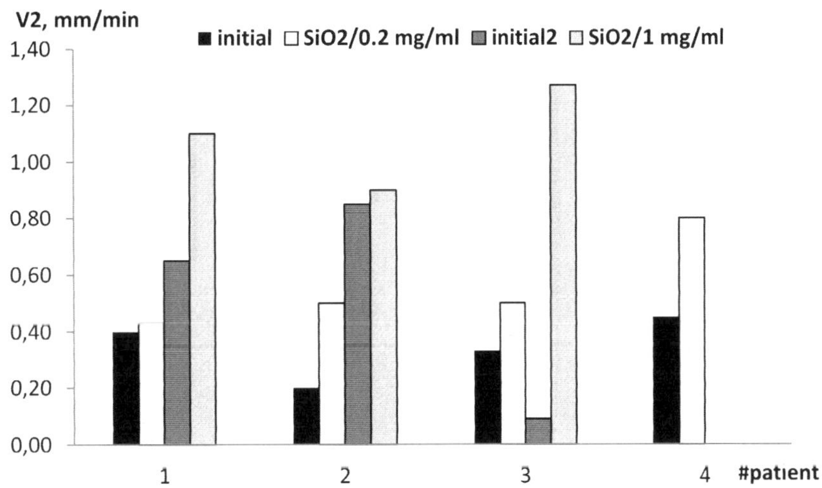

Figure 4. Changes of the erythrocytes sedimentation rate at the second stage after blood incubation with physiological solution of SiO_2 NP.

4. Conclusion

It was found that SiO_2 nanoparticles with concentration 1 mg/ml had the most significant influence on aggregation properties of erythrocytes. The obtained results are fundamental and can be used to develop methods to determine properties of blood cells under various affects and pathologies by atomic force microscopy.

Acknowledgments
The investigation was performed within the Programs of State Research "Energy systems, process and technologies", project 2.2.

References

[1] Sokolova I A *Regional bloodshed and microcirculation* 2010 **4** 4-26 (Russian edition)

[2] Baar S *J. Clin. Path.* 1967 **20** 239-43
[3] Choi J W and Pai S Hw *Clin. Lab. Sci.* 2002 **32** 393-8
[4] Baumler H, Neu B, Donath E and Kiesewetter H *Biorheology* 1999 **36** 439-42
[5] Melnikova G B, Kuzhel, N S, Tolstaya T N, Konstantinova E E, Drozd E S, Shishko O N, Mokhort T, Antonova N, Riha P, Kowalczuk A and Koseva N *Ser. Biomech.* 2015 **29** 12-19
[6] Johnson K L, Kendall K and Roberts A D *Proc. R. Soc. Lond.* 1971 **324** 301

Characterization of soft and heterogeneous surfaces by mechanical properties map obtained with atomic force microscopy

G B Melnikova, A S Petrovskaya, A A Makhanek and S A Chizhik

A. V. Luikov Heat and Mass Transfer Institute of NAS Belarus, Minsk, Belarus

galachkax@gmail.com

Abstract. In this research, mechanical properties of soft materials were estimated using atomic force microscopy methods. Different regimes of the nanoindentation process were used. Surface mapping and calculation by Jonson-Kendell-Roberts model allowed obtaining elastic modulus of the platelets heterogeneous surface, minimizing experimental deviations, and identifying the difference between the studied surfaces.

1. Introduction

In biomedical studies, the atomic force microscope (AFM) is used not only to visualize the topography of biological materials and cells, but also to determine their mechanical properties. One of the tasks is to select a correct model for calculating the elastic modulus. A review of the literature data on the nanoindentation of biological cells in the liquid phase by atomic force spectroscopy showed that there was no single technique for determining the elastic characteristics of the membranes of such cells. More experimental works on determining the elastic modulus of fixed biological cells in the liquid phase are based on the simplest model of contact interaction – Hertz or Johnson – Kendell – Roberts (JKR). In the present research, we did not take into consideration different types of influences on calculation of elastic modulus. Unlike the JKR, the Hertz model does not take into account the adhesion forces. But it is valid only for a perfectly elastic sample. Both models do not take into account the contribution of the viscous friction force that occurs during deformation of not fixed cells.

The elastic-viscous nature of the sample can be taken into account when analysing the nanoindentation data on the basis of the model developed by Ting [1]. It describes interaction between the tip of an axisymmetric probe and a flat elastic viscous sample both during the approach and retraction of the device working platform. Note that the Ting model does not take into account the adhesion forces and it is difficult to address them.

The general method for determining the viscoelastic characteristics of the material is to record the stress relaxation after the predetermined or known displacement (deformation of the sample) [2]. In this study, based on the atomic force spectroscopy, a force response was measured on time after the given depth of cell indentation. Analysis of the cells viscoelastic characteristics during measurements in liquid phase was carried out according to the Hertz model taking into account the finite thickness of the cell with assumption that it had a planar shape and the probe deformation could be neglected.

In the Hertz model approximation, under the assumption that the rate of deformation of the sample is constant, using the Voigt-Kelvin model, as well as the direct and inverse Laplace transforms, the authors of Ref. [3] obtained an expression for the magnitude of the interaction force between the probe

IOP Conf. Series: Materials Science and Engineering **443** (2018) 012023 doi:10.1088/1757-899X/443/1/012023

and the sample. But this model does not take into account adhesion forces. To solve this problem, we used the Tabor parameter. For example, if we calculate using JKR model we can take into consideration adhesion forces between cantilever and sample surfaces [4]. This model is more suitable for the polymeric materials as well as fixed and living biological cells [5].

One of the important problems is to determine local mechanical properties of heterogeneous surface by AFM [6]. As usual, objects are characterized by convexities and concavities. Characteristics may also vary in different points of studied objects. It is important to make an amendment for shape of the probe that is well described in F. Borodich's article [7]. Other corrections to be made are the angle, under which the presser on the convexity is applied, and the probe interaction with the concavity.

Another important problem is to obtain the statistically reliable data on the base of mechanical properties values distribution on the surface and their changes from cell to cell.

Therefore, the aim of our research is to establish availability of viscoelastic properties of fixed platelets membranes similarly with homogeneous materials polydymethylsiloxane using elastic properties map obtained by AFM method. Consideration of all these corrections, as well as elasticity maps constructing studies, allows us to minimize variance of elasticity for the surface under research.

2. Experimental

In our research we used two types of materials – fixed biological cells (heterogeneous surface) and polydimethylsiloxane (PDMS, homogeneous surface). PDMS was made by deposition a polymer solution onto planar substrate and then drying it in air.

Platelets were separated from stabilized whole blood by sodium citrate. Obtained cells precipitation was washed with phosphate buffered saline (PBS) 3 times. Then suspensions of cells were fixed with 0.5% buffer solution of glutaraldehyde during 30 min. Suspension twice washed by PBS and water precipitation with cells was used to make a smear on mica plates. The method of preparing cells for AFM researching was described in Ref. [8].

For research, we used an AFM device NT-206 (produced by MicrotestMashines, Belarus) with standard silicon probes of V-shaped type (produced by "Mikromacsh", Estonia) with stiffness 0.35 N/m (CSC 11) and 3 N/m (NSC 11) and curve radius of probes 10 nm. The stiffness of probe is indicated according to the technical passport of manufacturer. Elastic modulus was calculated by JKR model.

Process of point spectroscopy was made with different indentation times. This makes it possible to identify the viscous properties exhibited by the materials.

3. Results and discussion

The research surface of PDMS was characterized by homogeneous structure with low value of RMS, about 2.3 nm for scan sizes 4x4 μm^2 (Fig. 1a) and 5.5x5.5 μm^2 (Fig. 2a). In Figure 1, the lines and points of nanoindentation were demonstrated. Similar size of fixed platelets surface is interesting too, because it has the heterogeneity in the form cells (about 3 μm) on the mica plate surface (Fig. 3). Map of mechanical properties allowed identifying platelets on the surface. AFM images of research platelets are presented in Figure 4a.

It was found that with increasing of indentation time for PDMS in the range from 12 to 99 s, the elastic modulus did not change and constituted 2.6 MPa in average (Fig. 1b). It should also be taken into account that values of elasticity did not depend on probe stiffness or nanoindentation loads of probe on sample surface and remained constant (Fig. 2b). Depth of nanoindentation increased with increasing load from 80 to 106 nN. The obtained results confirm that this material does not have viscosity properties.

According to Figures 3 and 4, the platelets laid more in the form of conglomerates rather than single cells. The average elastic modulus of the cells is 163.5 MPa ± 15%. Deviations are due to various properties of the surface membranes platelets and include device error. During mapping mechanical properties and correlating with surface morphology, we can reject data not connected with the platelets properties (Fig. 4). So, after analysis by mapping we give less deviation, about 10%. On the map of elasticity modulus, one place with higher value was identified, due to hollow on the mica plate.

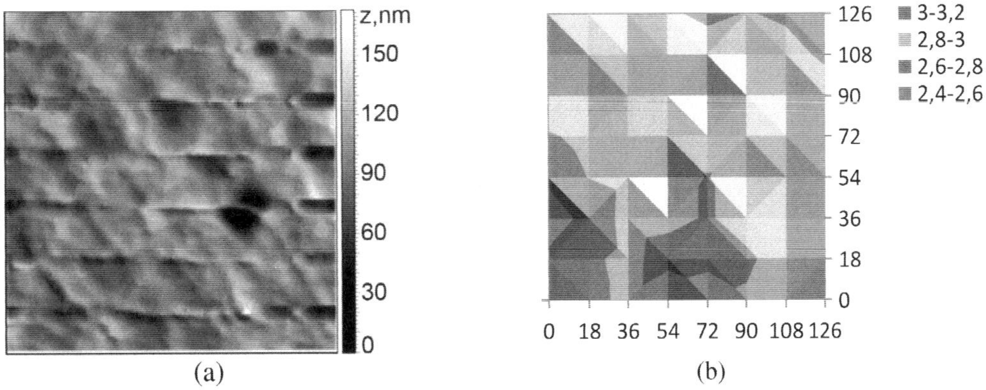

Figure 1. (a) Structure in Torsion and (b) elasticity modulus (MPa) of PDMS surface 4x4 μm² (CSC11 probe, time 61 s, load 106 nN).

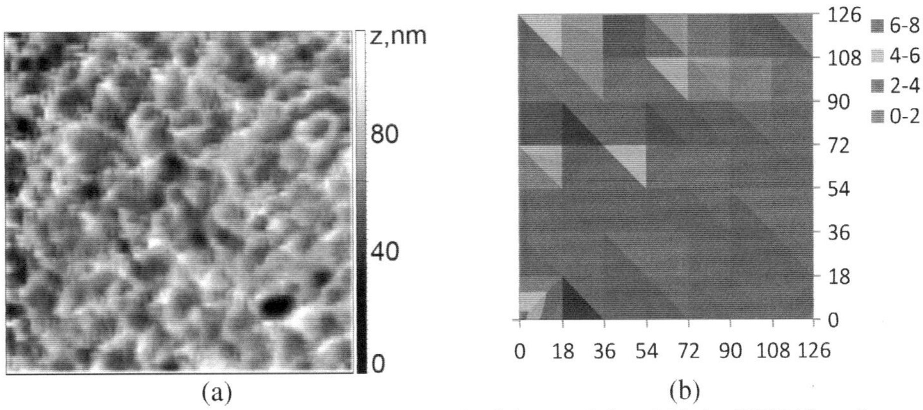

Figure 2. (a) Structure in Torsion and (b) elasticity modulus (MPa) of PDMS surface 5.5x5.5 μm² (NSC11 probe, time 37 s, load 106 nN).

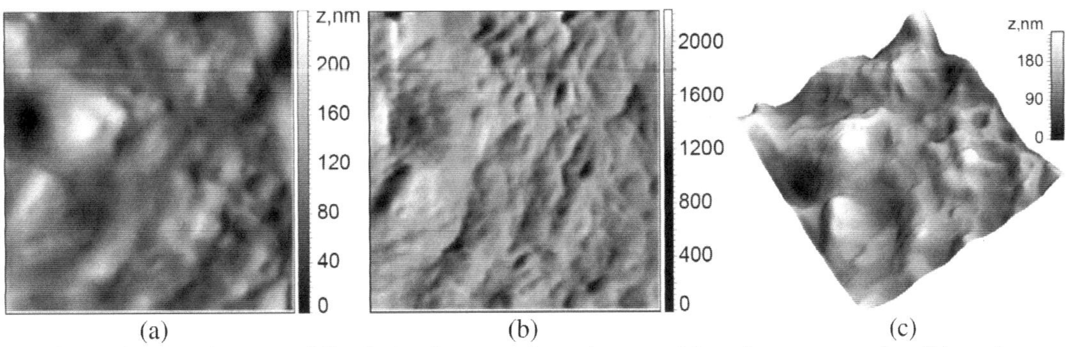

Figure 3. AFM-images of fixed platelets on mica substrate: (a) surface topography, (b) torsion, (c) 3D. Scan size 6.3x6.3 μm².

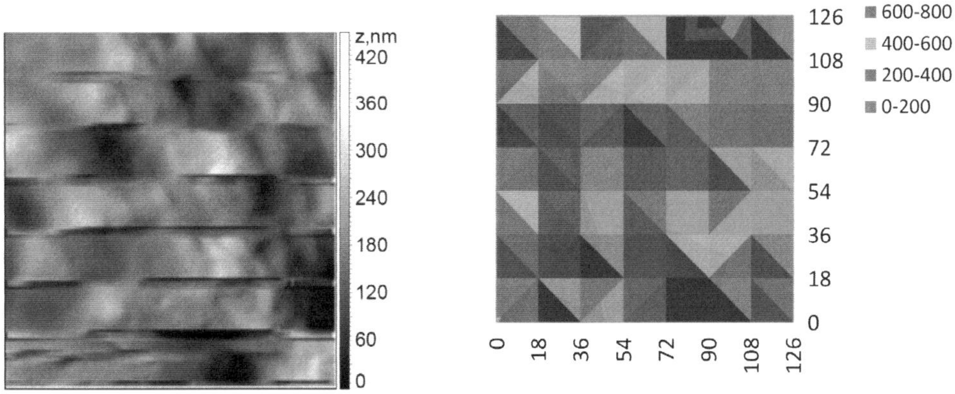

(a) (b)

Figure 4. (a) Structure and (b) elasticity modulus (MPa) of fixed platelets, surface 7x7 μm^2 using NSC 11A probe, load 40%.

The different indentation times in point were applied for fixed platelets too. It was found, that with increasing of nanoindentaion time the elastic modulus increased from 114.3 to 210.6 MPa. This is due to insignificant demonstration of platelets viscous properties. More optimal time of indentation is 70 s with load 80 nN.

4. Conclusion

In this research, mechanical properties of soft materials were estimated by AFM methods. Different regimes of the nanoindentation process were used. It was demonstrated that mapping mechanical properties allowed minimizing values deviation. It was found that fixed platelets had moderate viscous properties.

Acknowledgments

This research was supported by the grants of Belarusian Republican Foundation for Fundamental Research BRFFR No. F18V-007 and Programs of State Research «Konvergencia-2020», project 3.03.

References

[1] Darling E M, Zauscher S, Block J A and Guilak F 2007 *Biophys. J.* **92** 1784-91
[2] Gorshkova E, Pleskova E and Mikheeva E *Nanoindustry* 2012 **4** 50-3 (in Russian)
[3] Ting T C *J. Appl. Mech.* 1966 **33** 845-54
[4] Chizhik S A, Huang Z, Gorbunov V V, Myshkin, N K and Tsukruk V 1998 *Langmuir* **14** 2606-9
[5] Melnikova G B et al. *5thEurosummer school on biorheology and symposium on micro and nanomechanics and mechanobiology of cells, tissues and systems* 2015 33
[6] Zhu X, Siamantouras E, Liu K K and Liu X 2016 *J. Mech. Behav. Biomed. Mater.* **56** 77-86
[7] Borodich F *Adv. Appl. Mech.* 2014 **47** 225-336
[8] Melnikova G B, Kuzhel, N S, Tolstaya T N, Konstantinova E E, Drozd E S, Shishko O N, Mokhort T, Antonova N, Riha P, Kowalczuk A and Koseva N, *Ser. Biomech.* 2015 **29** 12-9

Domain kinetics during polarization reversal in 36° Y-cut congruent lithium niobate

E A Neradovskaia[1,2], M M Neradovskiy[1,2], A A Esin[2,3], M A Chuvakova[2], P Baldi[1], M P De Micheli[1], A R Akhmatkhanov[2,3], N Forget[4] and V Ya Shur[2,3]

[1] Institut de Physique de Nice, 06108, Nice, France
[2] School of Natural Sciences and Mathematics, Ural Federal University, 620000, Ekaterinburg, Russia
[3] Labfer Ltd., 620014, Ekaterinburg, Russia
[4] Fastlite Ltd., 06600, Antibes, France

neradovskaia.elizaveta@ya.ru

Abstract. We present experimental study of domain kinetics in 36° Y-cut congruent lithium niobate during polarization reversal by increasing external electric field using liquid electrodes. The motion of the domain walls has been imaged on the Z_s+ and Z_s- polar surfaces of the crystal simultaneously. The linear dependence of the threshold field on the field rate has been revealed. The threshold field in low field rate limit has been estimated as 32.8 ± 0.3 kV/mm, which is lower than predicted value. The switching process for polarization reversal in the constant field has dramatically accelerated with increase of external electrical field above threshold. The new approach, which accelerates significantly the *in situ* study of 3D domain kinetics in CLN crystals, has been demonstrated.

1. Introduction

The middle-infrared (mid-IR) wavelength range can be defined as a portion of electromagnetic spectrum varied from edge of near-infrared (2 μm) to tens of microns. Mid-IR spectral range has a number of so-called "water windows" representing light wavelengths, which can propagate through Earth's atmosphere without absorption by water vapour [1]. Moreover, most molecules have specific absorption lines in the mid-IR, affording a description that makes it possible to determine them by optical spectroscopy [2].

To create the optical frequency mixers adjusted to the ultra-short pulse sources (below 100 fs) and high average power (above 10 W) in mid-IR, it is necessary to obtain non-linear components with large aperture (about 1 cm^2). The periodically poled nonlinear optical crystals of lithium niobate (LN) and lithium tantalate (LT) are well fitted to the optical parametric amplification in the mid-IR due to their wide transparency range and high non-linearity, but the routine poling technique is not suitable for producing the components with appropriate large aperture [3].

In order to overcome this technological barrier, the "slanted poling" (axis-slant quasi-phase matching) has been elaborated [4]. The essence of the method is to produce periodical grating in a plate with slanted polar axis, specifically in 25° X-cut MgO-doped LN. In this work, we have investigated 36° Y-cut congruent LN (CLN) crystals, which are used for surface acoustic wave devices. The main advantages of these crystals are low price and great offer on the market.

2. Experimental

The studied samples represented CLN plates cut in such a way that the angle between normal to the crystal surface and Y-axis was 36°. The 0.5-mm-thick 3-inch wafers were polished from both sides to optical quality and cut on samples with sizes 10x15x0.5 mm^3. All experiments were performed at room temperature.

Polarization reversal was performed by application of external electric field pulses of two shapes: (a) rising field (Fig. 1b) and (b) constant field (Fig. 2b), using liquid electrodes (aqueous solution of lithium chloride). The top electrode 2 mm in diameter was located on the surface corresponding to positive direction of the slanted polar Z-axis (Z_s^+), and the continuous bottom electrode was located on the opposite side (Z_s^-).

Magnitude of the applied external electric field ranged from 30 to 35.2 kV/mm. *In situ* imaging of the domain structure mas made by polarizing microscope (Carl Zeiss LMA10, Germany) with magnification 3.2x and recorded by the high-speed CMOS camera (Photron Mini UX100, Japan) with film rate varied from 50 to 250 fps.

Resulting videos were analysed by methods of image processing realized in Wolfram Mathematica software.

3. Results and discussion

Two types of the switching experiments were realized by application of the single pulses of the external field of different waveforms: switching in rising field and in constant field.

The key parameter of polarization reversal in any ferroelectric is excess of the switching field (E_s) above the threshold value (E_{th}) defined as $\Delta E_{ex} = E_s - E_{th}$. It is known that the threshold field in Z-cut CLN is about 21 kV/mm [5]. Hence, the estimated threshold field of the 36° Y-cut CLN should be $E_{th}^N = E_{th}^Z/\text{Sin } 36° \approx 35.7$ kV/mm.

3.1. Switching in rising field

In order to measure dependence of E_{th} on the field rise rate (dE/dt) (Fig. 1a), several samples were switched by pulses with rising field (Fig. 1b) with dE/dt ranged from 0.2 to 1.3 kV/mm•s^{-1}. Maximum applied field was 35.2 kV/mm. The application of the pulses with higher field amplitude led to a breakdown. The threshold field was estimated as a field value, at which the first domain became visible. Linear fitting shows that the threshold field for quasi-static switching along the normal to the sample $E_{th}^N = 32.8 \pm 0.3$ kV/mm, which is lower than predicted value. This fact can be attributed to formation of the first domains at the electrode edges with higher field value due to a fringe effect.

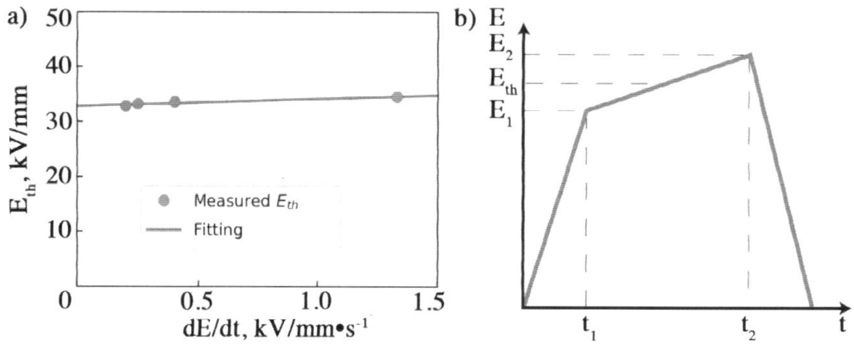

Figure 1. (a) Dependence of the threshold field (E_{th}) defined as a field value at the moment, when the first domain became visible on the field rise ($\Delta E/\Delta t$). Experimental points were fitted by the linear dependence. (b) The waveform of the rising field pulse. $\Delta t = t_2 - t_1$, $\Delta E = E_2 - E_1$.

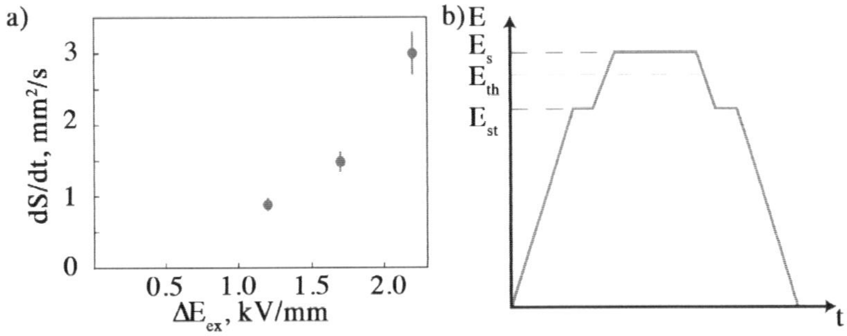

Figure 2. (a) Dependence of the switched area growth rate (dS/dt) on the excess of the switching field over the threshold value (ΔE_{ex}) for polarization reversal in constant field. (b) Waveform of the pulse used for switching in constant field. E_s - switching field, E_{st} - stabilization field, E_{th} - threshold field.

The kinetic maps [6] shown in Figure 3 were obtained by successive overlapping the domain wall positions recorded with frequency 5 Hz during polarization reversal by rising field pulse with $dE/dt = 0.2$ kV/mm•s^{-1}. The domain wall positions were detected by consistent application of an edge detection pipeline to the video frames. The edge detection pipeline included the following steps: 1) digital noise reduction by subtraction of the first frame (the image of the sample in the initial single domain state) and application of a curvature flow filter [7], 2) application of Canny Edge algorithm [8] to obtain the image with domain wall contrast only. The colour of each wall corresponds to the value of applied external field and the time, at which the wall was recorded. The final kinetic map represents a sum of the frames processed by the edge detection pipeline mentioned above. The maps demonstrate propagation of domain walls on Z_s^+ and Z_s^- surfaces. The clearly seen nonmonotonic change of the wall velocity can be attributed to pinning by defects and domain-domain interactions.

Dependence of the domain wall velocity along positive and negative projections of Y-axis on Z_s^+ surface on applied external electric field is shown in Figure 3c. The switching process always starts by appearance of isolated domains at the electrode edges. Two mechanisms are observed later: (1) slow jump like growth of the domains and (2) anisotropic wall motion after domain merging. For applied field just above the threshold (ΔE_{ex} ranged from 0.3 to 0.8 kV/mm), the walls stopped at the defects with local increasing of the threshold field (pinning centres). The walls restarted, when the applied field overcame the local threshold field value [9].

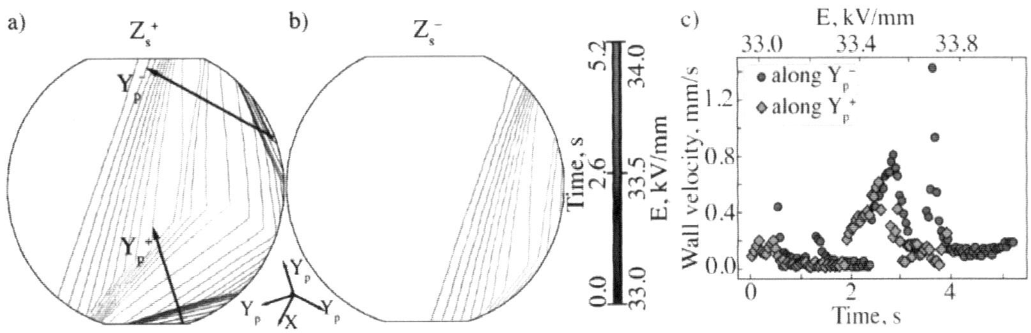

Figure 3. Kinetic maps of the domain structure evolution for polarization reversal in 36° Y-cut CLN by single pulse of rising field with field rate of 0.2 kV/mm•s^{-1} on (a) Z_s+ and (b) Z_s– surfaces. The domain wall colour corresponds to the time and applied external field. The time starts from appearance of the first visible domain in the field of vision. (c) The field dependence of the domain wall velocity along positive and negative projections of Y-axis shown by arrows on (a) on Z_s^+ polar surface.

It is known that the domain walls orientations in Z-cut CLN are defined by C_{3V} crystal symmetry. For equilibrium switching conditions characterized by effective screening of depolarization field, the domain walls are strictly oriented along Y-axis [10-13]. In our experiment, domain walls were oriented along projections of Y-axes at low fields ($\Delta E_{ex} \leq 0.8$ kV/mm) and their orientation on Z_s^+ and Z_s^- surfaces tended to X-axis with increasing of ΔE_{ex}. The obtained deviation from crystallographic directions is caused by increasing of the kink concentration at the wall [14,15].

3.2. Switching in constant field
The domain kinetics was studied in 36° Y-cut CLN in constant fields ranged from 34 to 35.2 kV/mm. The stabilization steps E_{st} = 30 kV/mm were added to the waveform before and after the switching step (Fig. 2b). The first stabilization step allowed preventing the crystal dielectric breakdown and the second one allowed minimizing the backswitching process [16,17].

Dependence of the switched area growth rate on the excess of the switching field above the threshold value (ΔE_{ex}) is shown in Figure 2a. It is seen that the relative increasing of the external electrical field in 0.0035 accelerated the polarization reversal three times.

4. Conclusion
The domain kinetics in 36° Y-cut CLN during polarization reversal by rising external electric field using liquid electrodes has been studied. The polarization reversal in tilted cut of CLN crystal gives information, which accelerates significantly the *in situ* study of 3D domain kinetics due to a possibility to observe domain walls motion on the Z_s^+ and Z_s^- polar surfaces of the crystal simultaneously. The kinetic maps containing complete information about wall motion have been used for analysis of data obtained during polarization reversal in rising field. Observed deviation of the domain walls from crystallographic axis is attributed to the increase of the kink concentration at the wall. The measured linear dependence of the threshold field on the field rate allows estimating the threshold field in low field rate limit as, $E_{th}^N = 32.8 \pm 0.3$ kV/mm which is lower than predicted value due to the fringe effect. The increase of external electrical field above threshold leads to a dramatic acceleration of the switching process for polarization reversal in the constant field. The obtained information will be used for further development of electric field poling technique in CLN crystals for high-power laser sources.

Acknowledgments
The equipment of the Ural Centre for Shared Use "Modern Nanotechnology" Ural Federal University was used.

References
[1] Lubin D 1994 *J. Climate* **7** 121-40
[2] Bertie J E and Lan Z 1996 *Appl. Spectrosc.* **50** 1047-57
[3] Ishizuki H and Taira T 2008 *Opt. Express* **16** 16963-70
[4] Ishizuki H and Taira T 2011 *Opt. Mater. Express* **1** 1376-82
[5] Shur V Ya, Rumyantsev E L, Nikolaeva E V, Shishkin E I, Batchko R G, Fejer M M, Byer R L and Mnushkina I 2002 *Ferroelectrics* **269** 189-94
[6] Shur V Ya, Esin A A, Alam M A and Akhmatkhanov A R 2017 *Appl. Phys. Lett.* **111** 152907
[7] Malladi R and Sethian J A 1995 *Appl. Math.* **92** 7046-50
[8] Canny J 1986 *IEEE Trans. Pattern Anal. Mach. Intell.* **8** 679–98
[9] Shur V Ya, Rumyantsev E L, Pelegov D V, Kozhevnikov V L, Nikolaeva E V, Shishkin E L, Chernykh A P and Ivanov R K 2002 *Ferroelectrics* **267** 347-53
[10] Shur V Ya 2006 *J. Mater. Sci.* **41** 199-210
[11] Shur V Ya, Gruverman A L and Rumyantsev E L 1990 *Ferroelectrics* **111** 123-31
[12] Shur V Ya 2006 *Ferroelectrics* **340** 3-16

[13] Shur V Ya, Rumyantsev E L, Nikolaeva E V, Shishkin E I, Batchko R G, Fejer M M and Byer R L 2001 *Ferroelectrics* **257** 191-202
[14] Shur V Ya, Akhmatkhanov A R and Baturin I S 2015 *Appl. Phys. Rev.* **2** 040604
[15] Lobov A I, Shur V Y, Baturin I S, Shishkin E I, Kuznetsov D K, Shur A G, Dolbilov M A and Gallo K 2006 *Ferroelectrics* **341** 109-16
[16] Batchko R G, Shur V Ya, Fejer M M and Byer R L 1999 *Appl. Phys. Lett.* **75** 1673-5
[17] Kolchina E A, Neradovskiy M M, Shikhova V A, Pelegov D V, Shur V Ya, Ivleva L I and Dec J 2016 *Ferroelectrics* **496** 149-56

IOP Conf. Series: Materials Science and Engineering **443** (2018) 012025 doi:10.1088/1757-899X/443/1/012025

Main results obtained in a series of animal experiments for the assessment of the organism's responses to metallic nanoparticles exposure

L I Privalova[1], M P Sutunkova[1], I A Minigalieva[1], V B Gurvich[1], O G Makeyev[2], I V Valamina[2], V Ya Shur[3], E V Shishkina[3], I V Zubarev[3], S N Solovyova[1], S V Klinova[1] and B A Katsnelson[1]

[1] Ekaterinburg Medical Research Centre for Prophylaxis and Health Protection in Industrial Workers, 620014, Ekaterinburg, Russia
[2] Central Research Laboratory of the Ural State Medical University, 620028, Ekaterinburg, Russia
[3] School of Natural Sciences and Mathematics, Ural Federal University, 620000, Ekaterinburg, Russia

privalovali@yahoo.com; privalova@ymrc.ru

Abstract. Nanoparticles (NPs) of Fe_3O_4 were produced by a chemical technique and NPs of Ag, Au, CuO, NiO, Mn_3O_4, PbO, ZnO, TiO_2, SiO_2, and Al_2O_3 – by laser ablation in water. In some experiments we compared particles of a given chemical composition having different diameters, while in others – equidimensional NPs of different metals or metal oxides (Me-NPs). We used two experimental models: a single intra-tracheal instillation of Me-NPs 24 h before the bronchoalveolar lavage procedure (collecting) and the repeated intra-peritoneal injections during 6-7 weeks in non-lethal doses. Besides, we carried out long-term inhalation experiments with NPs of Fe_2O_3, SiO_2, and NiO. We have demonstrated that NPs are much more noxious than their fine micrometric or even submicron counterparts and are usually the more toxic the smaller their dimensions within the nanoscale range. We have found also that toxicity of Me-NPs strongly depends on their chemical nature, solubility, and specific mechanisms of characteristic action of a given metal in any chemical form. Solubility of Me-NPs in biological milieus plays an important role in their toxicokinetics, which can prevail over that of the physiological mechanisms controlling their distribution, retention, and elimination. On the other hand, thanks to the high activity of these mechanisms, the organism is not defenceless against the impact of Me-NPs. The toxicity and even genotoxicity of Me-NPs can be significantly decreased by adequately composed combinations of some bioactive agents in innocuous doses.

1. Introduction

Nanoparticles (NPs) of metals and of their oxides are of special interest for risks assessment and management, because along with engineered NPs, there exists usually a substantial fraction of nanoscale ("ultrafine") particles of the same substances within the particle size distribution of condensation aerosols generated by arc-welding, metallurgical, and some chemical technologies. However, in such industrial aerosols chemically similar micrometre particles, including submicron ones having dimensions >100 nm, are usually present as well. This fact is confirmed by the distribution of particles by size in the submicron range on a filter sampled from the workplace air of a copper smelting and casting facility (Fig. 1) [1,2].

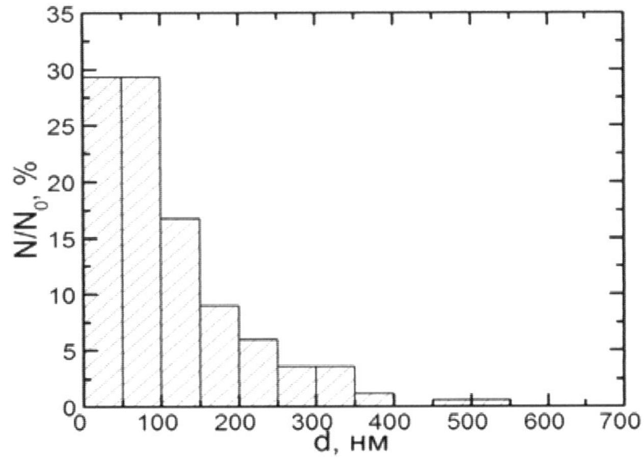

Figure 1. The percentage distribution of particles by size in the submicron range on a filter sampled from the workplace air of a copper smelting and casting facility. N - number of particles of a given diameter; N_0 - total number of particles.

The situation becomes even more complicated, if we take into consideration how rare in such industries the air is polluted by particles of one chemical composition. Indeed, most Me-NP-generating by metallurgical and welding technologies are bound to produce and do produce multi-component mixtures of chemically different particles of similar or dissimilar geometry. For instance, in arc-welding and alloyed steel production, one usually finds different combinations of iron, chromium, nickel, manganese, and silicon oxides in the form of NPs and microparticles (MPs), while in crude copper smelting and refining – those of copper, lead, cadmium, zinc, and arsenic oxides.

Thus, solving some general problems in relation to Me-NPs that nanotoxicology faced from the very beginning of its evolution as a special branch of the toxicological science was not only theoretically challenging, but also very urgent for the everyday practice of occupational health risks assessment and management. Starting from the very first nanotoxicological work in 2008-2009 [3,4] up to now, our team has been conducting *in vivo* animal experiments involving Me-NPs and has thus accumulated a certain wealth of relevant knowledge. This paper summarizes mostly our experience [3-5].

2. Materials and methods

Some suspensions of NPs were produced by laser ablation of metal targets (99.99% purity, 1-mm thick) placed on the bottom of a glass vessel with 5-30 mL of deionized water. NPs size distribution was obtained by a direct measurement using scanning electron microscopy. Suspension stability was characterized by zeta potential measured in a Zetasizer Nano ZS analyser (Malvern, UK) and it was high enough to increase the concentration by partial water evaporation at 50°C [2].

The experiments were carried out on outbred white male and female rats from our breeding colony 3-4 month old with an initial body weight of 150-220 g. Both the exposed and the control groups contained at least 12 animals. The rats were housed in conventional conditions (dry bulb temperature 20-22°C, relative humidity 50-60%), breathed unfiltered air, and were fed with standard balanced food.

The experiments were planned and implemented in accordance with the «International guiding principles for biomedical research involving animals» developed by the Council for International Organizations of Medical Sciences (1985) and were approved by the Bio-Ethics Committee of the Yekaterinburg Medical Research Centre for Prophylaxis and Health Protection in Industrial Workers.

In animal experiments we estimated toxicity of different metallic particles in nanometre and micrometre ranges using three experimental models:

IOP Conf. Series: Materials Science and Engineering **443** (2018) 012025 doi:10.1088/1757-899X/443/1/012025

- a single intra-tracheal instillation of Me-NPs 24 h before collecting the bronchoalveolar lavage to obtain a liquid sample for cytological and biochemical assessment [6];
- repeated intra-peritoneal injections during 6-7 weeks in non-lethal doses to assess the thus induced subchronic intoxication by a lot of functional, biochemical, and morphological indices;
- long-term inhalation experiments with NPs of Fe_2O_3, SiO_2, and NiO [6,7].

All comparative assessments were based on parallel experiments.

An intratracheal instillation of 1 ml of Me-NP or Me-MP water suspension (or of sterile deionized water from the same batch, without any particles) served as an experimental model for the response of the lower airways to particle deposition. To this point, cell population and some biochemical characteristics of bronchoalveolar lavage fluid (BALF) obtained 24 h after the instillation were studied.

The subchronic toxicity of the Me-NPs and some Me-MPs was investigated by means of repeated intraperitoneal (i.p.) injections of the same suspensions in sub-lethal doses 5 times a week during 5-7 weeks. NP inhalation by laboratory rodents is not an ideal model of real human exposures as it is often deemed to be [8], but the intraperitoneal animal model circumvents these interspecies differences. Dosing by injection is much more accurate, reliable, and reproducible compared to the more "natural" experimental methods, such as inhalation ones.

The rats were killed by decapitation and their blood was collected by exsanguination. The liver, spleen, kidneys, and brain were weighed. The assessed blood biochemical indices included the total serum protein, albumin, globulin, triglycerides, cholesterol, high and low density lipoproteins, bilirubin, ceruloplasmin, reduced glutathione (GSH), malondialdehyde (MDA), alkaline phosphatase, alanine and aspartate transaminases ALT, AST), catalase, gamma-glutamyltransferase, creatinine, and, in some experiments, thyrotropic hormone of hypophysis, thyroxin, triiodothyronine, follicle-stimulating and luteinizing hormones, progesterone, dehydroepiandrosterone, estradiol, and neuron-specific enolase.

The level of genomic DNA fragmentation as an index for the metals' in vivo genotoxicity was assessed using the Random Amplification of Polymorphic DNA (RAPD) test. The method is based on the fact that, unlike a fragmented DNA, which forms the so-called "comet tail" during agarose gel electrophoresis, a non-fragmented DNA has a very low degree of migration and virtually stays at the same place ("comet head"). The degree of migration is directly related to the extent of DNA fragmentation. To characterize the extent of damage to DNA tagged with tritium, we used "coefficient of fragmentation", i.e., the ratio of total radioactivity of all "tail" fractions to the radioactivity of the "head" [4]. To study structural and ultrastructural used changes in rats' viscera, we used optical and transmission electron microscopy.

3. Results and discussion

Figures 2 and 3 illustrate the dependences based on values of the neutrophil leukocytes to alveolar macrophages count ratio in broncho-alveolar lavage fluid, which is a very informative index for comparative cytotoxicity and pulmonary toxicity of particles deposited in the lower airways. We present the examples of the responses to iron oxide (magnetite) particles of different diameters and to gold and silver particles of different or of virtually equal dimensions [9].

Of a special significance is the affinity of metallic NPs to mitochondria with more or less prominent damage to these organelles; it was reported by many authors with the so-called oxidative stress resulting in generation of free radicals – one of the key primary mechanisms of the DNA damage [10]. Thus, NPs of silver and copper oxide are more likely to damage mitochondria than Au NPs. Figure 4 demonstrates that Ag NPs are also more genotoxic as compared with Au NPs [10].

If metallic NPs of a given composition are always the more cytotoxic the smaller their dimensions, the dependence of their organ-systemic chronic resorbtive toxicity on the diameter within nanoscale range is not so unique due to its intricate (complicated) and often contra-directional (antagonistic) influences of this parameter on toxicodynamics and toxicokinetics.

Unlike the effects on liver and other organs of the reticuloendothelial system (RES), the toxic action of metallic NPs on kidneys is probably less associated with their bio-persistency than with a quick dissolution of NPs in biological milieus.

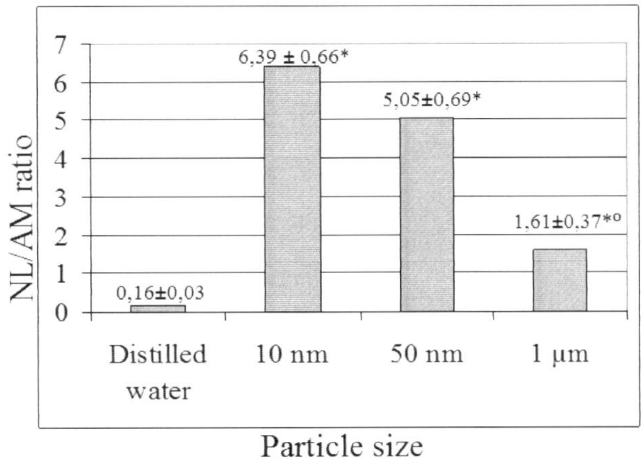

Figure 2. The neutrophil leukocytes to alveolar macrophages count ratio (NL/AM) in the BALF or rats 24 hr after i.t. instillation of Fe_3O_4 particles depending on their mean diameter. * statistically significant (P < 0,05) difference from controls; ° - the same from 10 nm.

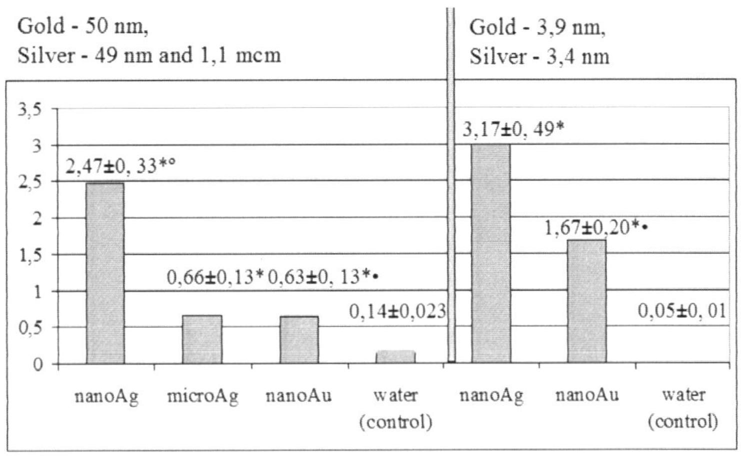

Figure 3. Index for comparative cytotoxicity of silver and gold particles. * statistically significant (P < 0,05) difference from controls, ° - the same from nanoAg, ▪ - the same from microAg.

Table 1. Morphometric indices for the state of rat's brain after repeat intraperitoneal injections of NiO or Mn_3O_4 NPs (X±s.e.) Note: * statistically significant difference from the control group (p < 0.05 by Student's t-test).

Index	Control	Mn_3O_4-NP	NiO-NP
Caudate nucleus			
Cells without nucleoli (%%)	30,50±2,77	69,90±1,79*+	47,36±2,45*
Hippocampus (CA 1)			
Cells without nucleoli (%%)	30,50±2,30	70,45±2,31*	35,8±2,21

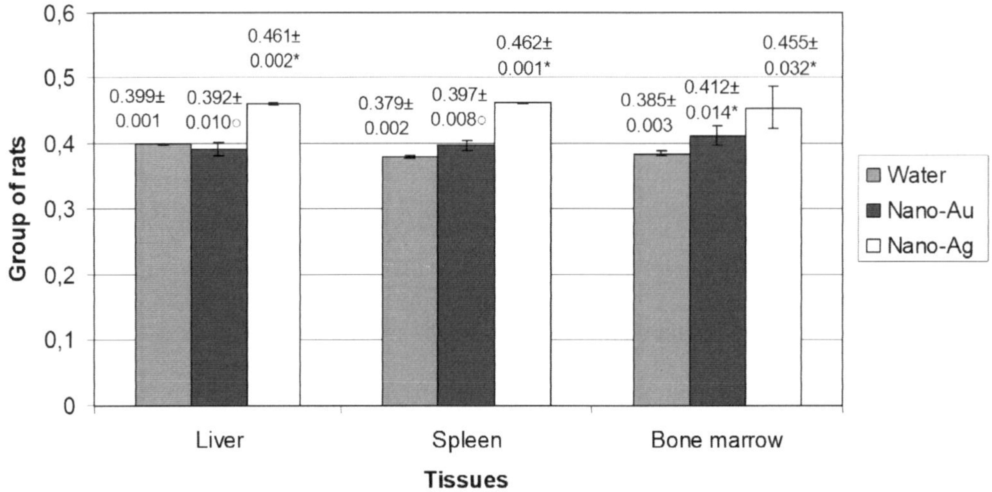

Figure 4. Coefficients of the genomic DNA fragmentation in rats exposed to subchronic administration of gold or silver equidimensional NPs (the RAPD-test), X±s.e Note: statistically significant difference * from the control group; ○ between the group receiving NG and the group receiving NS ($P<0.05$ by Student's t-test)

Speaking about the metal-associated specific differences between patterns of different NPs adverse health effects, we may give as an example a higher damage to the brain striatum (basal nuclei) and the hippocampus exerted by manganese oxide NPs as compared to those of nickel oxide ones (Table 1) [5].

Brain damage was induced also by the subchronic toxicity of copper oxide NPs [10]. Along with such damage to brain basal nuclei neurons, the toxicological syndrome comprised accumulation of CuO in the liver and brain, some decrease in the serum ceruloplasmin level, and anemia. All these phenomena are characteristic of Wilson's disease in humans associated with genetically determined impairment of copper metabolism. We believe that in our animal experiments this syndrome may be considered specific to the chronic toxicity of exogenous copper (in the form of NPs or ultrafine MPs).

However, the NPs neurotoxicity may be associated not only with mechanisms specific for a certain metal (such as manganese or copper) [6,11]. It is well known that any inhaled NPs having the smallest dimensions are able, after deposition in nasal cavity, to be transferred along the olfactory nerve fibres, thus bypassing the blood-brain barrier. Therefore, brain neurons may be damaged by them like any other cells due to the mechanisms common for all metallic NPs.

This statement is illustrated by Figure 5 showing scanning transmission electron microscopy (STEM) image of rat brains after 6 months' inhalation "nose-only" exposure to low concentrations of nickel oxide NPs. Neuron's myelin sheath contains NiO NPs. There are areas of demyelination in axon's sheath, where the electron-dense NPs are detected.

Figure 6 displays a STEM image of a rat alveolocyte containing NiO NPs in its cytoplasm after the type II inhalation exposure for 3 months. The emptied multilamellar bodies and NPs in cytoplasm are evident. The boxed area appears at higher magnification on the right and presents a large accumulation of NPs.

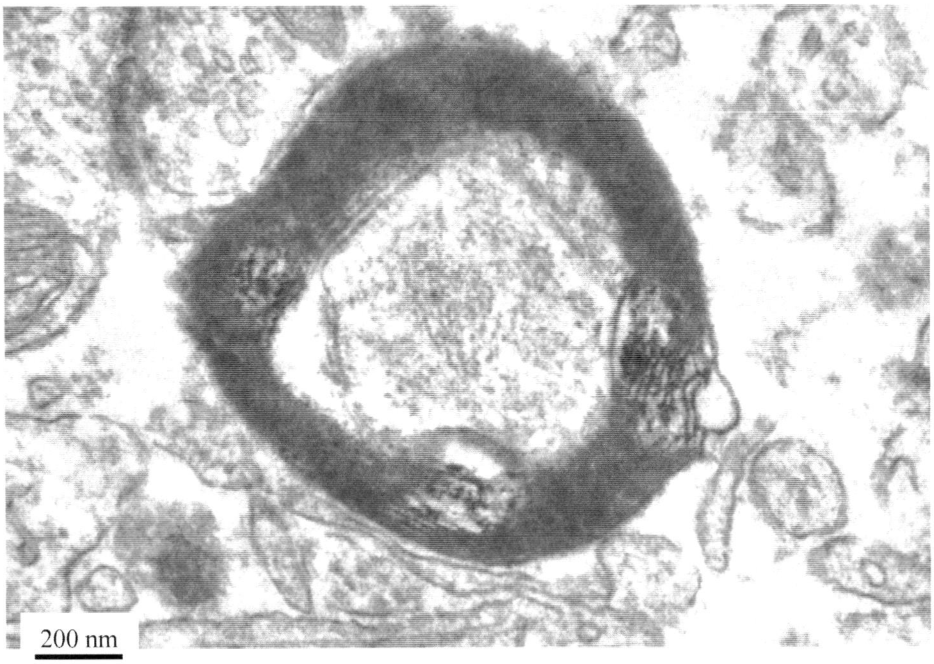

Figure 5. Brain by STEM. A neuron's myelin sheath containing NiO NPs.

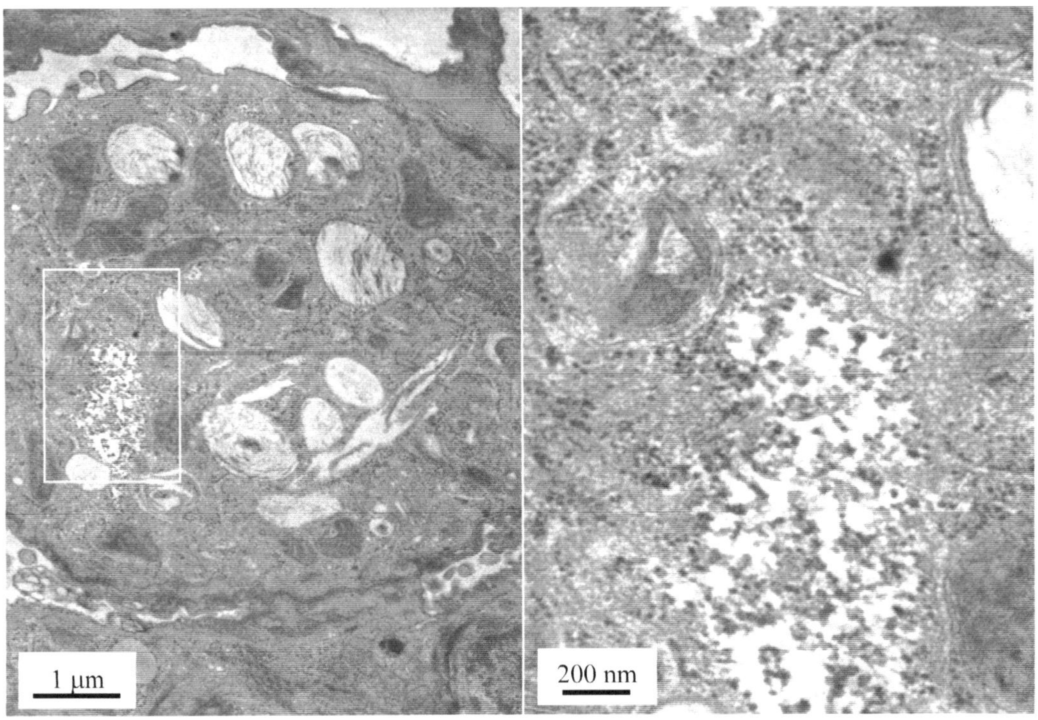

Figure 6. STEM image of type II alveolocyte with NiO NPs in its cytoplasm.

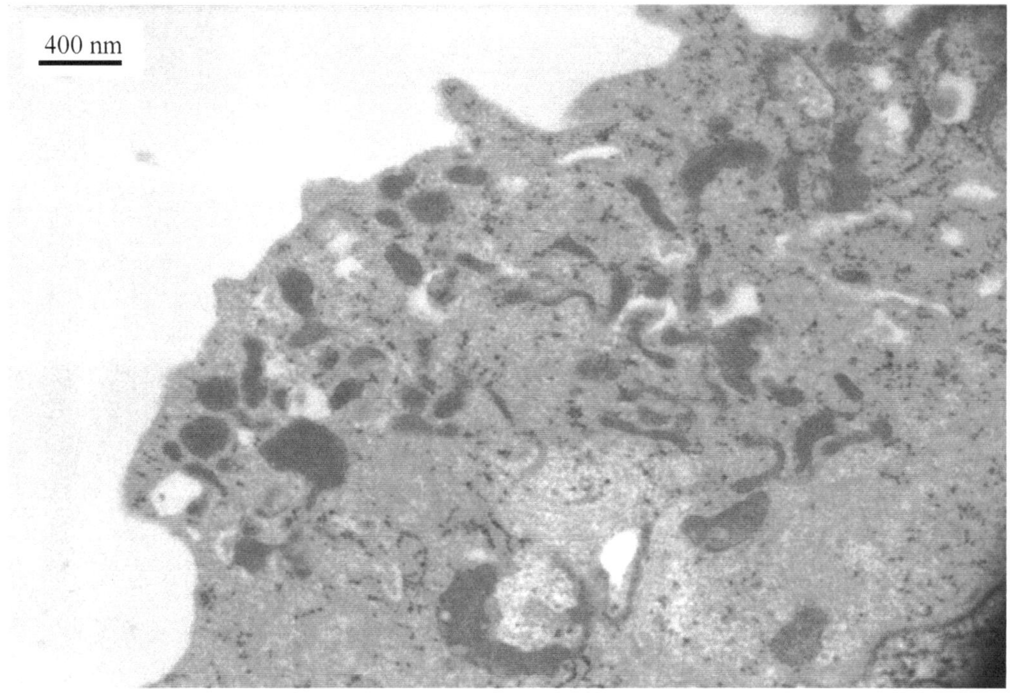

Figure 7. STEM image of an alveolar macrophage with NPs within cytoplasm.

Figure 7 displays a STEM image of a rat alveolar macrophage with NPs in its cytoplasm following the 6 months of inhalation exposure. NiO NPs are scattered in cytoplasm.

The coefficient of genomic DNA fragmentation in the blood nucleated cells (Table 2) is proved to be statistically significant and virtually equally increased in the rats exposed to all the NPs tested, but the background BPC administration significantly reduced the effect of the combined toxic exposure, while the BPC alone did not produce any significant influence on the control value.

NiO NPs show a distinct genotoxicity within every experimental timeline (of 3, 6, and 9 months), which later deepens as the exposure duration increases (Table 2). 6 months fragmentation coefficient is statistically significantly increased compared to that after 3 months of the exposure. 9 months fragmentation coefficient is statistically significantly higher than the 6 months one, while there is no increasing in the control group.

Table 2. Fragmentation coefficients of genomic DNA in nucleated blood cells of rats being chronically exposed to NiO NPs via inhalation, ($X \pm S_x$). *NB*: * marks the estimated values that statistically significant ($P<0{,}05$) differ from the control group.

The nucleated cells of peripheral blood	**The duration of inhalational exposure**					
	3 months		**6 months**		**9 months**	
	Control	**NiO NPs**	**Control**	**NiO NPs**	**Control**	**NiO NPs**
	0,4229± 0,0008	0,4480± 0,0017*	0,4247± 0,0006	0,5332± 0,0031*	0,4244± 0,0005	0,5447± 0,0036*

Table 3. Fragmentation coefficients of genomic DNA in rats following their chronic inhalation exposure to NiO NPs and / or the Bioprotective complex administration (BPC), (X±Sx). * marks the estimated values that statistically significant (P<0,05) differ from the control group; + marks the estimated values that statistically significant (P<0,05) differ from the NiO NPs group.

Nucleated cells of peripheral blood	Duration of inhalational exposure - 3 months			
	Control	**NiO NPs**	**NiO NPs + BPC**	**BPC**
	0,4229±0,0008	0,4480±0,0017*	0,4264±0,0008*+	0,4219±0,0003

As is shown in the Table 3, BPC administration in NiO NPs inhalation exposure has taken the fragmentation coefficient down (to about the normal values).

Our research has demonstrated that metallic NPs are much more noxious than their fine micrometric or even submicron counterparts. At the same time, cytotoxicity, organ-systemic toxicity, and in vivo genotoxicity of NPs having a given geometry strongly depend on their chemical nature and on the properties associated with the latter (the solubility included) both quantitatively and qualitatively due to specific mechanisms of action characteristic of a given metal in any chemical form.

Even with the high activity of defence mechanisms, metallic NPs are one of the most dangerous occupational and environmental hazards due to their extremely high toxicity and an apparently imminent genotoxicity. However, even if the remarkably low levels are actually harmless, they are too difficult to maintain to guarantee workers' occupational safety. That is why we believe an additional protective system is needed, the so-called biological prophylaxis designed to enhance the organism's resistance to general and specific mechanisms of nanotoxicity. The idea emerged from our long-term experience with a successful bioprotection of the organism against a number of other toxicants, including some mineral microparticles.

Bioprotectors acting by different toxicokinetic and toxicodynamic mechanisms proved to be most effective when administered not separately but in combinations (the so-called "bioprophylactic complexes" or, abbreviated, BPC). In all the studies we have found that, as expected, the toxicity and even genotoxicity of metallic NPs could be really decreased against the use of certain BPCs compositions. A lot of bioprophylactic complexes are of a similar composition and, we believe, similar BPCs should comprise most of the same components when further looking for bioprotection against other metallic NPs.

Concerning bioprotection against the adverse effects of metallic NPs, up to now we have substantiated theoretically and tested experimentally four BPCs protecting against:

- nano silver,
- nano copper oxide [4],
- a combination of nano nickel oxide and nano manganese oxide [6],
- a combination of nano copper oxide, nano zinc oxide, and nano lead oxide [3],
- a combination of nano titanium dioxide, nano aluminum oxide, and nano silicon dioxide [12].

The above-described results demonstrate the attenuating effects of the bioprotectors on the subchronic systemic toxicity of Me-NPs. In all the studies we found that, as expected, the toxicity and even genotoxicity of metallic NPs could be really decreased against the background of adequately composed BPCs (Fig. 8) [10].

Based on our results, we have obtained the Patents Rights Documents, related to the Bioprotection against toxicity of different NPs: silver, copper oxide, nickel and manganese oxides.

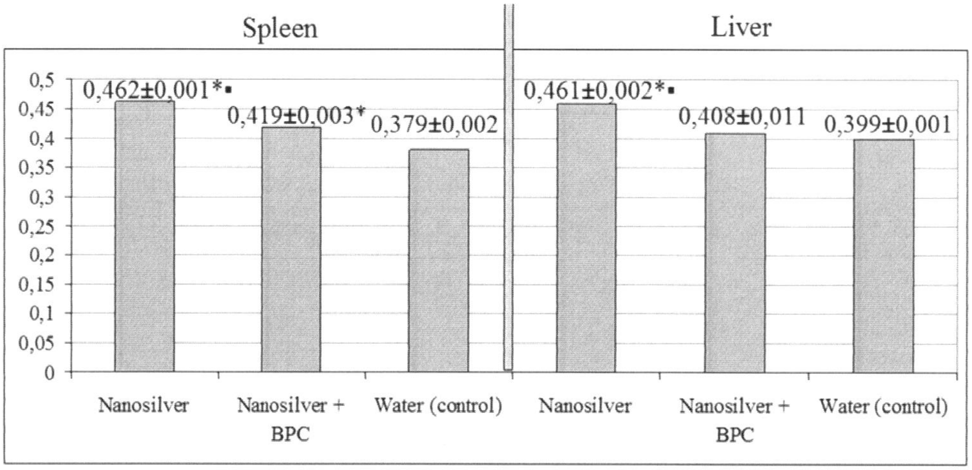

Figure 8. Coefficient of fragmentation of the genomic DNA in spleen (left) and liver (right) cells of rats exposed to nanosilver without (1st columns) or with (2nd columns) bioprotectors; 3rd columns – control (X±Sx). There is statistically significant difference from the control group ($P<0.05$) and from the group nAg+bioprotectors. * statistically significant ($P < 0,05$) difference from controls; ▪ - the same from nanoAg + BPC.

4. Conclusion

The toxic effects of metal and metal oxide NPs are much higher than those of their micrometric counterparts, given the same exposure pathways and similar chemistry, even of minimal (including submicron) sizes. For a given size, the toxicity of NPs depends on their chemical nature and the related properties, including solubility.

It was assumed that at some low level of potentially dangerous exposure, a balance between a toxic's adverse biological activity and the organism's natural defensive mechanisms could prevent the development of any identifiable disease or condition or even mild subclinical anomalies.

It was shown that the toxicity and even genotoxicity of metallic NPs could be significantly decreased using adequately composed Bioprotective Complexes.

Acknowledgements

The equipment of the Ural Centre for Shared Use "Modern Nanotechnology" Ural Federal University was used. The work was partially supported by Government of the Russian Federation (Act 211, Agreement 02.A03.21.0006) and the state task of Ministry of education and science of the Russian Federation (No. 3.4993.2017/6.7).

References

[1] Katsnelson B A, Privalova L I, Gurvich V B, Kuzmin S V, Kireyeva E P, Minigalieva I A, Sutunkova M P, Loginova N V, Malykh O L, Yarushin S V, Soloboyeva J I and Kochneva N I 2014 *J. Environ. Protect.* **5** 1435-49

[2] Katsnelson B A, Privalova L I, Sutunkova M P, Gurvich V B, Loginova N V, Minigalieva I A, Kireyeva E P, Shur V Ya and Shishkina E V 2014 *Int. J. Nanomed.* **10** 3013-29

[3] Minigalieva I A, Katsnelson B A, Panov V G, Privalova L I, Varaksin A N, Gurvich V B, Sutunkova M P, Shur V Y, Shishkina E V, Valamina I E, Zubarev I V, Makeyev O H, Meshtcheryakova E Y and Klinova S V 2017 *Toxicology* **380** 72-93

[4] Privalova L I, Katsnelson B A, Loginova N V, Gurvich V B, Shur V Ya, Valamina I E, Makeyev O H, Sutunkova M P, Minigalieva I A, Kireyeva E P, Rusakov V O, Tyurnina A E, Kozin R V, Meshtcheryakova E Y, Korotkov A V, Shuman E A, Zvereva A E and Kostykova S V 2014 *Int. J. Mol. Sci.* **15** 12379-406

[5] Privalova L I, Katsnelson B A, Loginova N V, Gurvich V B, Shur V Y, Beikin Y B, Sutunkova M P, Minigalieva I A, Shishkina E V, Pichugova S V, Tulakina L G and Beljayeva S V 2014 *Int. J. Mol. Sci.* **15** 21538-53

[6] Minigalieva I A, Katsnelson B A, Privalova L I, Sutunkova M P, Gurvich V B, Shur V Ya, Shishkina E V, Valamina I E, Makeyev O H, Panov V G, Varaksin A N, Grigoryeva E V and Meshtcheryakova E Y 2015 *Int. J. Mol. Sci.* **16** 22555-83

[7] Katsnelson B A, Degtyareva T D, Minigalieva I A, Privalova L I, Kuzmin S V, Yeremenko O S, Kireyeva E P, Sutunkova M P, Valamina I I, Khodos M Y, Kozitsina A N, Shur V Y, Vazhenin V A, Potapov A P and Morozova M V 2011 *Int. J. Toxicol.* **30** 60-7

[8] Fröhlich E and Salar-Behzadi S 2014 *Int. J. Mol. Sci.* **15** 4795-822

[9] Katsnelson B A, Privalova L I, Sutunkova M P, Tulakina L G, Pichugova S V, Beykin J B and Khodos M J 2012b *Bull. Exp. Biol. Med.* **152** 627-31

[10] Katsnelson B A, Privalova L I, Gurvich V B, Makeyev O H, Shur V Ya, Beikin Ya B, Sutunkova M P, Kireyeva E P, Minigalieva I A and Loginova N V 2013 *Int. J. Mol. Sci.* **14** 2449-83

[11] Katsnelson B A, Panov V G, Minigaliyeva I A, Varaksin A N, Privalova L I, Slyshkina T V and Grebenkina S V 2015 *Toxicology* **334** 33-44

[12] Minigalieva I A, Katsnelson B A, Privalova L I, Sutunkova M P, Gurvich V B, Shur V Y, Shishkina E V, Valamina I E, Makeyev O H, Panov V G, Varaksin A N, Bushueva T V, Sakhautdinova R R, Klinova S V, Solovyeva S N and Meshtcheryakova E Y 2018 *Int. J. Mol. Sci.* **19** 837

IOP Conf. Series: Materials Science and Engineering **443** (2018) 012026 doi:10.1088/1757-899X/443/1/012026

Adaptive time-frequency analysis of signals in AFM

V Pukhova

Department of Photonics, Saint Petersburg Electrotechnical University "LETI", Saint Petersburg, 197376, Russia

vmpukhova@etu.ru

Abstract. The probe-sample interactions during the AFM experiments represent a case of a multi-frequency behaviour, since they are characterized by the simultaneous excitation of cantilevers eigenmodes and/or their harmonics. Analysis of the data obtained as a result of the interaction between the probe and the sample is of a particular interest not only from a practical point of view, i.e. obtaining information about the surface of the sample, but also this is an interesting methodological task, since the instantaneous processes resulting from the interaction of the probe with the sample surface carry a huge amount of information. To obtain this information, it is necessary to provide an analysis method that is capable of characterizing all of the excited cantilevers frequencies simultaneously and without averaging. In other words, it is necessary to obtain the complete spectral response of the cantilever interacting with the sample surface, i.e. information about the dynamics of cantilever behaviour in time. The aim of the present work is to show the possibilities of time-frequency analysis techniques in analysing the signals of AFM experiments.

1. Introduction

Atomic Force Microscopy (AFM) [1] experiments are aimed at gaining information about the sample surface, not only depicting its topography, but also revealing its physical properties. In AFM experiments, cantilevers are used to probe the sample, and thus information about the sample surface is coded in its dynamics. Cantilevers are micromechanical oscillators and the dynamics of the cantilevers has a multi-frequency nature, inasmuch as it consists of simultaneously excited cantilevers eigenmodes and/or their harmonics originated while the probe interacts with the surface distinctive features. All this information can be obtained only if the behaviour of the cantilever is properly described and analysed. This task requires accurate and adequate signal analysis.

The main difficulty in solving this task is that the dynamics of the cantilever is not only of a multi-frequency nature, but also that each of the frequencies combining the resulting signal is changing in time, and sometimes abruptly. Hence, in terms of the signal processing, the AFM signals are non-stationary ones. The analysis of such signals demands for special thoroughness that can be provided if the signal analysis is adaptive.

2. Signal analysis approaches

Commonly, analysis of the signals in AFM is carried out by applying Fourier transform (FT) to the signal. This well-known approach is used due to its high accuracy and analysis speed [2]. However, despite its many benefits, FT provides a full spectral analysis only when stationary signals are investigated, which are caused by its working principle implying displaying of the integrated spectrum of the studied signal. This results in a loss of information about cantilever dynamics in particular time moments, which ultimately may lead to incorrect conclusions about the system behaviour. To avoid such situation, a time-frequency analysis needs to be implemented. The most powerful time-frequency analysis approach is a wavelet transform (WT) method [2].

Wavelet transform is a known method for image processing and recently it has found its applications in data analysis, when non-stationary signals need to be studied [3-4]. The distinctive difference between FT and WT is an availability of the time scale in the last technique. This possibility stems from the working principle of the WT, in which scaled and delayed wavelets are used unlike cosines infinitely spread on time scale from minus to plus infinity as in FT. It is worth to notice that the only basis functions in FT are cosines of various amplitudes, frequencies, and phases, while in WT basis functions are of different shapes and should be chosen in accordance with the form of the analysed signal. The correct choice of the basis wavelet function or so called mother wavelet allows outperforming standard FT and, hence, this task need to be solved individually in each case [5]. WT has already been successfully implemented to study experimental data in dynamic AFM experiments and for those tasks the Morlet wavelet has been used [6-9] since it has superior time-frequency resolution [10].

Morlet wavelet is formed by multiplying a complex exponential with fixed carrier frequency by a Gaussian shape envelope [2]:

$$\psi(t) = \frac{1}{\sqrt[4]{\pi c^2}} e^{-\frac{t^2}{2c^2}} e^{idt} \tag{1}$$

where $\psi(t)$ is the mother wavelet, c and d are positive parameters controlling the carrier frequency of the mother wavelet and its time spread, i.e. time-frequency resolution. The product of these parameters is a dimensionless parameter $G_S = cd$, known as Gabor shaping factor. It is worth to notice that Morlet wavelet satisfies the admissibility criterion.

Mother wavelet in WT provides a source function, which is scaled and delayed to obtain so called daughter wavelets, and to obtain the WT spectrum of the analysed time signal; the projection of the signal onto daughter wavelets in the entire time-frequency range is calculated.

The WT provides adaptive time-frequency resolution of the signal spectrum due to its working principle. However, it also satisfies the uncertainty principle, which implies that the higher the frequency resolution, the lower the time resolution.

In most AFM data analysis tasks, the joint use of FT and WT fulfils the requirements imposed on the accuracy of the mapping of signal spectral characteristics. However, some tasks require a higher frequency resolution or higher signal to noise ratio than conventional wavelet transform analysis can provide. In these cases, the approach can be used that in the present work is referred to as up-chirplet and down-chirplet transform, (U)CT and (D)CT, respectively. The principal concept of the chirplet transform has been previously introduced [11-13]; however, without consideration of the difference between the up-chirplet and down-chirplet transform and up to present days it is rarely used in real experimental data analysis tasks.

Up-chirplet /down-chirplet transform can be seen as WT, where the mother wavelet is obtained by multiplying a complex exponential with linearly varying carrier frequency by a Gaussian shape envelope [11]:

$$\psi(t) = \frac{1}{\sqrt[4]{\pi c^2}} e^{-\frac{t^2}{2c^2}} e^{i(a+bt)t} \tag{2}$$

where $\psi(t)$ is the mother chirplet, a and b are parameters defining the starting and final carrier frequencies and chirpyness, i.e. rate of the carrier frequency change. The parameters a and b could be positive or negative, and their signs and values also define whether the carrier frequency increasing (up-chirplet) or decreasing (down-chirplet). Using chirplet instead of wavelet as the mother function allows adding a shear in frequency within existing time-frequency windows, thus improving the frequency resolution. However, this is true only in the case if the choice of the parameters a and b has been carried out in such a way that the mother chirplet resembles the analysed signal.

Illustrations of the basis functions used in FT, WT, (U)CT, and (D)CT methods are presented in Figure 1.

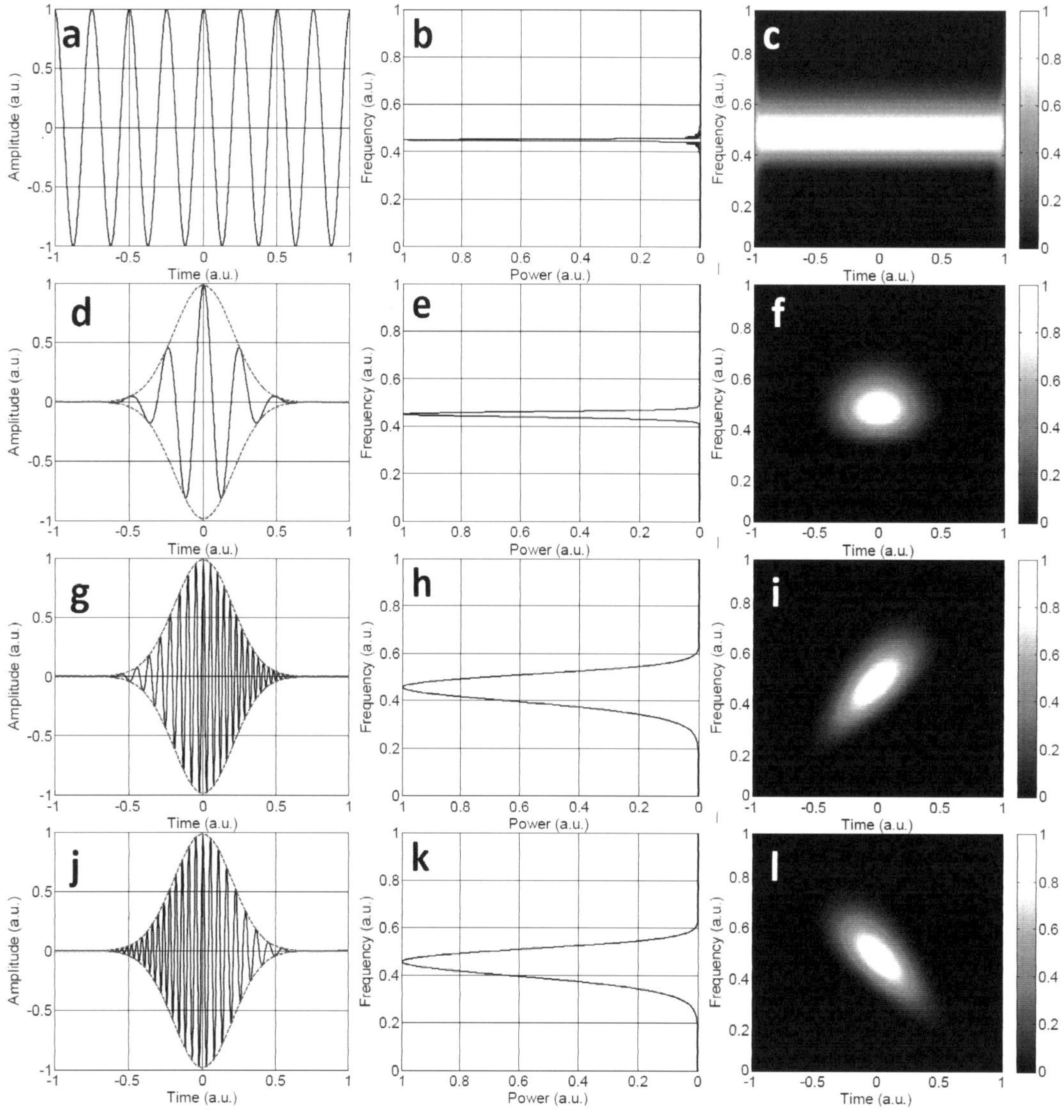

Figure 1. Basis functions used in signal analysis methods. (a,d,g,j) Temporal representations of basis functions real parts used in FT, WT, (U)CT, and (D)CT methods, respectively. Blue lines – Gaussian shape envelopes. (b,e,h,k) Spectral contents of signals (a,d,g,j), respectively. (c,f,i,l) Time-frequency representations of signals (a,d,g,j), respectively. All scales are normalized and given in arbitrary units.

3. Adaptive analysis of experimental AFM data

To show an example of the approach to data analysis discussed in the previous section, a force curve of dynamic atomic force spectroscopy experiment is analysed (Fig. 2). The investigated experimental force curve (Fig. 2a) has been acquired in ambient conditions (room temperature 296 K, relative humidity 55%, atmospheric pressure) with thermally driven rectangular Au-coated silicon cantilever. Its first free flexural eigenmode is equal to 12.5 kHz and spring constant is equal to 0.15 N/m. Spring constant has been calculated by using the thermal tune method applied to the first flexural eigenmode [14]. The experiment itself has been performed on HOPG surface with constant approach velocity 0.8 nm/ms during the entire dynamic force spectroscopy experiment. Calculated inverse optical lever sensitivity [15] in the present experiment is equal to 190.4 nm/V.

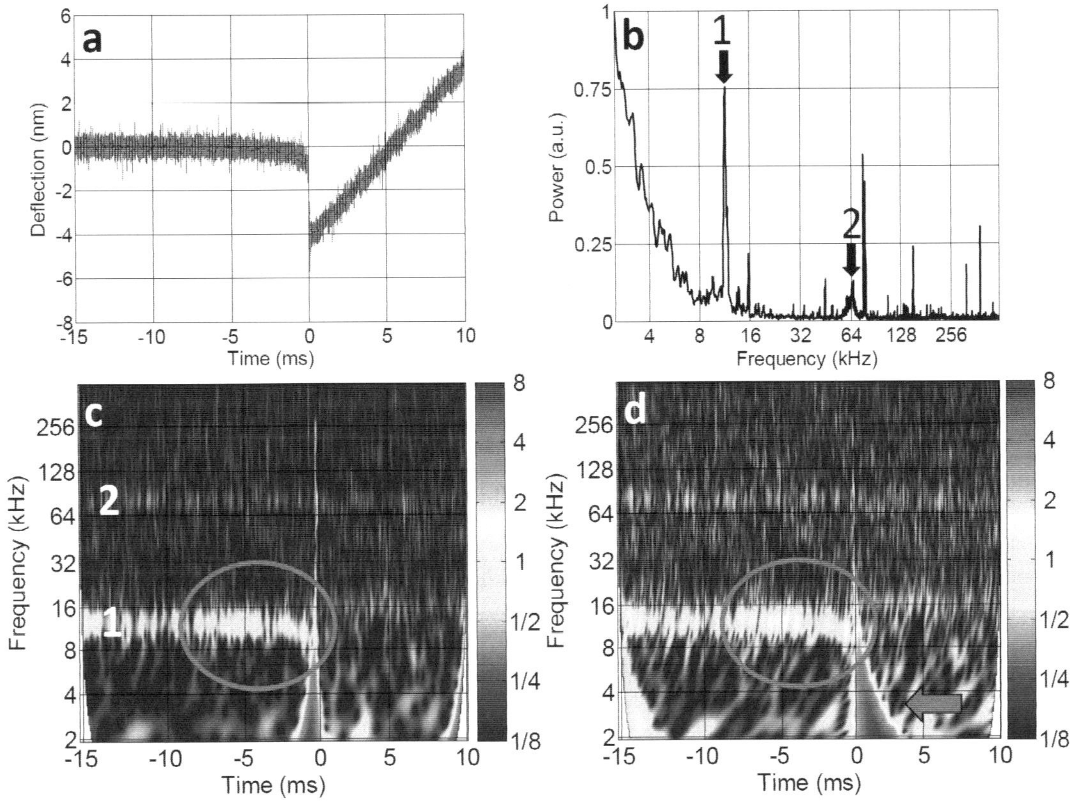

Figure 2. Analysis of the force curve obtained in dynamic atomic force spectroscopy experiment: (a) Cantilever deflection as a function of time registered by a standard optical beam deflection system. (b) FT of the signal (a). (c) WT of the signal (a). (d) Down-chirplet transform of the signal (a). The first and the second flexural eigenmodes are labelled as 1 and 2. Red ellipse allocates an area of particular interest. Red arrow indicates an area, where the applied approach causes artificial shear in time-frequency space. (c,d) Side lobes on both sides marked by white colour are edge effects of the applied transforms. (a,c,d) Zero time corresponds to the jump-to-contact transition. (b,c,d) Frequency axes and colour scales are presented in base-2 logarithm scales.

The analysis of the force curve obtained in dynamic atomic force spectroscopy experiment is carried out in several consecutive steps, namely FT, WT, and chirplet transform (Fig. 2).

The first transform that is applied to the acquired signal is FT, which gives an overall understanding of the signal spectrum (Fig. 2b). The system dynamics is not available from the Fourier power spectrum, since it shows spectrum integrated over 25 ms. Obtained Fourier power spectrum allows observing few resonance peaks, and in order to properly attribute these spectral components to the eigenmodes of the cantilever, the WT analysis can be carried out.

The WT analysis of the force curve shown in Figure 2a is presented in Figure 2c. In this case, the mother wavelet is Morlet wavelet (see section 2). On the WT spectrum, the behaviour of the cantilever during the experiment is unfolded in time. It is clearly seen that the behaviour of the cantilever before and after the jump-to-contact (JTC) transition is different and while interacting with the sample surface the evident spectral broadening around the first flexural mode and short-term excitations of the higher flexural modes appear. The detailed description of these phenomena could be found here [6, 8].

Comparing both FT and WT spectra of the signal, it is possible to confidently identify flexural eigenmodes of the cantilever. For the given system, the value of the first flexural mode is equal to 12.5 kHz and the second flexural mode is equal to 77.8 kHz, in Figure 2b,c they are labelled as 1 and 2.

IOP Conf. Series: Materials Science and Engineering **443** (2018) 012026 doi:10.1088/1757-899X/443/1/012026

In most research tasks, such joint FT and WT analysis provides enough information about the studied system; however, in some cases additional information is required or it is necessary to enhance visibility of particular area on wavelet transform spectrum. In such cases, it is useful to apply a complementary analysis. One of such analyses is up-chirplet/down-chirplet transform (see section 2).

As an example, consider the area within the red ellipse in Figure 2c. In the allocated area the down shifting of the first flexural eigenmode is clearly visible. This phenomenon appears due to the acting of the tip-sample force gradient on a cantilever when its tip approaches the sample surface. The data from this time-frequency range could be used to calculate instantaneous values of the tip-sample force gradient acting on a cantilever. Since the frequency is down shifting in the encircled area, it is reasonable to apply (D)CT analysis to improve the frequency resolution. Down-chirplet transform of the signal is shown in Figure 2d. The improvement of the frequency resolution, and thus visibility of the spectrum, is obvious in the allocated area.

(D)CT/(U)CT analysis could be very useful in particular tasks, but it should be applied carefully. For example, consider area in Figure 2d indicated by a red arrow pointing left. In this time-frequency range, the spectral broadening around the first flexural mode due to the JTC appears, but since no frequency down shifting takes place in this region (Fig. 2c), the down-chirplet of the applied (D)CT does not resemble this part of the analysed signal, and thus application of the (D)CT causes artificial shear in time-frequency space. This effect is particularly evident at low frequencies and should be taken into account every time while applying (U)CT/(D)CT to avoid possible misinterpretation of the spectrum.

4. Conclusion

AFM experimental data analysis is not a trivial task; thus, its solution requires a signal analysis method that can be adaptive to allow detail representing of the signal spectral response. Approach proposed in the present work implies a joint use of a few consistently applied data analysis techniques. It is worth to begin with application of the Fourier transform to the signal under investigation, which gives a range of frequencies for the given system and a general understanding about the spectral content of the signal. Then, to correctly attribute the spectral components of the obtained Fourier spectrum, the wavelet transform analysis is applied to the observed range of frequencies. This analysis displays evolution of each spectral component of the signal in time, which allows registering distinctive nonlinearities of the signal to be studied in details. However, it often happens that such signal parts have poor visibility due to the fact that they are usually changing their properties in a very short time interval. To enhance the resolution of these signal parts on a time-frequency plane, it could be beneficial to additionally use a complementary analysis based on the up-chirplet or down-chirplet transform described in the present work. If duly executed, such approach to the signal analysis gives the most comprehensive understanding about the system behaviour in the time-frequency space.

References

[1] Binnig G, Quate C C and Gerber C 1986 *Phys. Rev. Lett.* **56** 930

[2] Mallat S 1999 *A Wavelet Tour of Signal Processing* (San Diego: Academic Press)

[3] Pukhova V, Gorelova E, Ferrini G and Burnasheva S 2017 *Proc. IEEE Conf. of Russian Young Researchers in Electrical and Electronic Engineering (EIConRus)* 715-8

[4] Pukhova V, Kustov T and Ferrini G 2018 *Proc. IEEE Conf. of Russian Young Researchers in Electrical and Electronic Engineering (EIConRus)* 1141-5

[5] Chopra S and Marfurt K 2015 *Proc. SEG Int. Exposition and 85th Annual Meeting* 2957-61

[6] Pukhova V, Banfi F and Ferrini G 2013 *Nanotechnology* **24** 505716

[7] Pukhova V, Banfi F and Ferrini G 2014 *Beilstein J. Nanotechnol.* **5** 494-500

[8] Pukhova V, Banfi F and Ferrini G 2015 *Nanotechnology* **26** 175701

[9] Pukhova V and Ferrini G 2017 *IOP Conf. Ser.: Mat. Sci. Eng.* **256** 012004

[10] Deng Y, Wang C, Chai L and Zhang Z 2005 *Appl. Phys. B* **81** 1107-11

[11] Mann S and Haykin S 1991 *Proc. Vision Interface* 205-12

[12] Hu G and Zhu F 2011 *Circuits and Systems* **2** 107-11

[13] Peng Z, Meng G, Chu F, Lang Z, Zhang W and Yang Y 2011 *IEEE Trans. Instrum. Meas.* **60** 3222-9
[14] Butt H J and Jaschke M 1995 *Nanotechnology* **6** 1-7
[15] Higgins M J, Proksch R, Sader J E, Polcik M, Mc Endoo S, Cleveland J P and Jarvis S P 2006 *Rev. Sci. Instrum.* **77** 013701

IOP Conf. Series: Materials Science and Engineering **443** (2018) 012027 doi:10.1088/1757-899X/443/1/012027

Some notes about scanning probe microscopy, nanoengineering and methods of quantum mechanics

A E Rassadin[1], **T S Sazanova**[1,2], **A V Stepanov**[1,3] **and L A Fomin**[1,4]

[1] Nizhny Novgorod Mathematical Society, Nizhny Novgorod, 603950, Russia
[2] Nizhny Novgorod State Technical University n.a. R.E. Alekseev, 603950, Nizhny Novgorod, Russia
[3] Chuvash State Agricultural Academy, 428000, Cheboksary, Russia
[4] Institute of Microelectronics Technology and High Purity Materials RAS, 142432, Chernogolovka, Moscow region, Russia

brat_ras@list.ru

Abstract. In the presented paper, a research program for new vision of nanoengineering has been suggested. In the framework of this program, growth of a solid state surface under the influence of external source of sputtering particles is considered as the problem of optimal control by distributed parameter system, the source being a control. As one of cornerstones in this approach, general theory of the one-dimensional KPZ-equation in approximation of small angles under the action of the spatially-inhomogeneous source has been developed. The complete mathematical model of this process with the Dirac delta function source of external particles illustrates this theory, Green's function for this system having been calculated exactly. Scheme of physical realization of delta-functional source by means of series of carbon nanotubes has been discussed. Opportunity of combined characterization of solid state surface by both atomic force microscopy and scattering data for electromagnetic waves has been demonstrated. The height of the sample has been constructed with help of the Cole-Hopf substitution from single eigenfunction of discrete spectrum being an example of the surface shape under investigation.

1. Introduction

From 1981 to 2018, scientific community all over the world has been witnessing intensive development of scanning probe microscopy [1]. Nowadays, various modes of scanning probe microscopy have covered different spheres of contemporary material science – from the use of magnetic force microscopy for the research of epitaxial ferro- and antiferromagnetic structures [2] to the application of atomic force microscopy for the study of chitosan's and chitosan-based copolymers' mechanical properties [3]. In particular, the increase of precision for measurements using atomic force microscopy proves to give us an opportunity of verification of different models for growth of solid state surfaces [4].

At present, the most popular model of such kind is the Kardar-Parisi-Zhang equation (KPZ) [5]:

$$\frac{\partial H}{\partial t} = c \cdot \sqrt{1 + (\nabla H)^2} + \nu \cdot \nabla^2 H + U(\vec{x}, t) \qquad (1)$$

This phenomenological equation describes temporal evolution of the height $H(\vec{x}, t)$ of solid state surface under sputtering of a substance on it, $\vec{x} = (x_1, x_2)$ being two-dimensional vector and ∇ being two-dimensional gradient. The first term on the right-hand side of the KPZ-equation demonstrates that the growth of the surface occurs only along the local normal $\vec{n}(\vec{x},t)=\left(\frac{-\nabla H}{\sqrt{1+(\nabla H)^2}},\ \frac{1}{\sqrt{1+(\nabla H)^2}}\right)$ to it, rate c of this growth being constant. The second term on the right-hand side of the equation (1) corresponds to the surface diffusion of sputtering substance, ν being a diffusion coefficient. Moreover, this term regularizes the input equation preventing a gradient catastrophe, which might be generated by the presence of the first term only. The third term $U(\vec{x},t)$ is the external source of sputtering particles. It is obvious that at arbitrary moment of time $t\geq 0$ this function ought to be nonnegative everywhere: $U(\vec{x},t)\geq 0$. As a rule, the Cauchy problem for the equation (1) is studied:

$$H(\vec{x},0)=h_0(\vec{x}),\qquad \vec{x}\in D\subset R^2, \tag{2}$$

initial condition being corresponded to initial shape of the surface under consideration.

For instance, the KPZ-equation proves to be highly adequate to the physical experiment during the simulation of manufacturing process of multilayer mirrors and gratings employed in X-ray optics [6]; therefore, verification procedure for model (1) of the surface growth with support of this one by means of atomic force microscopy is not required. The situation gives rise to the problem of uncertainty under choice of direction of further epistemological movement. Nevertheless, there is the way to overcome this obstacle by considering the external source of particles $U(\vec{x},t)$ as a control. At last, in this case the equation (1) ought to be added by demand of minimization of some functional of height $H(\vec{x},t)$ and control $U(\vec{x},t)$:

$$F[H,U]=\int_0^T\int_D f(H(\vec{x},t),U(\vec{x},t))\cdot d^2x\cdot dt\rightarrow \min \tag{3}$$

Conditions (1)-(3) form the problem of optimal control by distributed parameter system [6]. Namely, starting from the initial shape $h_0(\vec{x})$ by means of control $U(\vec{x},t)$ it is required to achieve minimum of functional (3) on the fixed interval of time $[0,T]$. For example, one can set up the problem of the closest fitting of fixed shape $H_*(\vec{x})$ by height $H(\vec{x},T)$ at the moment of time $t=T$ in $L^2(D)$-metric:

$$\int_D [H(\vec{x},T)-H_*(\vec{x})]^2 d^2x\rightarrow \min \tag{4}$$

We stress that on the interval $[0,T]$ of time function $H(\vec{x},t)$ must obey to the equation (1). In other words, it is the problem of nanoengineering.

The first obvious solution of this problem is to apply the infinite-dimensional generalization of Pontryagin's maximum principle [8]. In order to reduce the KPZ-equation to infinite-dimensional dynamical system in the space l_2 of square-summable sequences of complex numbers, let us expand functions $H(\vec{x},t)$ and $U(\vec{x},t)$ on an orthonormal basis $\{\Phi_n(\vec{x})\}_{n=0}^{n=\infty}$ in separable Hilbert space $L^2(D)$:

$$H(\vec{x},t)=\sum_{n=0}^{\infty}h_n(t)\cdot\Phi_n(\vec{x}),\quad U(\vec{x},t)=\sum_{n=0}^{\infty}u_n(t)\cdot\Phi_n(\vec{x}) \tag{5}$$

denumerable sets of Fourier coefficients $h_n(t)$ and $u_n(t)$ representing dynamical variables in l_2 and controls therein, respectively. But the substitution of the first Fourier-series expansion from (5) into the equation (1) ends in failure, because the strong nonlinearity of this equation does not allow obtaining any visible result. This circumstance demonstrates that to do the step forward, one must simplify the KPZ-equation in the framework of the conjecture about correctness of the inequality $|\nabla H|<<1$ using the so-called approximation of small angles [5]. The result of this simplification is equal to:

$$\frac{\partial H}{\partial t}=c+\frac{c}{2}\cdot(\nabla H)^2+\nu\cdot\nabla^2 H+U(\vec{x},t) \tag{6}$$

Further, let us introduce a new unknown function $\varphi(\vec{x},t)$ as follows:

$$H(\vec{x},t) = c \cdot t + \frac{2 \cdot \nu}{c} \cdot \ln \varphi(\vec{x},t) \tag{7}$$

The expression (7) is known to be the Cole-Hopf substitution [5]. By means of this formula, one can reduce the nonlinear equation (6) to the Cauchy problem for the next one:

$$\frac{\partial \varphi}{\partial t} = \nu \cdot \nabla^2 \varphi + \frac{c \cdot U(\vec{x},t)}{2 \cdot \nu} \cdot \varphi, \qquad \varphi(\vec{x},0) = \exp\left[\frac{c \cdot h_0(\vec{x})}{2 \cdot \nu}\right] \tag{8}$$

The linear parabolic equation (8) is simpler than the nonlinear equation (6).

For the functional (4) it is interesting to consider two-stage control:

$$U(\vec{x},t) = \begin{cases} Q(\vec{x},t), & 0 \le t \le T \\ 0, & t > T \end{cases} \tag{9}$$

where positive value T being fixed.

The formula (9) describes the situation of free evolution of the system (6) under $t > T$. If $t > T$, then the second term on the right-hand side of the equation (8) disappears and this equation is transformed into a trivial two-dimensional diffusion equation:

$$\frac{\partial \varphi}{\partial t} = \nu \cdot \nabla^2 \varphi \tag{10}$$

Let us assume that on the interval of time $[0,T]$ there is a control $Q_*(\vec{x},t)$, under which global minimum of the functional (4) would be achieved that is $H(\vec{x},T) = H_*(\vec{x})$ almost everywhere in the region D. Hence, further evolution of height can be described by means of Green's function of the equation (10) [9] and the Cole-Hopf substitution (7):

$$H(\vec{x},t) = c \cdot (t-T) + \frac{2 \cdot \nu}{c} \cdot \ln\left(\int_{R^2} \exp\left[-\frac{(\vec{x}-\vec{\xi})^2}{4 \cdot \nu \cdot (t-T)} + \frac{c \cdot H_*(\vec{\xi})}{2 \cdot \nu}\right] \cdot \frac{d^2 \xi}{4 \cdot \pi \cdot \nu \cdot (t-T)}\right) \tag{11}$$

For example, if the fixed shape $H_*(\vec{x})$ is equal to:

$$H_*(\vec{x}) = \frac{2 \cdot \nu}{c} \cdot \ln\left[\left(1 + m_0 \cdot \cos\frac{2 \cdot \pi \cdot x_1}{a}\right) \cdot \left(1 + m_0 \cdot \cos\frac{2 \cdot \pi \cdot x_2}{a}\right)\right] \tag{12}$$

then under $t > T$ the formula (11) gives us [10]:

$$H(\vec{x},t) = c \cdot (t-T) + \frac{2 \cdot \nu}{c} \cdot \ln\left[\left(1 + m(t) \cdot \cos\frac{2 \cdot \pi \cdot x_1}{a}\right) \cdot \left(1 + m(t) \cdot \cos\frac{2 \cdot \pi \cdot x_2}{a}\right)\right] \tag{13}$$

where $0 < m_0 < 1$, $m(t) = m_0 \cdot \exp(-4 \cdot \pi^2 \cdot \nu \cdot t / a^2)$ and a is the lattice constant.

The explicit solution (13) may be interpreted as surface of crystal with cubic symmetry growing in the direction of one of the primitive translation vectors. At each moment of time $t > T$, this function is homothetic to the initial condition (12) for it. Graph of function (12) is presented in Figure 1.

It is well known that besides Pontryagin's maximum principle [8], there exist quite a few alternative approaches to solving problems of optimal control by distributed parameter systems ([7, 11-13] and references therein). On the other hand, to solve the problem defined by conditions (2), (4), (6), and (9) under $t \in [0,T]$, it is necessary to study a number of model situations in order to find the most convenient phase space for this problem. That is why in accordance with the classical sample [8] let us restrict area of our research by the next simplifications. Namely, we shall consider only surfaces with cylindrical generatrix (Fig. 2) and stationary controls depending on longitudinal coordinate only.

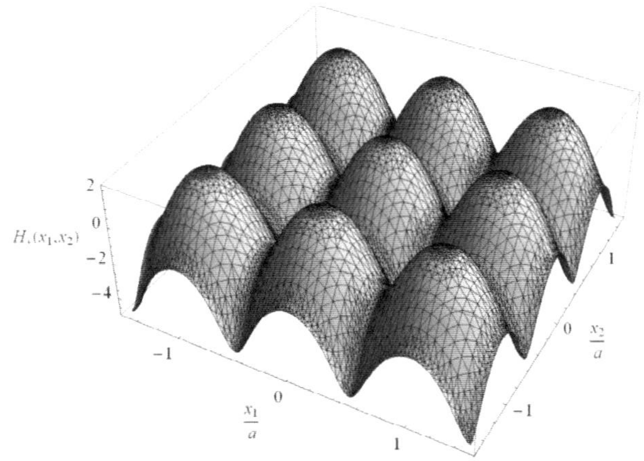

Figure 1. An example of the initial shape of the surface with convenient expression for its free evolution.

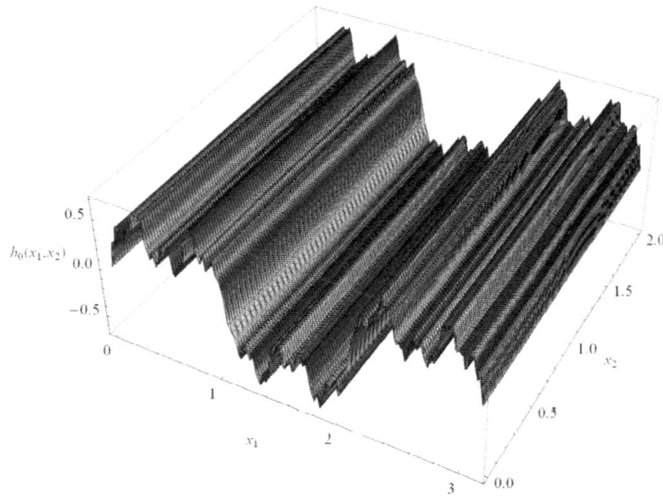

Figure 2. An example of a surface with a cylindrical generatrix.

The rest of the article is organized as follows: in Section 2 we demonstrate, how to construct a general solution of the one-dimensional KPZ-equation in approximation of small angles under the action of the spatially-inhomogeneous source $Q(x)$, which vanishes on infinity: $\lim\limits_{x\to\pm\infty} Q(x) = 0$. Section 3 deals with the case, when the source profile is proportional to the Dirac delta function: $Q(x) = \frac{4 \cdot \nu^2 \cdot q}{c} \cdot \delta(x)$. In section 4 we consider both the simplest solution of the KPZ-equation with the delta-functional source and the approach to measurement of surface diffusion coefficient ν along conducting surfaces by means of electromagnetic waves scattering. Final section is devoted to the discussion of results elaborated and perspectives of further investigations.

2. General theory of one-dimensional KPZ-equation with a spatially-inhomogeneous source

The aim of this section is to compare a number of methods to find the exact solution of the one-dimensional KPZ-equation in approximation of small angles with a time-independent external source $Q(x)$. In this case, the Cauchy problem for equation (6) is reduced to:

$$\frac{\partial H}{\partial t}=c+\frac{c}{2}\cdot\left(\frac{\partial H}{\partial x}\right)^2+\nu\cdot\frac{\partial^2 H}{\partial x^2}+Q(x)\,,\quad H(x,0)=h_0(x)\,,\qquad x\in R \tag{14}$$

The Cole-Hopf substitution (7) transforms the equation (14) to the next one:

$$\frac{\partial\varphi}{\partial t}=\nu\cdot\frac{\partial^2\varphi}{\partial x^2}+\frac{c\cdot Q(x)}{2\cdot\nu}\cdot\varphi\,,\qquad \varphi(x,0)=\exp\left[\frac{c\cdot h_0(x)}{2\cdot\nu}\right] \tag{15}$$

This linear second-order parabolic partial differential equation might be solved in the framework of the newest technique developed in Ref. [14]. Exact solution of Eq. (15) can be expressed as follows:

$$\varphi(x,t)=\lim_{n\to\infty}\left[\hat{S}^n\left(\frac{t}{n}\right)\cdot\varphi(x,0)\right] \tag{16}$$

where

$$\hat{S}(t)=ch^2\left[2\cdot\sqrt{\nu\cdot t}\cdot\frac{\partial}{\partial x}\right]+\frac{c\cdot t}{2\cdot\nu}\cdot Q(x) \tag{17}$$

is the so-called space shift operator [14]. However, application of Eq. (16) under $Q(x)\neq 0$ is very hard, because calculation of sequential degrees of the operator (17) under increase of exponent n in Eq. (16) becomes more awkward. On the other side, the exact solution of equation (15) can be represented as:

$$\varphi(x,t)=\int_{-\infty}^{+\infty}G(x,\xi;t)\cdot\varphi(\xi,0)\cdot d\xi \tag{18}$$

where $G(x,\xi;t)$ is Green's function of this equation, which can be found by means of the Feynman-Kac formula [15, 16]:

$$G(x,\xi;t)=\int_{C(t,x-\xi)}\exp\left[\frac{c}{8\cdot\nu^2}\cdot\int_0^t Q(x+\eta(\tau))\cdot d\tau\right]\cdot d_{W(t,x-\xi)}\eta \tag{19}$$

Path integral (19) is calculated on functional space $C(t, x-\xi)$ of continuous on interval $[0,\tau]$ functions $\eta(\tau)$ obeying to the boundary conditions: $\eta(0) = 0$ and $\eta(t) = x-\xi$. The symbol $d_{W(t,\,x-\xi)}\eta$ designates the so-called conditional Wiener measure [15, 16]. But calculation of Green's function according to the Eq. (19) is quite difficult too; that is why let us find the solution of the linear Eq. (15) in the form:

$$\varphi(x,t)=\exp(-\nu\cdot\varepsilon\cdot t)\cdot\psi(x) \tag{20}$$

It is easy to check that a new unknown function $\psi(x)$ obeys to the stationary Schrödinger equation:

$$\hat{\mathrm{H}}\cdot\psi=\varepsilon\cdot\psi \tag{21}$$

with the Hamiltonian

$$\hat{\mathrm{H}}=-\frac{d^2}{dx^2}+W(x) \tag{22}$$

where the role of potential energy is played by the function $W(x)\equiv-\frac{c}{2\cdot\nu^2}\cdot Q(x)$. Due to nonnegativity and vanishing on infinity of the source $Q(x)$, the Hamiltonian (22) possesses both the discrete spectrum of negative eigenvalues ε_n $(n=\overline{0,N-1})$ and the continuous spectrum of positive eigenvalues ε [17]. Green's function $G(x,\xi;t)$ as a kernel of operator of evolution $\exp(-\nu\cdot t\cdot\hat{\mathrm{H}})$ for (15) is equal to:

$$G(x,\xi;t)=\sum_{n=0}^{N-1}\exp(-\nu\cdot\varepsilon_n\cdot t)\cdot\psi_n(x)\cdot\psi_n^*(\xi)+\int_0^{+\infty}\exp(-\nu\cdot\varepsilon\cdot t)\cdot\psi(x,\varepsilon)\cdot\psi^*(\xi,\varepsilon)\cdot d\varepsilon \tag{23}$$

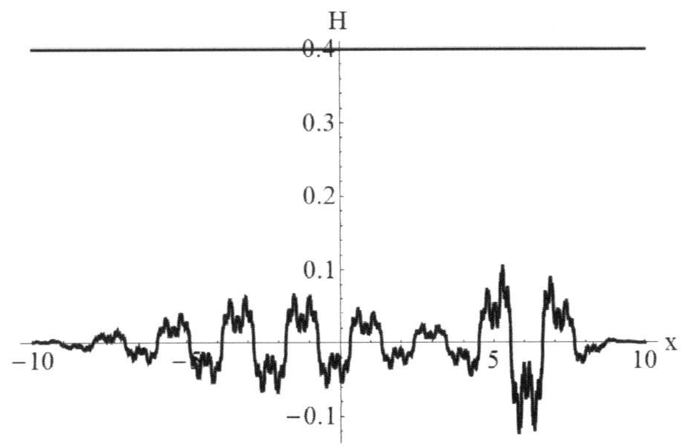

Figure 3. Initial shape (from below) and constant fixed shape (from above).

In the formula (23), N eigenfunctions of discrete spectrum $\psi_n(x)$ and eigenfunctions $\psi(x,\varepsilon)$ of continuous spectrum ought to be normalized according to standard rules of quantum mechanics [17]:

$$\int_{-\infty}^{+\infty} \psi_n^*(x)\cdot\psi_m(x)\cdot dx = \delta_{nm},\ \int_{-\infty}^{+\infty} \psi_n^*(x)\cdot\psi(x,\varepsilon)\cdot dx = 0,\ \int_{-\infty}^{+\infty} \psi^*(x,\varepsilon)\cdot\psi(x,\varepsilon')\cdot dx = \delta(\varepsilon-\varepsilon') \quad (24)$$

Moreover, integrals (24) mean that the phase space $L^2(R)$ of the system (15) is decomposed into direct sum of subspace V^d with basis from eigenfunctions of the discrete spectrum and subspace V^c with basis from eigenfunctions of the continuous spectrum: $L^2(R) = V^d \oplus V^c$.

Having found Green's function from the expression (23), we can return to investigation of the height $H(x,t)$ using formulae (18) and (7). In the case under consideration, the functional (4) is reduced to:

$$\int_{-\infty}^{+\infty} [H(x,T) - H_*(x)]^2 dx \to \min, \quad (25)$$

and as a fixed shape one can choose the function $H_*(x) = h_T \equiv const$ corresponding to uniformly flat surface. This situation is presented in Figure 3.

If on the interval of time $[0,T]$ there is a control under which global minimum of the functional (25) would be achieved that is $H(x,T) = h_T$ almost everywhere on R then under $t > T$ evolution of the system (14) with the source $Q(x) \equiv 0$ obeys to the next very simple expression:

$$H(x,t) = h_T + c\cdot(t-T) \quad (26)$$

The formula (26) claims that under $t > T$ there is no roughness on the surface under investigation.

3. Surface with cylindrical generatrix under the action of the Dirac delta function source of sputtering particles

It this section, general theory developed in the previous section is applied to the source proportional to the Dirac delta function: $Q(x) = \frac{4\cdot\nu^2\cdot q}{c}\cdot\delta(x)$. In geometry with cylindrical generatrix this source can be physically realized by means of the channelling of slow atomic particles along the series of carbon nanotubes [18] situated on the straight line $x = 0$. This configuration is illustrated by Figure 4.

Thus, in this case Eq. (15) is equal to:

$$\frac{\partial\varphi}{\partial t} = \nu\cdot\frac{\partial^2\varphi}{\partial x^2} + 2\cdot\nu\cdot q\cdot\delta(x)\cdot\varphi \quad (27)$$

and the stationary Schrödinger equation (21) transforms to the following:

$$-\frac{d^2\psi}{dx^2} - 2 \cdot q \cdot \delta(x) \cdot \psi = \varepsilon \cdot \psi \tag{28}$$

This equation is known to describe the quantum mechanical problem about the delta-functional well [19]. In accordance with [19] under $x \neq 0$ the equation (28) is reduced to the following one:

$$-\frac{d^2\psi}{dx^2} = \varepsilon \cdot \psi \tag{29}$$

which ought to be added by boundary conditions at $x = 0$ [19]:

$$\psi(0+0) = \psi(0-0) \equiv \psi(0) \ , \quad \frac{d\psi}{dx}(0+0) - \frac{d\psi}{dx}(0-0) + 2 \cdot q \cdot \psi(0) = 0 \tag{30}$$

It is easy to obtain from Eqs. (29) and (30) that in this system there is the unique state of the discrete spectrum:

$$\psi_0(x) = \sqrt{q} \cdot \exp(-q \cdot |x|) \tag{31}$$

corresponding to eigenvalue $\varepsilon_0 = -q^2$ [19]. Graphs of potential well $W(x) = -2 \cdot q \cdot \delta(x)$ and the normed ground state (31) in dimensionless coordinate are presented in Figure 5.

But our further construction sharply differs from this one in Ref. [19] because we take into consideration the next projection operator $\hat{\mathrm{P}}$ acting on arbitrary wave function $\psi(x)$ as follows:

$$\hat{\mathrm{P}} \cdot \psi(x) = \psi(-x) \, . \tag{32}$$

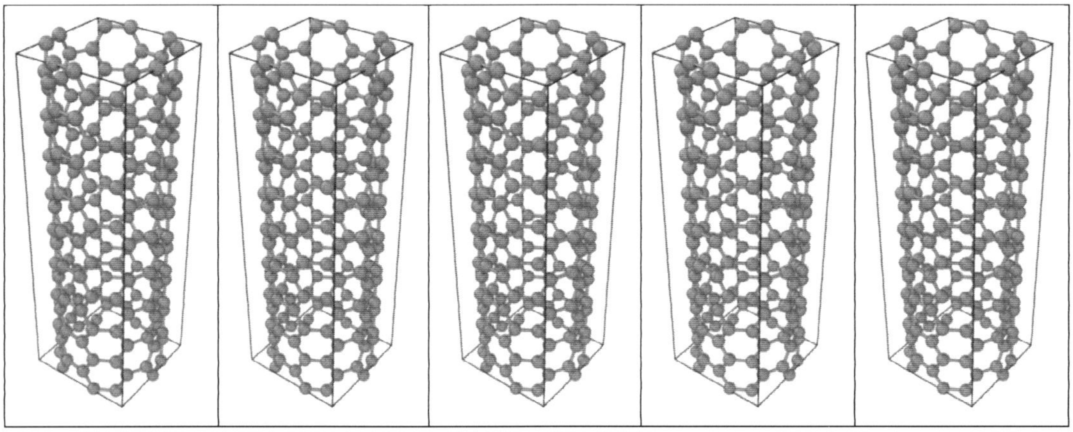

Figure 4. Series of carbon nanotubes as realization of delta-functional source.

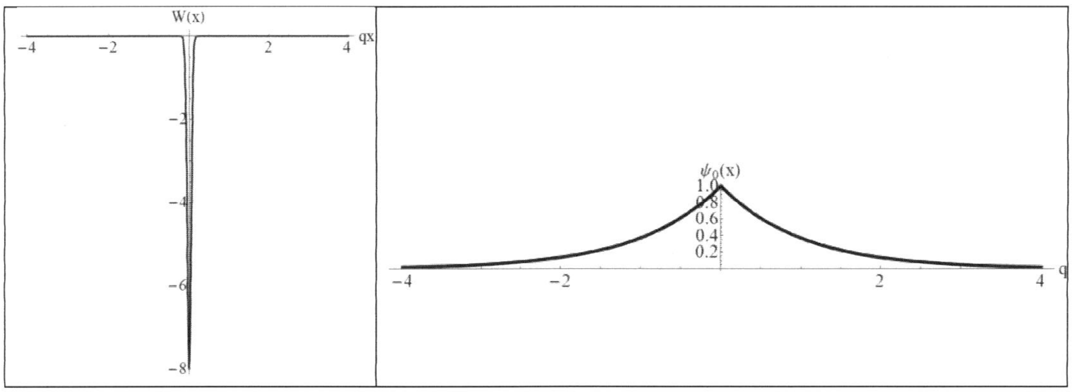

Figure 5. The delta potential well (on the left) and the ground state (on the right).

The Hermitian operator (32) is the well-known parity operator with eigenvalues P = ±1 [17, 19].

The Hamiltonian of the problem $\hat{\mathrm{H}} = -\dfrac{d^2}{dx^2} - 2\cdot q\cdot\delta(x)$ commutes with the parity operator: $[\hat{\mathrm{H}},\hat{\mathrm{P}}]=0$. Hence, for both of them there are common eigenfunctions [17, 19]: $\hat{\mathrm{H}}\cdot\psi_k^{\pm} = k^2\cdot\psi_k^{\pm}$ and $\hat{\mathrm{P}}\cdot\psi_k^{\pm} = \pm\psi_k^{\pm}$. Using Eqs. (29) and (30), one can establish that:

$$\psi_k^{+}(x) = \frac{1}{\sqrt{\pi}}\cdot\frac{k\cdot\cos(k\cdot x) - q\cdot\sin(k\cdot|x|)}{\sqrt{k^2+q^2}}, \qquad \psi_k^{-}(x) = \frac{1}{\sqrt{\pi}}\cdot\sin(k\cdot x) \tag{33}$$

Both of eigenfunctions (33) are normed on the Dirac delta function on k:

$$\int_{-\infty}^{+\infty}\psi_k^{\pm}(x)\cdot\psi_{k'}^{\pm}(x)\cdot dx = \delta(k-k') \tag{34}$$

and correspond to the same eigenvalue $\varepsilon = -k^2$. Graphs of functions (33) in dimensionless coordinate under $k = \pi\cdot q$ are presented in Figure 6. Application of the general formula (23) for the delta potential well gives us the following expression for Green's function of the equation (27):

$$G(x,\xi;t) = \exp(\nu\cdot t\cdot q^2)\cdot\psi_0(x)\cdot\psi_0(\xi) + \int_0^{+\infty}\exp(-\nu\cdot t\cdot k^2)\cdot[\psi_k^{+}(x)\cdot\psi_k^{+}(\xi) + \psi_k^{-}(x)\cdot\psi_k^{-}(\xi)]\cdot dk \tag{35}$$

It is easy to check that for functions (31) and (33) the next relation is true:

$$\psi_0(x)\cdot\psi_0(\xi) + \int_0^{+\infty}[\psi_k^{+}(x)\cdot\psi_k^{+}(\xi) + \psi_k^{-}(x)\cdot\psi_k^{-}(\xi)]\cdot dk = \delta(x-\xi) \tag{36}$$

Therefore, Green's function (35) obeys to the standard initial condition: $G(x,\xi;0) = \delta(x-\xi)$.
The result of calculation of Green's function in accordance with Eq. (35) is equal to:

$$G(x,\xi;t) = \frac{1}{2\cdot\sqrt{\pi\cdot\nu\cdot t}}\cdot\exp\left[-\frac{(x-\xi)^2}{4\cdot\nu\cdot t}\right] + \exp(\nu\cdot t\cdot q^2)\cdot[G_1(x,\xi;t) + G_2(x,\xi;t) + G_3(x,\zeta,t)] \tag{37}$$

where

$$G_1(x,\xi;t) = q\cdot\exp(-q\cdot|x| - q\cdot|\xi|) \tag{38}$$

$$G_2(x,\xi;t) = -\frac{\pi\cdot q}{4}\cdot[g(x+\xi,t) + g(-x-\xi,t) + g(x-\xi,t) + g(\xi-x,t)] \tag{39}$$

$$G_3(x,\xi;t) = \frac{\pi\cdot q}{4}\cdot[g(|x|-|\xi|,t) + g(|\xi|-|x|,t) + g(|x|+|\xi|,t) - 3\cdot g(-|x|-|\xi|,t)] \tag{40}$$

Functions (39)-(40) express via the next auxiliary function connected with the complementary error integral:

$$g(\zeta,t) = \exp(q\cdot\zeta)\cdot erfc\left(\frac{\zeta}{2\cdot\sqrt{\nu\cdot t}} + q\cdot\sqrt{\nu\cdot t}\right) \tag{41}$$

Moreover, functions (38)-(40) are invariant under the action of the abelian point group C_4 on the plane (x,ξ).

Typical graph of Green's function (37) under fixed time t is presented in Figure 7.

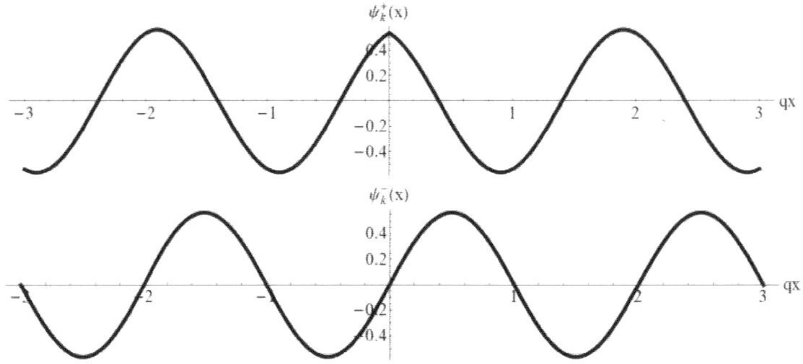

Figure 6. Eigenstates of continuous spectrum for the delta potential well corresponding to different eigenvalues of the parity operator: P = +1 from above and P = -1 from below.

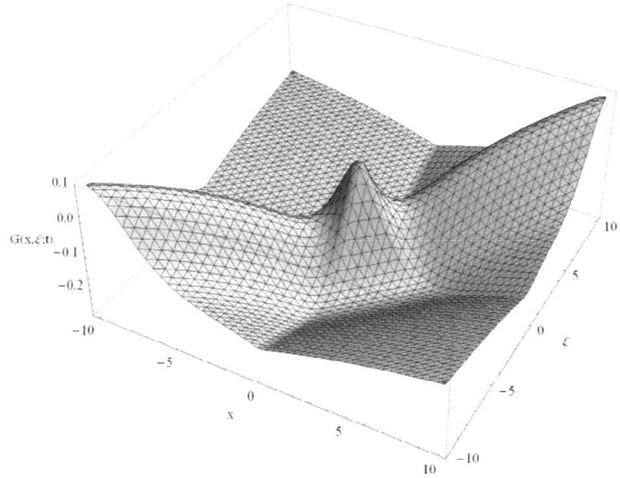

Figure 7. Green's function of linear parabolic equation with Dirac delta function source.

4. Wedge shape and measurement of surface diffusion coefficient

This section is devoted to analysis of the simplest solution of the KPZ-equation with delta-functional source and a corollary from it, which is important for physical experiment.

The identity (36) is known to claim that functions (31) and (33) form complete system in the Lebesgue space $L^2(R)$. Further, this fact in common with formulae (34) and the expressions:

$$\int_{-\infty}^{+\infty} \psi_k^+(x)\cdot\psi_{k'}^-(x)\cdot dx = 0\ ,\quad \int_{-\infty}^{+\infty} \psi_k^\pm(x)\cdot\psi_0(x)\cdot dx = 0 \tag{42}$$

It means that the phase space $L^2(R)$ of the infinite-dimensional system (27) can be expanded into the following direct sum: $L^2(R) = V_0 \oplus V_+^c \oplus V_-^c$, where $V_0 \subset L^2(R)$ is its subspace constructed from the single function (31) of the discrete spectrum of 'energy' ε and $V_\pm^c$ are linear subspaces of $L^2(R)$ corresponding to the continuous spectrum of this one. Bases of these subspaces V_+^c and V_-^c consist from functions $\psi_k^+(x)$ and $\psi_k^-(x)$, respectively.

It is quite hard to deal with eigenfunctions of continuous spectrum, that is why we shall consider the one-dimensional subspace V_0 only. In this case, initial condition for the equation (27) is equal to:

$$\varphi(x,0) = c_0 \cdot \psi_0(x) \quad (c_0 > 0) \tag{43}$$

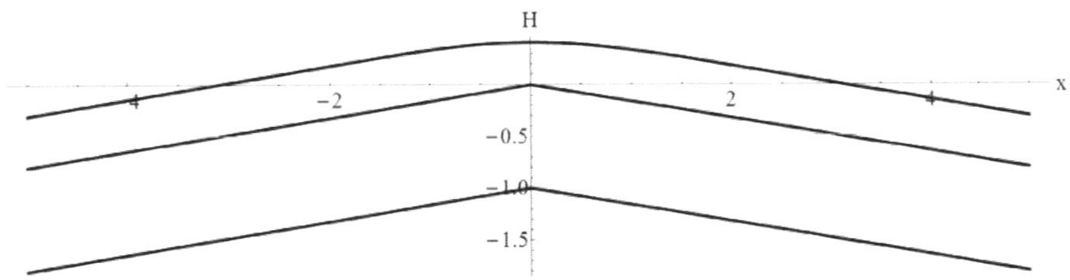

Figure 8. Temporal evolution of wedge corresponding to the single state of discrete spectrum: initial shape (from below), shape at $t = T$ (in the middle) and shape under $t > T$ (from above).

and the general solution of the equation (27) can be extracted from the formula (20):

$$\varphi(x,t) = c_0 \cdot \psi_0(x) \cdot \exp(\nu \cdot q^2 \cdot t) \tag{44}$$

Applying the Cole-Hopf substitution (7) in opposite direction from (44) we find that:

$$H(x,t) = \left(c + \frac{2 \cdot \nu^2 \cdot q^2}{c} \right) \cdot t + h_0(x), \tag{45}$$

where dependence

$$h_0(x) = \ln(c_0 \cdot \sqrt{q}) - \frac{2 \cdot \nu \cdot q}{c} \cdot |x| \tag{46}$$

is the initial shape of the surface corresponding to initial condition (43) for the auxiliary function (44).

Using formulae (45)-(46) it is easy to check that $\tan \beta \equiv \left| \frac{\partial H(x,t)}{\partial x} \right| = \frac{2 \cdot \nu \cdot q}{c}$. Hence, for validity of small angles approximation for the KPZ-equation (14) intensity of the Dirac delta function must satisfy to the next inequality: $q << c/\nu$. Thus, in this case initial profile (46) is the wedge with the angle near its edge close to π. And according to the expression (45), temporal evolution of the initial condition (46) reduces to the motion of the wedge in vertical direction with constant velocity $V = c + \frac{2 \cdot \nu^2 \cdot q^2}{c}$. This situation is presented in Figure 8.

If under $t > T$ source of external particles is switched off ($Q(x) \equiv 0$), then the shape of the surface is the following:

$$H(x,T) = h_T - \frac{2 \cdot \nu \cdot q}{c} \cdot |x|. \tag{47}$$

For further considerations it is convenient to assume that $h_T = 0$. One can achieve it by choosing $c_0 = \frac{1}{\sqrt{q}} \cdot \exp\left[-c \cdot T \cdot \left(1 + \frac{2 \cdot \nu^2 \cdot q^2}{c^2} \right) \right]$ in the expression (43). In this case, the one-dimensional variant of Eq. (11) gives us the next law of growth of the surface under $t > T$:

$$H(x,t) = c \cdot (t - T) \cdot \left[1 + \frac{2 \cdot \nu^2 \cdot q^2}{c^2} \right] + \frac{2 \cdot \nu}{c} \cdot \ln\left[\frac{g(x,t-T) + g(-x,t-T)}{2} \right] \tag{48}$$

where $g(x,t)$ is the auxiliary function (41). Graph of the function (48) under the fixed time t is presented in Figure 8. One can observe that under $t > T$ smoothing of edge of the wedge takes place. If sputtering substance is ideal conductor (at $t = T$), there is a possibility to measure coefficient of surface diffusion ν (it is supposed that values c and q are known from other measurements).

Let us consider linearly polarized simple harmonic incident electromagnetic wave with complex amplitude $A \cdot \exp[-i \cdot k \cdot r \cdot \cos(\varphi - \varphi_0)]$, r and φ being polar coordinates in plane xz and $k = 2\pi/\lambda$ being wave vector of the wave expressed via its wavelength λ. This wave falls on the wedge (47) from the direction $(\cos\varphi_0, \quad 0, \quad \sin\varphi_0)$ and gives rise to the wave of horizontal polarization:

$$\vec{E}_H = (0, \quad E(r,\varphi), \quad 0) \tag{49}$$

with the complex amplitude:

$$E(r,\varphi) = A \cdot [u(k \cdot r, \varphi - \varphi_0) - u(k \cdot r, \varphi + \varphi_0 + 2 \cdot \beta)] \tag{50}$$

and to the wave of vertical polarization:

$$\vec{E}_V = \left(\frac{1}{i \cdot k \cdot r} \cdot \frac{\partial H(r,\varphi)}{\partial \varphi}, \quad 0, \quad \frac{1}{i \cdot k} \cdot \frac{\partial H(r,\varphi)}{\partial r} \right), \tag{51}$$

with the complex amplitude of magnetic field:

$$H(r,\varphi) = A \cdot [u(k \cdot r, \varphi - \varphi_0) + u(k \cdot r, \varphi + \varphi_0 + 2 \cdot \beta)]. \tag{52}$$

expressing via the following auxiliary function (see Ref. [20] and references therein):

$$u(\rho,\phi) = \frac{2\pi}{\alpha} \cdot \left[\frac{J_0(\rho)}{2} + \sum_{l=0}^{\infty} \exp\left(-\frac{\pi \cdot i}{2} \cdot \frac{\pi \cdot l}{\alpha} \right) \cdot J_{\frac{\pi \cdot l}{\alpha}}(\rho) \cdot \cos\frac{\pi \cdot l \cdot \phi}{\alpha} \right] \tag{53}$$

In the formula (53) $\alpha = \pi + 2 \cdot \beta$ is the angle of the wedge and $J_\nu(\rho)$ is Bessel function.

Further, one can both measure and calculate using expressions (49)-(53) the specific cross section of this wave:

$$\frac{d\sigma}{d\varphi} = \frac{|\vec{E}|^2}{A^2} \cdot r \tag{54}$$

After that, one can extract the value of the angle of the wedge and, therefore, the value of surface diffusion coefficient v from comparison of experimental and theoretical data.

We underline that in expression (54) the electric field strength $\vec{E}$ ought to contain reflected wave only. It is obvious that at $t = T$ series of carbon nanotubes from Figure 4 must be removed from zone of experiment. To simplify formulae (49)-(53), it is suitable to apply both the Sommerfeld ray asymptotics and the Pauli ones [20].

5. Conclusion

In this work, the newest research program has been claimed; namely, we have proposed the point of view on nanoengineering as the problem of optimal control by distributed parameter system. We have illustrated this vision by the KPZ-equation (1) describing the growth of a solid state surface, external source of sputtering particles being a control.

We have carried out a number of preliminary investigations in the framework of the suggested research program. In particular, we have constructed the exact solution for the Cauchy problem (14) for the one-dimensional KPZ-equation in approximation of small angles with the source $Q(x) = \frac{4 \cdot v^2 \cdot q}{c} \cdot \delta(x)$ of external particles:

$$H(x,t) = c \cdot t + \frac{2 \cdot v}{c} \cdot \ln\left(\int_{-\infty}^{+\infty} G(x,\xi;t) \cdot \exp\left[\frac{c \cdot h_0(\xi)}{2 \cdot v} \right] \cdot d\xi \right) \tag{55}$$

where $G(x,\xi;t)$ Green's function (37)-(41).

Success in obtaining of the result (55) gives us a confidence that by methods of quantum mechanics [17, 19] we shall construct the exact solution of the Cauchy problem (14) with the following source:

$$Q(x) = \frac{4 \cdot v^2}{c} \cdot \sum_{n=0}^{N-1} q_n \cdot \delta(x - n \cdot b). \quad (56)$$

Function (56) can be physically realized by means of the passing of slow atomic particles through the 'comb' with step b of N series of carbon nanotubes presented in Figure 4. We underline that intensities q_n of flows of such particles can be different in different series. Also we stress that carbon nanotubes start to decay at temperature about 1500 K [18]; therefore, these nanostructures can be applied under different kinds of epitaxial procedures.

The suggested in previous section method of surface diffusion coefficient determination is highly realistic, because in work [21] the possibility of the channelling of atomic particles with different masses – from He to Xe – and with energies less than 0.5 keV is theoretically established, metals being among these particles. Moreover, usage of electromagnetic waves scattering data is very fruitful idea, because atomic force microscopy usually presents the shape of the surface under investigation at the beginning and at the end of epitaxial process [4]. But for values like (54) real time measurement can be organized. This information about states of the observed system's motion may be very important especially for reconstruction of unknown source [7].

It is necessary to emphasize that a wide range of nanotechnological problems can be solved as problems of optimal control by distributed parameter system. For instance, in paper [22] it has been demonstrated that for a multilayer Ge/Si(001) heterostructure with vertically aligned Ge nanoclusters the nonuniform spatial elastic strain distribution gives rise to a three-dimensional potential well for electrons in the strained Si layers surrounding Ge quantum dots. In this situation, elastic strain distribution ought to be considered as a control. And the depth of the potential well ought to be considered as a functional, its maximization resulting in maximization of efficiency of the radiative recombination for the nanostructure [22]. Another example concerns magnetic field controlled domain wall nucleation, pinning and depinning effects in ferromagnetic nanowires [23]. Great attention is paid to these effects to realize magnetic logical cells, which are able to perform main logical operations [23].

On the other hand, scanning probe microscopy gives an opportunity both to create and to investigate nanostructures of different types for various applications [24-26], because nowadays the same device can work as scanning probe lithographer, as atomic force microscope, as magnetic force microscope, and so forth [24-26].

At last, let us remind that Eigler and Schweizer were the first to use a scanning tunneling microscope tip to arrange 35 individual xenon atoms on a single-crystal nickel surface, famously spelling out the letters 'IBM' [27]. Thus, nanoengineering and scanning probe microscopy have been closely relating since invention of the second one. And at present, powerful development of functional analysis and computer science is setting the qualitatively new level of this interrelation.

Acknowledgements
The authors prepared this article during working on the project № 18-08-01356-a of Russian Foundation for Basic Research.

References

[1] Ma Z, Zhao M, Qu Z, Gao J, Wang F, Shi Y, Qin L and Liu J 2017 *J. Nanosci. Nanotechnol.* **17** 2213-34
[2] Mikhailov G M, Fomin L A and Chernykh A V 2017 *Materials* **10** 1156
[3] Sazanova T S, Otvagina K V and Vorotyntsev I V 2018 *Polym. Test.* **68** 350-8
[4] Kulikov D A, Potapov A A, Rassadin A E and Stepanov A V 2017 *IOP Conf. Ser.: Mater. Sci. Eng.* **256** 012026
[5] Kardar M, Parisi G and Zhang Y C 1986 *Phys. Rev. Lett.* **56** 889-92
[6] Goray L I and Lubov M N 2014 *J. Surf. Invest.* **8** 444-55
[7] Korotkii A I and Mikhailova D O 2013 *Proc. Steklov Inst. of Math.* **280** 98-118

[8] Pontryagin L S, Boltyanskii V G, Gamkrelidze R V and Mishchenko E F 1986 *Selected works, v. 4, Classics of Soviet Mathematics, The mathematical theory of optimal processes* (Gordon & Breach Science Publishers)
[9] Vladimirov V S 1971 *Equations of Mathematical Physics* (Marcel Dekker Inc.)
[10] Rassadin A E, Stepanov A V and Fomin L A 2018 *Adv. Sci.* **1** 28-33 (in Russian)
[11] Fursikov A V 2001 *Sb. Math.* **192** 593-639
[12] Obradovich O, Potapov M M and Razgulin A V 1992 *Comput. Math. & Math. Phys.* **32** 1075-83
[13] Vasil'ev F P 1970 *USSR Comput. Math. & Math. Phys.* **10** 171-190
[14] Remizov I D 2018 *Appl. Math. Comput.* **328** 243-6
[15] Kac M 1949 *Trans. Amer. Math. Soc.* **65** 1-13
[16] Feynman R P 1953 *Phys. Rev.* **91** 1291-1301
[17] Landau L D and Lifshitz E M 1977 *Quantum Mechanics: Non-Relativistic Theory* (Pergamon Press)
[18] Stepanov A V 2015 *J. Surf. Invest.* **9** 789-98
[19] Galitsky V M, Karnakov B M and Kogan V I 1992 *Problems on Quantum Mechanics* (Nauka) (in Russian)
[20] Ufimtsev P Ya 2014 *Fundamentals of the Physical Theory of Diffraction* (Wiley-IEEE)
[21] Dedkov G V 2002 *Surf. Coat. Tech.* **158-159** 75-80
[22] Yakimov A I, Bloshkin A A and Dvurechenskii A V 2008 *Phys. Rev. B* **78** 165310
[23] Ermolaeva O L and Mironov V L 2018 *IEEE Trans. Magn.* **54** 1-8
[24] Huang Z Q, Jiang J J, Hua Y M, Li C Z, Wagner M, Lewerenz H J, Soriaga M P and Anfuso C 2015 *Energy Environ. Focus* **4** 260-77
[25] Schwager P, Bulter H, Plettenberg I and Wittstock G 2016 *Energy Technol.* **4** 1472-85
[26] Denning D, Guyonnet J and Rodriguez B J 2016 *Int. Mater. Rev.* **61** 46-70
[27] Eigler D M and Schweizer E K 1990 *Nature* **344** 524-6

IOP Conf. Series: Materials Science and Engineering **443** (2018) 012028 doi:10.1088/1757-899X/443/1/012028

The study of supramolecular structure of asphaltenes by atomic force microscopy

T I Sharipov, M Yu Dolomatov and R Z Bakhtizin

Department of Physical Electronics and Nanophysics, Bashkir State University, 450076, Ufa, Russia

sha-t@yandex.ru

Abstract. The article is devoted to the actual problem of studying the supramolecular structure of oil asphaltenes. They are the prospective structural materials for molecular electronics and nanotechnology. At present, the properties of asphaltenes at the supramolecular level are studied by various physical and physicochemical methods, as well as by methods of mathematical modelling. The aim of the work is research of asphaltenes supramolecular structure by atomic force microscopy (AFM), comparison of experimental data with the calculations of computer simulation, and high resolution visualization of asphaltene crystallites. The topography of asphaltenes monomolecular layer is observed by AFM. Separate objects of disc-shaped form – crystallites obtained as a result of the association of individual molecules of asphaltenes are clearly visible. In addition, some structural parameters are calculated.

1. Introduction

Recently, much attention has been paid to the works, in which organic substances are used as materials for electronics. This is due to the unique properties of elements of organic electronics, which exhibit semiconductor and even metal properties, and the size of these elements can be comparable to the size of an organic molecule [1-3].

Asphaltenes are the prospective structural materials for molecular electronics. It is known that asphaltenes in oil dispersed systems are in the form of supramolecular structures ranging in size from 1 to 1000 nm, which allows us to consider asphaltenes as a promising object of nanotechnology. At present, the properties of asphaltenes at the supramolecular level are studied by various physical, physicochemical methods and methods of mathematical modelling [4-7].

Asphaltene is the high molecular weight component of crude oil, consisting of 90-95% of hydrocarbon. Besides carbon, hydrogen, oxygen, and nitrogen, the asphaltene molecule also includes sulphur [8]. Asphaltenes have an increased tendency to association and formation of liquid crystal or supramolecular structures. In 1961, T. Yen proposed the so-called burst model of the asphaltenes structure of "plate to plate" type [9]. At present, two models of asphaltenes aggregation – the "continental" type and the "archipelago" type structure are generally recognized [10,11]. The structure of the "archipelago" is unlikely for asphaltenes extracted from the residue of thermal cracking, because chemical bonds binding polycyclic fragments of the molecule are destroyed in the process of cracking. Considering the tendency of polycyclic molecules to associate, an important question is to determine the degree of molecules association in solution. Asphaltenes were also studied by mass spectrometry and field ionization [12], mass spectrometry with photoionization [13], and laser desorption mass spectrometry [14]. By means of physical methods, concentrations, at which the asphaltenes structure changes in the solution, were determined [15].

It was found by Schneider et al. [16] using the method of fluorescent correlation spectroscopy, that solutions of oil asphaltenes in toluene at concentrations up to 10^{-4} wt. (0.1 g/l) manifested themselves as true solutions. Mullins et al. [17], as well as Freed et al. [18] found that with an increase in the concentration of more than 10^{-4} wt. stable nanoaggregates of 8 - 10 molecules were formed. With a mass fraction of more than 5 10^{-3} wt. (5 g/l), further formation of clusters of nanoaggregates took place. At concentrations above 5 10^{-2} wt. (50 g/l) a gel-like grid was formed. Thereby, the degree of molecular association increases depending on the concentration of the solution, which is important to consider during microscopic studies.

The aim of the work is study of asphaltenes supramolecular structure by atomic force microscopy (AFM), comparison of experimental data with the calculations of computer simulation, and high resolution visualization of asphaltene crystallites.

2. Materials and methods

The asphaltenes isolated from thermal cracking residues by Soxhlet extractor were used as the research object. Average molar mass number of asphaltenes according to the optical spectroscopy is 1200 a.m.u. The solution concentration for film preparing is 0.1 g/l. Under these conditions, we should expect a weak molecules association in solution.

Mica was chosen as the substrate because of its atomically smooth surface. We have used AFM Ntegra-Prima (NT-MDT, Zelenograd, Russia) for the high resolution imaging of surface topography and calculation of structure parameters of asphaltene molecules. The scanning of the samples was carried out by the HA_NC cantilevers (TipsNano, Russia) with the curvature radius of 10 nm, tip's cone angle of 30°, and force constant of 12 N/m. Under the ambient conditions and with the correct concentration of the solution, a single molecule of asphaltene can be visualized.

The initial task was to select the optimal concentration of asphaltenes in toluene, a working solution for subsequent preparation of samples suitable for AFM study. The working solutions with a concentration of 0.1 g/l were prepared by calculating the mass ratio of asphaltene in the toluene volume. Langmuir-Blodgett technique was used to obtain a monomolecular film of asphaltenes. After trying each of the four standard Langmuir-Blodgett techniques, we used up-draw process of the mica substrate through the film, whereby we obtained the highest quality AFM images. The deposited asphaltenes thin film was dried in the Petri dish for 60 minutes at the temperature of 24°C. The final sample was studied in the air atmosphere in AFM tapping mode.

3. Results and discussion

To demonstrate the reproducibility of the sample preparation technique, two samples prepared by one method were scanned. The topography of the monomolecular layer of asphaltenes was obtained by AFM (Figs. 1a, 2a). Individual disc-shaped objects are visible on them.

For detailed analysis, the cross-sections have been made along the lines marked in Figures 1a and 2a. Each cross-section was performed on three separate objects. According to the first cross-section profile (Fig. 1b), it can be calculated that the first object has a height of 1.5 nm and a width of 33 nm, the second one – 1.35 and 34 nm, and the third one – 1.4 and 31 nm, respectively. According to the second cross-section profile (Fig. 2b), it can be calculated that the first object has a height of 1.7 nm and a width of 36 nm, the second one – 1.3 and 36 nm, and the third one – 1.4 and 34 nm, respectively.

Thus, the results of AFM studies show disc-shaped structures constructed from asphaltene molecules associates. Analysing and comparing the sizes of objects and their location on the surface with the results of the work of other researchers [17,18], we came to the conclusion that individual disc-shaped objects are clusters of asphaltenes and their placement on the surface in a nanoaggregate state can be described by the disk-like aggregation model. When comparing our experimental results with the results of study and simulation by Ese et al. [19] and Korzhov et al. [20], it can be assumed that the investigated aggregates have a disk-shaped structure and these structures are similar to discotic liquid crystals.

IOP Conf. Series: Materials Science and Engineering **443** (2018) 012028 doi:10.1088/1757-899X/443/1/012028

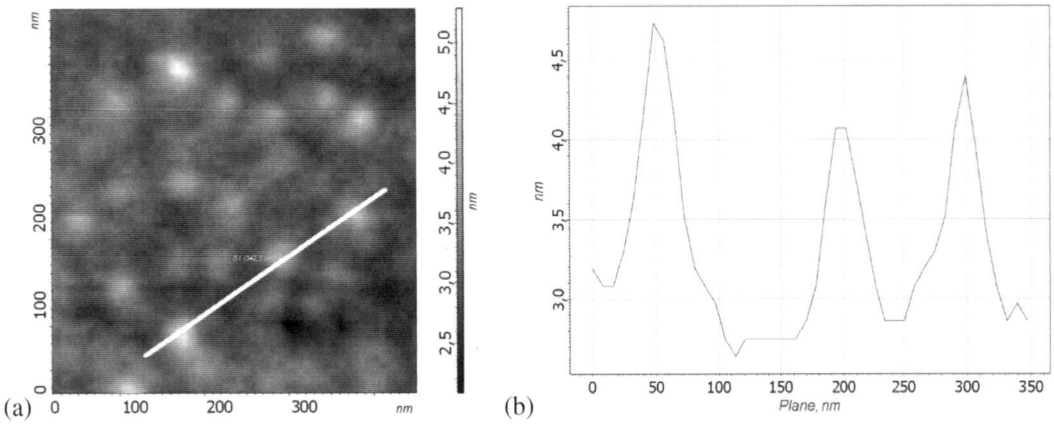

Figure 1. (a) AFM topography image of asphaltenes film (1 sample), (b) cross-section profile along the line marked at (a).

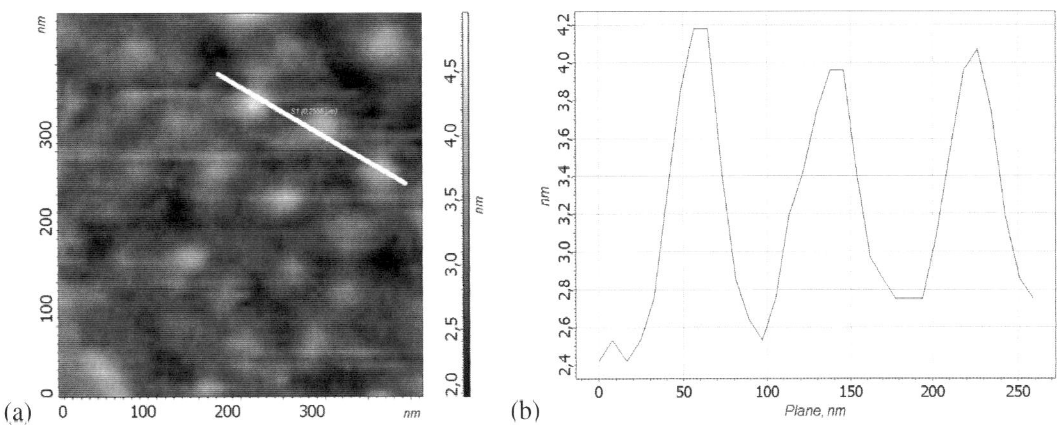

Figure 2. (a) AFM topography image of asphaltenes film (2 sample), (b) cross-section profile along the line marked at (a).

Table 1. Structural characteristics of crystallites.

Structural characteristics	Sample 1	Sample 2
Average crystallite height, nm	1,4±0,1	1,5±0,1
Average crystallite width, nm	33	35
Interlayer distance*, nm	0,35	0,35
Average number of layers in the crystallite, pcs.	4	4
Average number of molecules in a layer, pcs.	75	81
Average number of molecules in the crystallite, pcs.	300	324
Molar mass, amu	360000	388800

* - calculated by the simulation method

According to the results of previous works on the modelling of asphaltene molecules, the height of one molecule is approximately 0.1 nm, and the length is about 1 nm [21]. Therefore, knowing the height and length of one molecule, we can calculate some parameters that are presented for clarity in Table 1.

Based on the data of Table 1, it follows that the crystallite parameters of sample 1 and sample 2 have approximately the same quantitative values.

4. Conclusion

The results of the interpretation of AFM studies indicate the existence of supramolecular structures of the discotic type formed by associated molecules of asphaltenes. We have found that individual disc-shaped objects are asphaltenes crystallites and their location on the substrate surface can be described by the disk-like aggregation model. The average number of molecular structures in the crystallite is 304-324 pcs. The average number of layers in the crystallite is 4. Within each layer there are 76-81 molecules. The molar mass of the asphaltene crystallite was found. It is 364800-388800 amu. The obtained results correspond to the data of studies performed by other physical methods.

Acknowledgments

This work was supported by the Russian Foundation for Basic Research under grant No. 17-42-020616 and Republic of Kazakhstan under project No. AP05132165.

References

[1] Rambidi N G 2007 *Nanotekhnologii i molekulyarnyye komp'yutery.* M.: FIZMATLIT – 256
[2] Ratner M A 2013 *Nat. Nanotechnol.* **8** 378-81
[3] Sharipov T I and Bakhtizin R Z 2017 *IOP Conf. Series: Mat. Sci. Eng.* **195** 012002
[4] Unger F G and Andreyeva L N 1995 *Fundamental'nyye aspekty khimii nefti. Priroda smol i asfal'tenov.* – Novosibirsk: Nauka. – 192
[5] Ganeeva Yu M, Yusupova T N and Romanov G V 2011 *Russ. Chem. Rev.* **80** 993-1008
[6] Shutkova S A, Dolomatov M Yu, Bakhtizin R Z, Telin A G, Shulyakovskaya D O, Kharisov B R and Dezortsev S V 2012 *Bashkir chemical journal* **19** 220
[7] Bakhtizin R Z, Sharipov T I, Nurakhmetov T N, Kuterbekov K A, Salikhodga Dj M, Dolomatov M Yu and Kainarbaj A Dg 2017 *Nauchnyj jurnal Fizika (Fizika ilimij jurnaly)* **1** 79-84
[8] Sergiyenko S R, Taimova B A and Talalayev Ye I 1979 *Vysokomolekulyarnyye neuglevodorodnyye soyedineniya nefti.* M.: Nauka - 17
[9] Yen T F, Erdman J G and Pollack S S 1961 *Anal. Chem.* **33** 1587-94
[10] Sabbah H, Morrow A L, Pomerantz A E and Zare R N 2011 *Energy Fuels* **25** 1597-604
[11] Dmitriev D E and Golovko A K 2010 *Chemistry for Sustainable Development* **18** 171-80
[12] Boduszynski M W 1981 *Chemistry of Aspholtenes*. Washington. – Am. Chem. Soc. 119-35
[13] Merdrignac I, Desmazieres G, Terrie P, Delobet A and Laprevote O 2004 *Int. Conf. on Heavy Organics Deposition.* – Mexico. – P. 265
[14] Hortal A R, Martinez-Haya B, Lobato M D, Pedrosa J M and Lago S 2006 *J. Mass Spectrom.* **41** 960-8
[15] Andreatta G, Bostrom N and Mullins O C 2007 *Asphaltenes, Heavy Oils, and Petroleomics* Springer, New York, NY 231-57
[16] Schneider M, Andrews B, Mitra-Kirtley S and Mullins D 2007 *Energy & Fuels* **21** 2875-82
[17] Mullins O, Betancourt S, Cribbs M, Creek J, Dubost F, Andrews A and Venkatarmanan L 2007 *Energy & Fuels* **21** 2785-94
[18] Freed D, Lisitza N, Sen P and Song Y-Q 2007 *Asphaltenes, Heavy Oils, and Petroleomics* Springer, New York, NY 279-300
[19] Ese M-H, Sjoblom J, Djuve J and Pugh R 2000 *Colloid Polym. Sci.* **278** 532-8
[20] Korzhov Yu V and Orlov S A 2016 *Izvestiya Tomskogo polytekhnicheskogo universiteta. Injiniring georesursov* **327** 62-74
[21] Dolmatov M Yu, Shutkova S A and Dezortsev S V 2012 *Neftekhimiya* **52** 299-303

IOP Conf. Series: Materials Science and Engineering **443** (2018) 012029 doi:10.1088/1757-899X/443/1/012029

Probe microscopy and other methods in the investigation of aluminium alloys

O O Shcherbakova[1], T I Muravyeva[1], Yu A Bobrov[2], D L Zagorskiy[1,3] and I V Shkalei[1]

[1] Ishlinsky Institute for Problems in Mechanics RAS, 119526, Moscow, Russia
[2] NT-MDT, Zelenograd, Russia
[3] Gubkin Russian State University of Oil and Gas, 119991, Moscow, Russia

shcherbakovaoo@mail.ru

Abstract. The work is devoted to the study of the surface of anti-friction aluminium alloys. The influence of alloying elements and heat treatment on the physical and mechanical properties of the surface, its thermal conductivity, and electrical conductivity was estimated. It is shown that the addition of a number of elements (Pb, Bi, Cd, and In) to the base alloy Al-5Si-4Cu and subsequent heat treatment lead to an increase in physical and mechanical properties (so, the greatest hardness value has an alloy with cadmium). Thermal conductivity at the macrolevel was estimated from the measurement of electrical conductivity; it is shown that doping and heat treatment in most cases bring to a little change of these parameters. A complex method of microscopy (scanning electron microscopy with elemental analysis and scanning probe microscopy) was used to study the surface of the samples. Various modes of scanning probe microscopy were used: topography, spreading currents, and surface temperature were measured independently. It is shown that the values of electrical conductivity and thermal conductivity measured at the microlevel are correlated well enough. A significant heterogeneity in the surface thermal conductivity of the cast sample is found, which correlates with the corresponding phase composition of the impurities.

1. Introduction

It is known that aluminium alloys have a wide range of applications. One of the directions is their use as bearing materials in mechanical engineering. This is due to their lower cost compared to expensive bronze, which is often used in these units. At the same time, aluminium alloys have performance properties comparable to those of bronze [1-4]. This is achieved by alloying several elements, including low-melting metals (in particular, tin, lead, bismuth, cadmium, etc.), which makes it possible to improve the characteristics of the material and vary them in a wide range. The principle of operation of bearings based on these alloys is that under unfavourable friction conditions, a soft phase component (low-melting elements) is released onto the surface of the solid matrix, which plays the role of a lubricant that protects the base material from destruction.

This work is devoted to the study of alloys of the Al-Si-Cu-X system, where X is a low-melting element. Based on the calculations and results obtained earlier [5-7], a base composition based on Al-5% Si-4% Cu was taken.

The structure (primarily, the surface topography) was studied by electron and probe microscopies. The study of properties at the macrolevel was carried out by standard methods – Brinel hardness tests and electrical conductivity. Probe microscopy was also used to study the complex properties of the surface at the micro level – electrical conductivity and thermal conductivity. Note that the thermal conductivity is the most important characteristic of the alloy, which determines its working properties. However, it is often difficult to measure this value. It is known, however, that in many cases this parameter correlates with conductivity, which is easy to measure. Based on this, the tasks of this work were also independent measurements and establishing a correlation between these parameters.

The general aim of the work was to study the influence of individual elements and heat treatment on the structure and properties of the alloys.

2. Materials and methods

2.1. Alloys

The model alloys based on the basic composition of the Al-5% Si-4% Cu system were studied. The elements were chosen for the following reasons: copper was introduced for hardening, and its 4% concentration was taken close to the limiting solubility of this metal in an aluminium solid solution. Silicon, used to improve casting properties, is taken in an amount of 5%; while achieving the best balance between mechanical and technological properties.

At the first stage of the work, low-melting metals Bi, Cd, In, and Pb were added to the alloy base and the influence of each of these elements on the mechanical, heat-conducting, and tribological properties of the material was investigated. At the second stage, similar alloys based on the base system were investigated, but with the addition of tin taken in an amount of 4%; this concentration corresponds to the content of this element in branded aluminium antifriction alloys. Selected fusible elements, which proved to be the best at the first stage of the experiment, were added to this system (Al-5%Si-4%Cu-4%Sn).

The obtained samples were examined in the cast (initial) state, and also after heat treatment (HT). It is known that the best for mechanical properties is the globular morphology and it is necessary to strive for maximum spheroidization of all the particles that make up the composition [8]. The structures of alloys of similar compositions were studied by the authors in previous works [6,7,9]. Thus, it was shown that for analogous alloys, the optimal mode of maintenance was heating to 500°C followed by quenching in water. This regime leading to an improvement in the structure - the spheroidization of phases and the dissolution of copper in an aluminium solid solution - has been chosen in the present study.

2.2. Physical and mechanical properties

With the use of modern experimental methods, the physical and mechanical properties of the alloys of the Al-Si-Cu-X system, in particular, their hardness and electrical conductivity, were studied. The Brinell hardness was measured using the standard method in accordance with GOST 9012-59 on the Nemesis 9000 universal hardness tester with the following parameters: ball diameter 2.5 mm, load on the indenter was 612.9 N, and holding time 20 s. The electrical conductivity of the samples was investigated with a VE-26NP vortex structuroscope. The principle of operation of a structuroscope is based on the method of eddy currents using the phase method for processing the signal of an overhead eddy current transducer. The measuring range of the device is 5-62 MSm/m. The relative error is 2%.

2.3. Microscopy

Investigation of topography: electron microscopy was carried out on a scanning electron microscope (SEM) "Quanta-650" with an X-ray spectral microanalyzer EDAX. The regime of secondary electrons with an accelerating voltage of up to 25 kV was used. Samples were studied before and after HT. Studies at the nanoscale were carried out using various methods of scanning probe microscopy (SPM). The study of topography was carried out using the Smart SPMTM microscope (manufactured by AIST-NT). The tapping mode (resonance frequency 250 kHz) was used, cantilevers (probes) of the fpN10 series with the tip curvature radius of 20 nm were used. A special feature of the work was the investigation of the same selected areas of the surface by both methods of microscopy – SEM and

SPM. For doing it, i.e. for reliable identification of the investigated regions, the "reference method" was used – the investigated region of the surface was accurately labelled with marks [9].

Investigation of the properties of the surface of alloys was carried out using the Ntegra Prima SPM (NT-MDT). Preliminary measurement of the surface relief was carried out in a contact and semi-contact modes. At the same time, a map of the surface spreading current and the thermal conductivity of the samples was measured.

When measuring the spreading current between the sample surface and the probe, a constant voltage of up to 10 V was supplied and a current was measured for different points of the surface.

The thermal conductivity was estimated with the aid of a scanning thermal microscopy (STM) using a special probe, the end of which was a resistor to which a voltage of a predetermined value was applied. The temperature of the probe was measured by the change in its resistance built into the arm of the Winston bridge. To determine the thermal conductivity of the surface, the probe is first heated to a certain temperature, which changes when interacting with the surface. In conditions of a sufficiently slow scan, when transient processes can be neglected, the change in probe temperature will vary inversely with the thermal conductivity of the material in the region of its contact with the surface. Thus, the temperature difference recorded in this case provides information on the heat-conducting properties of the samples.

3. Results and discussion

For all samples, the surface structure and its change as a result of HT were studied. The influence of HT on the physical and mechanical properties of alloys was also evaluated.

In addition, at the second stage of research for alloys containing tin, new methods of studying surface properties, including additional modes of SPM, were used. In this case, the electrical conductivity and thermal conductivity were measured independently.

It should be noted that at the first stage the thermal conductivity was estimated from the electrical conductivity measured at the "macro level" using the known dependence; at the same time, their correlation was assumed on the basis of the literature data. At the second stage, these parameters were measured by SPM modes independently at the micro level and their correlation was shown.

3.1. Microscopic examination of the surface of Al-Si-Cu-X system

The complex method of microscopic studies [9] was used to study the initial surface of experimental aluminium alloys, both in the cast and HT state. The use of SEM made it possible to study the structure of experimental alloys and to estimate the effect of HT on its change. Figure 1 demonstrates the obtained SEM images of the initial surface of the samples (with additives of low-melting elements) in the cast (virgin) and HT state.

Studies have shown (Fig. 1) that all alloys are close in structure and contain in their composition two similar phases: compound Al_2Cu and eutectic Al-Si. They are distinguished by low-melting phase components – Pb, Bi, Cd, and In. It was also found that heat treatment led to the spheroidization of the phase constituents of the alloys.

The complex method of microscopic research used in the work included the joint use of various methods of electron and probe microscopy. Thus, the same surface regions were studied by the SEM and SPM methods. For this purpose, special markers (benchmarks) were applied to the surface, which made it possible to isolate and analyse the necessary area. This approach made it possible to obtain topographic images with elemental composition (both over the entire surface and in selected regions), and also to estimate the surface relief. Such a complex technique was applied to the study of all experimental alloys. As an example, the results obtained for a sample of the Al-5Si-4Cu-1Cd alloy in a cast (initial) and HT state are presented in Figure 2: they are given as SEM images with element analysis and SPM images, with corresponding 3D images. Table 1 shows the chemical composition of the selected regions.

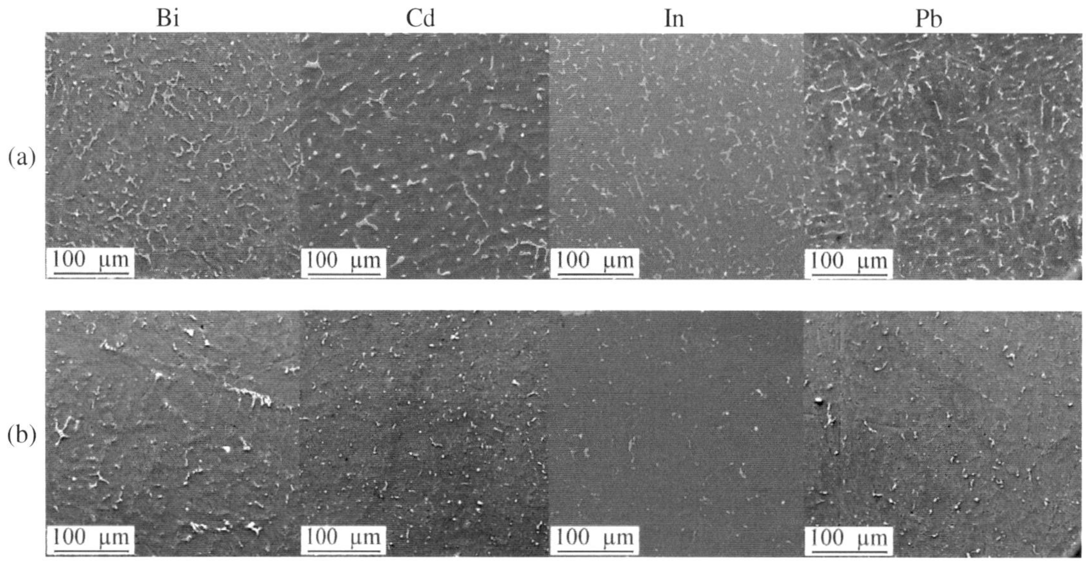

Figure 1. SEM images of the initial surface of new model aluminium alloys with additives (indicated above): (a) in the cast; (b) in the heat-treated state.

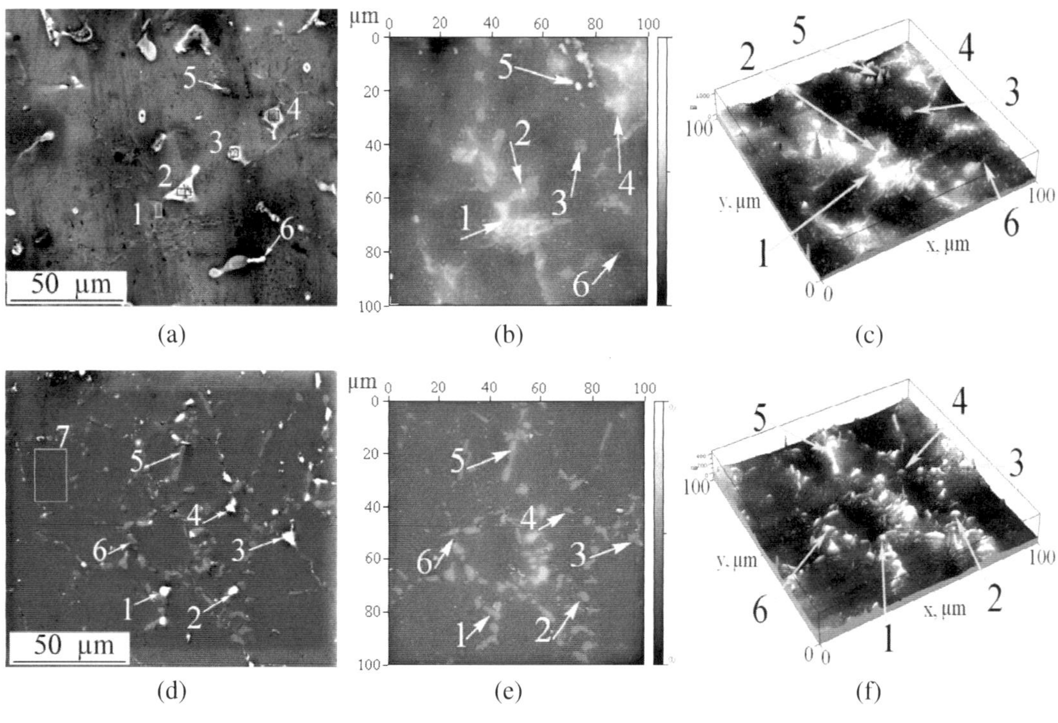

Figure 2. Image of the surface of the alloy Al-5Si-4Cu-1Cd: (a), (d) SEM images; (b), (e) 2D SPM topography image; (c), (f) 3D SPM topography image; (a), (b), (c) cast and (d), (e), (f) after HT.

Table 1. Chemical composition of the initial structure of the sample Al-5Si-4Cu-1Cd.

Sample	№ studied area	Chemical composition (% mass.)			
		Al	Si	Cu	Cd
Al–5Si–4Cu–1Cd cast	**1** (Al-Si)	40.60	57.33	1.60	0.47
	2 (Al_2Cu)	55.06	0.79	43.33	0.82
	3 (Al_2Cu)	59.29	1.28	38.58	0.85
	4 (Al_2Cu)	56.48	1.25	41.83	0.44
	5 (Al-Si)	39.30	55.14	2.16	3.40
	6 (Soft phase)	21.37	-	1.41	77.22
	7 (Matrix)	97.16	1.25	1.01	0.58
Al–5Si–4Cu–1Cd after HT	**1** (Soft phase)	24.61	1.22	1.79	72.38
	2 (Soft phase)	23.49	9.43	1.24	65.84
	3 (Al_2Cu)	54.34	0.77	44.71	0.18
	4 (Al_2Cu)	55.88	0.44	36.28	7.40
	5 (Al-Si)	64.25	32.40	2.92	0.43
	6 (Al-Si)	59.26	37.71	2.46	0.57
	7 (Matrix)	95.46	0.34	3.84	0.36

The obtained results indicate that the SEM and SPM images of the surface correlate well and substantially complement each other. We note that on SEM images, silicon in the matrix is seen very poorly (this is due to the proximity of atomic numbers of silicon and aluminium), while the use of SPM makes it possible to obtain additional information [9], so these elements are well separated on the SPM-picture. So, the joint use of two methods makes it possible to estimate the parameters of the relief of irregularities on the surface and the spatial geometry of the alloy.

Additional information on the phase composition of the multicomponent alloy was obtained by mapping. Figure 3 shows the distribution of chemical elements of the Al-5Si-4Cu-1Cd alloy in a cast and HT state.

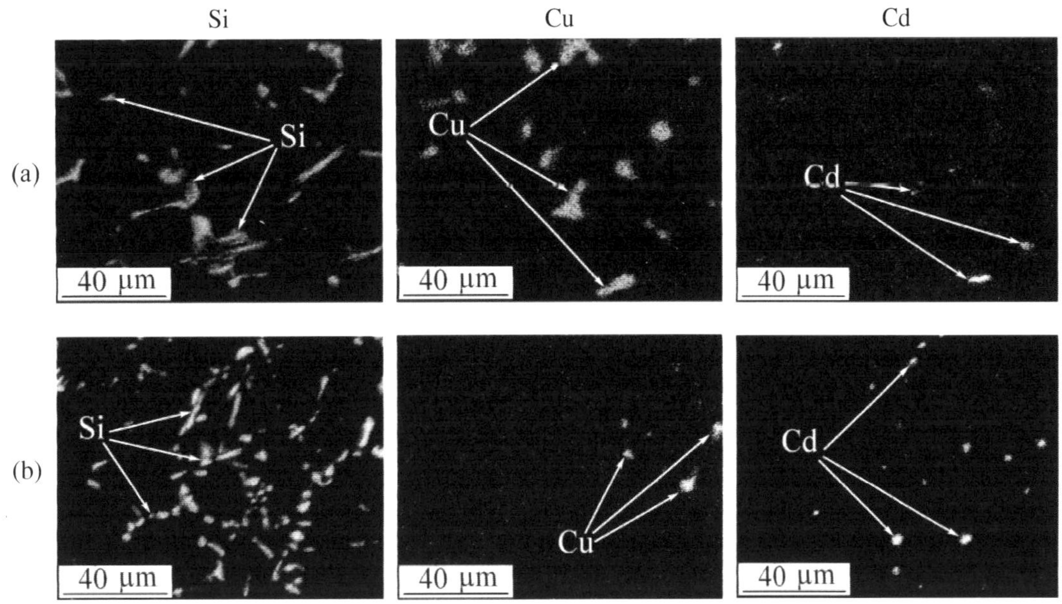

Figure 3. Map of the individual elements distribution on the surface of the alloy Al-5Si-4Cu-1Cd: (a) in the cast, (b) in the HT state.

The location of the alloying elements could be estimated from Figure 3. Thus, during the HT, the silicon phase becomes more globular, and its amount varies insignificantly compared with the cast state. The soft phase (Cd) is spheroidized after annealing (i.e., assumes a rounded shape). The number of Al_2Cu phase inclusions becomes significantly less than in the cast state, which is due to their partial dissolution (in Al-matrix). It is also indicated by the analysis of the elemental composition in Table 3 (the content of copper in the matrix increases to 3.84% by mass).

Thus, it is shown that HT at 500°C, followed by quenching in water, changes the structure of the alloy. The joint use of SEM and SPM methods allows qualitative evaluating structural and phase changes.

3.2. Investigation of physico-mechanical properties of Al-Si-Cu-X system

The strength properties of aluminium alloys are usually indicated by the value of hardness, and thermal conductivity is one of the most important characteristics for antifriction alloys. This is due to the need for maximum heat dissipation from contact surfaces during friction. To determine the thermal conductivity, measurements were made of the specific electric conductivity of alloys, since there is a correlation between these characteristics. Using the equation $y = 5.64 \cdot x + 23.43$, where x is the experimental value of the SES, given in Refs. [10,11], the value of the thermal conductivity of the four-component alloys was calculated. To compare and evaluate the effect of each fusible element separately on the mechanical and thermal conductive properties, the obtained values of the physical and mechanical properties of the investigated alloys were compared with the values of the base alloy Al-5% Si -4% Cu.

Table 2 shows the measured values of the hardness of alloys, the conductivity, and the corresponding values of the thermal conductivity of alloys in the cast (initial) state and after HT.

Table 2. Physical and mechanical properties of the investigated alloys in different states.

Alloy	Hardness, HB		Electrical Conductivity, MSm/m		Thermal conductivity, Wt/(m·K)	
	Cast	After HT	Cast	After HT	Cast	After HT
Al–5Si –4Cu	71.1 ± 0.8	77.9 ± 0.1	23.4	22.9	155.4	152.6
Al–5Si –4Cu–1Bi	69.2 ± 0.8	101.6 ± 0.8	20.7	20.5	140.2	139.1
Al–5Si–4Cu–1Cd	68.7 ± 0.8	141.7 ± 0.9	20.5	19.8	139.1	135.1
Al–5Si–4Cu–1In	67.7 ± 0.1	124.1 ± 1.8	20.4	19.9	138.5	135.7
Al–5Si–4Cu–1Pb	68.9 ± 1.4	111.2 ± 0.5	20.1	19.7	136.8	134.5

Hardness. According to the results given in Table 2, it can be seen that the addition of low-melting metals into the Al-5% Si-4% Cu alloy does not greatly affect the hardness values of the samples in the cast (initial) state, which is approximately 70 HB. After HT for all alloys, the hardness increases; this is primarily due to the dissolution of nonequilibrium excess phases containing copper. It should be noted that the addition of low-melting elements in a small amount to experimental alloys subjected to maintenance increases the hardness values to 100-140 HB, which is 40 to 75% higher than that of the base alloy. The greatest value of hardness after HT (142 HB) has an alloy with cadmium, and the lowest one has an alloy with bismuth (102 HB). This difference is probably due to the various effects of these additives on the process of nucleation and isolation of the hardening phase in aging (θ-Al_2Cu).

Electrical conductivity and thermal conductivity. Table 2 also shows the effect of HT on the change in the specific electrical conductivity (SEC). It is known that SEC of aluminium increases with decreasing impurities [8]. In our case, SEC is reduced by the elements that enter the solid solution after the HT (such alloying additives as Cu and Si). Low-melting metals are practically insoluble in Al, but they to a small extent reduce the value of the SEC. Proceeding from the assumption of the existence of a correlation between electrical conductivity and thermal conductivity, the addition of alloying elements has the same effect on the aluminium alloy – reducing the value of thermal conductivity. It should be noted that all experimental alloys have a higher electrical conductivity, and, consequently, thermal conductivity than bronze BrO4TS4S17 (the value of the SEC is 11.7 MSm/m). To a large extent, this is explained by the fact that the electrical conductivity of copper is significantly reduced, when there is even a very small amount of impurities.

3.3. Complex microscopic investigation of the surface of initial samples of the Al-Si-Cu-Sn-X system

In the second part of the work, the alloy of Al–5% Si–4% Cu–4% Sn system with addition of tin was taken as the basis, to which selected fusible elements were added based on the results obtained. The presence of a low-melting component in the investigated alloys corresponds to the content of this additive in branded aluminium alloys and allows us to consider that an alloy of the composition Al-5%Si-4%Cu-4%Sn (up to 0.5% Pb, Bi, and Cd) should have a sufficiently high level of operational (tribological) properties. Lead and bismuth, as well as tin, in the alloy serve to reduce friction and provide resistance to scoring under dry friction or lubrication conditions. In addition, at the first stage it was found that cadmium could affect the hardness. The results of hardness measurements are shown in Table 3.

Table 3. Results of the hardness test of alloys under study

Alloy	Hardness, HB	
	Cast	After HT
Al–5Si –4Cu–4Sn–(0,5Bi-0,5Pb-0,5Cd)	72.7 ± 0.9	129.2 ± 1.6
Al–5Si –4Cu–4Sn–(0,5Bi-0,5Pb)	71.2 ± 1.5	125.4 ± 1.2

It can be seen that the HT significantly increases the strength and hardness. Cadmium in this case also slightly increases the hardness of the alloy.

Using the example of Al-5Si-4Cu-4Sn (0.5Bi-0.5Pb) sample (without cadmium), the effect of HT on the structure as well as the thermal and conductive properties of the alloy was also evaluated. The results of comparative microscopic studies of an alloy of this composition – cast (initial) and after HT (500 °C), are shown in Figure 4.

It can be seen that the methods used complemented each other: as in the first stage, the SPM made it possible to visualize elements that are difficult to distinguish on SEM images, and also to estimate the spatial geometry of individual phase components [9]. It can be seen that, just as in the first case, the spheroidization of particles, the phase constituents of the alloy, takes place at HT.

To study the conductive properties of the surface, new SPM techniques have been applied at this stage. Of particular interest here is the possibility of directly measuring the thermal conductivity map and correlating it with a map of the independently measured electrical conductivity of the samples.

Figure 5 shows the obtained images of the same sample in the cast and HT state: the relief images and the temperature map are shown.

The higher temperature of the probe corresponds to the lighter areas in the image and qualitatively indicates their lower thermal conductivity. In Figure 5, it is possible to isolate the relief areas (phase 1), at which the probe temperature is virtually the same as the matrix (solid phase inclusions - Al_2Cu or Al-Si), i.e. their thermal conductivity is the same. In the areas indicated by 2 (soft phase component Sn-Pb-Bi), the probe has the highest temperature, i.e. in these places there is the greatest thermal resistance at heat exchange. In other areas, the temperature of the probe is intermediate, but higher than the temperature of the matrix.

On the HT sample, the probe temperature distribution is more homogeneous (Fig. 5d). A higher temperature corresponding to a lower thermal conductivity is observed only on impurities localized mainly at the boundaries of the phase-Al-Si, the inclusions of which have an oval shape. The thermal conductivity of the Al-Si phase does not differ much from the thermal conductivity of the matrix. Another phase (Al_2Cu) is not observed, since at HT, copper dissolves in the Al matrix.

A comparison of the thermal conductivity maps and the surface conductivity of the cast sample is shown in Figure 6. The same section was scanned at first in the thermal microscopy mode, and then in the surface conductivity mode (current measurement).

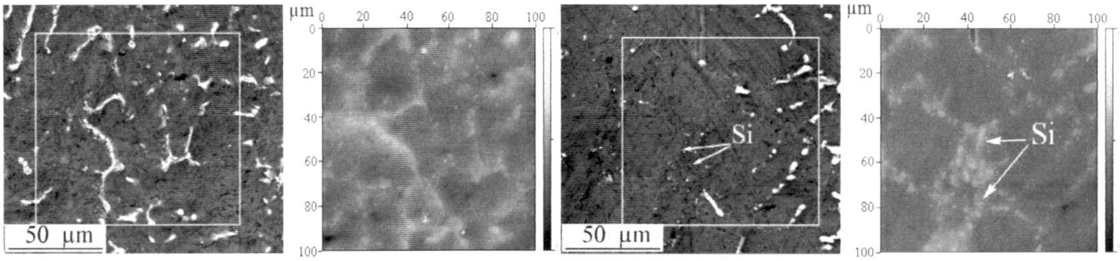

(a) (b) (c) (d)

Figure 4. The surface image of the Al-5Si -4Cu-4Sn- (0.5Bi-0.5Pb) alloy: the SEM image of (a) cast, (c) after the HT; Squares indicate the areas of the SPM study; SPM image (2D) (b) cast, (d) after HT.

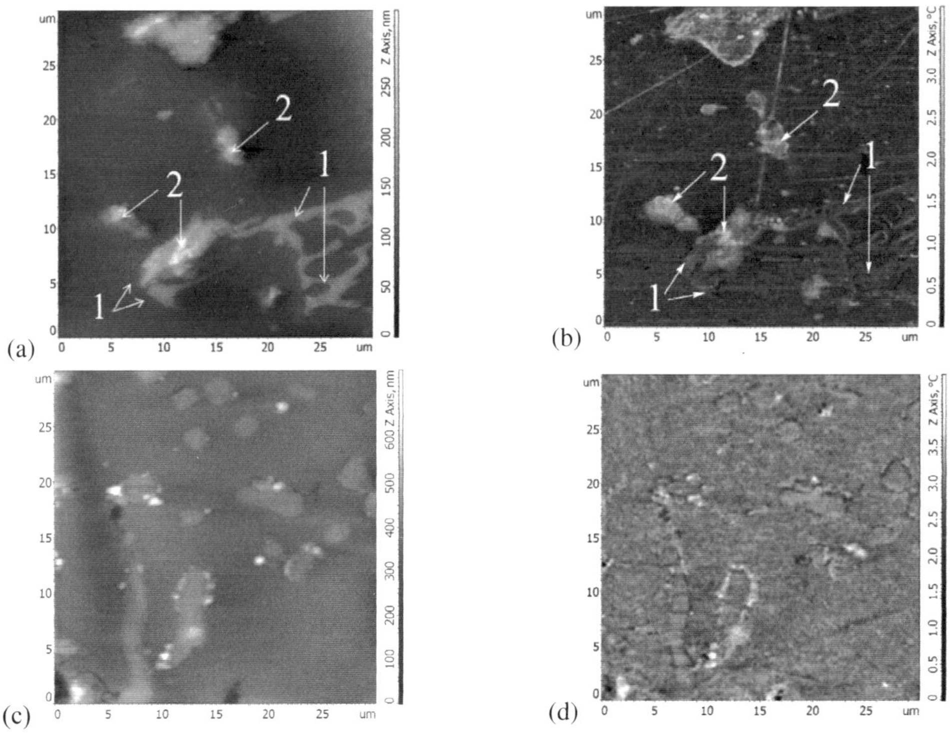

Figure 5. SPM images of the topography of the sample Al-5Si-4Cu-4Sn- (0.5Bi¬-0.5Pb): (a) cast and (c) HT alloys; a map of the deviation of the probe temperature from the nominal value: (b) cast and (d) HT alloys.

A comparison of Figures 6b and 6c shows that there is no complete correspondence between the thermal conductivity of the surface and its electrical conductivity. Indeed, there are sections, which thermal conductivity and electrical conductivity are smaller (denoted by 1) or roughly equal to (denoted by 2) those of the matrix, which qualitatively agrees with the Wiedemann-Franz law for metals. However, there are areas, which thermal conductivity is much lower than the thermal conductivity of the matrix, but the electrical conductivity is much higher (region 3). There may be several reasons for this. Thus, a special feature of the measurements of the samples under study is that the aluminium surface is covered by a nonconducting oxide, the thickness of which most likely depends on the local composition of the sample. Therefore, it is difficult to obtain high-quality images in the contact scanning regime and to correlate the heat-conducting properties of the surface with its electrical conductivity.

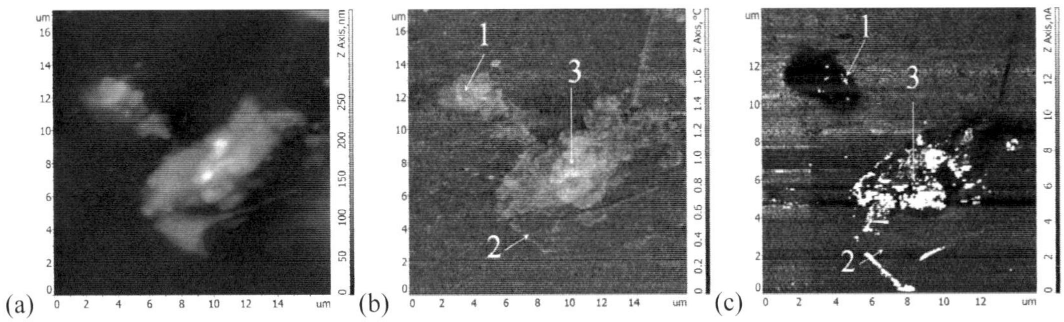

Figure 6. SPM images of the surface of the cast sample Al-5Si-4Cu-4Sn-(0.5Bi-0.5Pb): (a) topography (17x17 μm^2); (b) map of the probe temperature deviation from the nominal value (17x17 μm^2); (c) is a spreading current distribution map (18x18 μm^2).

In addition, the heat exchange depends on the contact area of the probe and the surface, which can lead to the influence of the relief on the measured probe temperature, for example, at the extremum points of the relief. Also, the reason for this behaviour can be related to the peculiarity of the phase composition of individual sections. In general, this question requires more detailed study.

However, in most cases, for the measured samples there is a correlation between electrical conductivity and thermal conductivity. All these results indicate the prospects of using the electrical methods for studying the surface properties of alloys.

4. Conclusion

It has been found that the addition of even a small amount of low-melting elements to the base alloy Al-5Si-4Cu and subsequent heat treatment have a significant effect on the mechanical properties of the material. Physico-mechanical properties increase for the alloy having cadmium with the highest hardness value. The thermal conductivity during alloying changes insignificantly; the addition of elements somewhat reduces this index.

Study of the surface by the complex method of scanning probe microscopy has shown that the separately measured values at the microlevel of the electrical conductivity and thermal conductivity correlate qualitatively.

Substantial inhomogeneity of the surface thermal conductivity of the cast sample is established, which correlates with the corresponding phase composition of impurities.

It is shown that the temperature treatment of the sample leads to a more uniform surface thermal conductivity; the phases with a lower thermal conductivity are concentrated at the boundaries of phases with a higher thermal conductivity.

On the whole, the results obtained show that the use of "electric methods" and scanning thermal microscopy is promising for studying the surface properties of alloys.

Acknowledgments

This work was supported by the Russian Scientific Foundation, project no. 14-19-01033-P (study of topography and elemental composition of the surface) and grant of the President of the Russian Federation MK-871.2018.8 (no. AAAA-A18-118080290023-0) (study of electrical properties of the surface). The authors are grateful to N.A. Belov (MISiS) for providing samples.

References

[1] Mironov A E, Gershman I S, Ovechkin A V and Gershman E 2016 *J. Frict. Wear* **37** 31-5
[2] Mironov A E, Gershman I S, Ovechkin A V and Gershman E 2015 *J. Frict. Wear* **36** 334-9
[3] Patent RF №2571665
[4] Mironov A E, Gershman I S, Gershman E I and Zheleznov M M 2017 *J. Frict. Wear* **38** 67-72
[5] Belov N A, Mikhailina A O, Alabin A N and Stolyarova O O 2016 *Met. Sci. Heat Treat.* **58** 195-201

[6] Belov N A, Stolyarova O O, Muraveva T I and Zagorskiy D L 2016 *Phys. Met. Metallogr.* **117** 579-87
[7] Sachek B Y, Mezrin A M, Muravyeva T I, Stolyarova O O, Zagorskiy D L and Belov N A 2015 *J. Frict. Wear* **36** 103-11
[8] Zolotorevskiy V S and Belov N A *Metallovedenie liteinyh aluminievyh splavov* – M.: Iz. dom MISiS, 2005. – 375.
[9] Stolyarova O O, Muravyeva T I, Zagorskiy D and Belov N A 2016 *Phys. Mesomech.* **19** 104-13
[10] Belov N A, Akopyan T K, Gershman I S, Stolyarova O O and Yakovleva A O 2017 *J. Alloys Compd.* **695** 2730-9
[11] Yakovleva A O, Belov N A, Bazlova T and Shkalei I V 2018 *Phys. Met. Metallogr.* **119** 35-43

IOP Conf. Series: Materials Science and Engineering **443** (2018) 012030 doi:10.1088/1757-899X/443/1/012030

Complex microscopy investigation of special purpose elastomers

O O Shcherbakova[1], T I Muravyeva[1], R V Gainutdinov[2], I V Shkalei[1], N N Petrova[3], E N Timofeeva[3] and D L Zagorskiy[1,2]

[1] Ishlinsky Institute for Problems in Mechanics RAS, 119526, Moscow, Russia
[2] Shubnikov Institute of Crystallography of Federal Scientific Research Centre "Crystallography and Photonics" of Russian Academy of Sciences", 119333, Moscow, Russia
[3] Ammosov North-Eastern Federal University, 677000, Jakutsk, Russia

shcherbakovaoo@mail.ru

Abstract. New types of elastomers (rubbers) for use in extremal conditions of north were proposed and investigated. New type of filler – carbon nanotubes were added to a basic elastomer composition in different proportions. Surface parameters of the samples were studied before and after tribological tests. Electron microscopy was used to study the relief and elemental composition of the surface and section (after the tests). Probe microscopy made it possible to investigate the adhesion and elastic properties of the surface at the nanoscale. Unmodified samples and samples with a small amount of nanotubes are remarkably changing in topography and elemental composition after tribological tests: the amount of carbon increased, while amounts of nitrogen, sulphur, and zinc decreased. Modification of the basic composition by nanotubes significantly changes the properties of rubber. Cross-sections showed that composition of the depth became more stable. Modified samples have a great adhesion strength and Young's modulus.

1. Introduction

Rubbers and other elastomers are widely used in many areas of the national economy, and development of new composites is an important task. In particular, the actual problem is fabrication of the materials operating in the climatic conditions of the north [1-2]. These conditions are often close to extremal. So, the materials must have high frost resistance, strength, and resistance to aggressive media. To achieve these goals, the rubber is modified with various fillers [3-4]. Recently, nanofillers, for example, carbon nanotubes (CNTs), which have a unique structure and properties, are increasingly being used as modifiers [5]. Thus, it is known that the addition of CNTs to a rubber composite often leads to a significant improvement of its properties [6-7]. However, the influence of the CNTs on the structure and the main properties of the resulting composite have not been studied in details to date.

In the present work, a series of samples with different contents of CNTs have been studied at the nanoscale using scanning electron microscopy (SEM) and scanning probe microscopy (SPM). The influence of tribological tests on the surface change was also studied. The aim of the work was to study the effect of the addition of CNTs on the structure and properties of elastomers based on epichlorohydrin rubber.

2. Materials and methods

2.1. Materials

In the present work, samples from epichlorohydrin-based composites (EPCG), which are intended to be used as seals in machine units operating at negative temperatures, were investigated. Samples of rubbers were made using Brabender plasticizer equipment. The components were mixed at 50°C for 15 minutes, then vulcanized on a hydraulic press GT-7014-H10C. The modifiers of rubber were carbon nanotubes (CNTs), added to the base composition in different percentages. For the subsequent study, samples of five different contents were used (Table 1).

Table 1. Composition of the samples under study: the amount of CNTs in the mass parts is given by 10 parts by weight of EPCG.

Sample number	**1**	**2**	**3**	**4**	**5**
CNT (m.p.)	0	0.3	1.0	2.0	10.0

2.2. Tribological tests

The operation regime of elastomers was modelled by tribological tests conducted on UMT-2 friction machine under the following conditions: contact pressure varied from 0.1 to 0.3 MPa, the sliding velocity varied from 1 to 100 mm/s, and the bulk temperature of the samples varied from -25 to 22°C. The description of tribological tests is presented in other works [8]. In the present work, the results obtained after tribological tests with the maximum parameters of pressure and velocity are presented.

2.3. Surface investigation

The surface of the obtained samples was studied by various methods of microscopy before and after tribological tests. The topography and elemental composition of the surface were examined on a Quanta 650 SEM using secondary and back-scattered electrons detectors with EDAX analytical equipment. The work was carried out at the low vacuum conditions, which allowed studying dielectric rubber without preliminary metallization of its surface.

NTEGRA Prima SPM (NT-MDT, Russia) was used to study the topography, adhesion, and viscoelastic properties of the sample surface at the nanoscale. HA-NC (NT-MDT) probes were used having stiffness ~ 3.5 N/m, a resonance frequency ~ 140 kHz, and a tip radius $\leq$ 10 nm. The heat stabilization studies were carried out in a clean climatic zone "TRACKPORE ROOM-05", which ensured the maintenance of temperature and humidity of the air environment with high accuracy.

Measurements of adhesion forces and elasticity (Young's modulus) were carried out by analysing the dependences of force on distance obtained in the regime of power spectroscopy [9]. Another method used, so-called force modulation method, consists in studying the damping of oscillations in a sample: it is known that attenuation is determined by the elastic properties of the surface [10]. The scanner with the sample performs vertical periodic oscillations at the resonant frequency of the scanner (6.2 kHz in the present work). The probe, which is in contact with the sample surface, registers the oscillations of the sample. In this case, the amplitude of oscillations of the probe-sample system is recorded, which depends on the rigidity of the sample surface.

3. Results

3.1. Electron microscopy

The most important characteristic is the changing of the elastomers surface during operation. At the initial stage of work, the surface of the samples before the tests was investigated in the regimes of secondary electrons and back-scattered electrons. The results obtained for the first sample are shown in Figure 1.

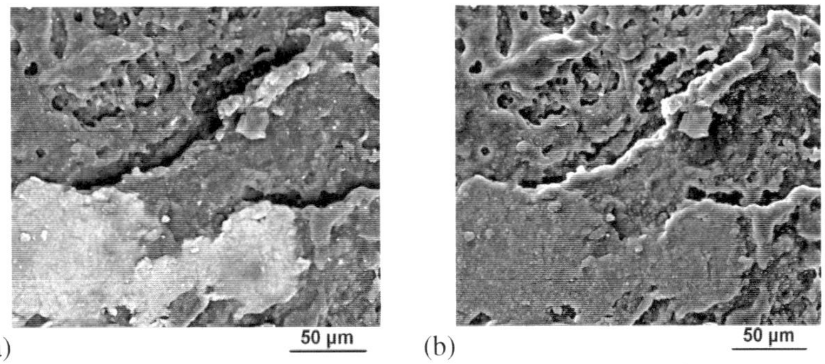

(a) (b)

Figure 1. SEM image of the surface of sample 1 regions after tribological tests: (a) in back-scattered electrons regime, (b) in secondary electrons regime.

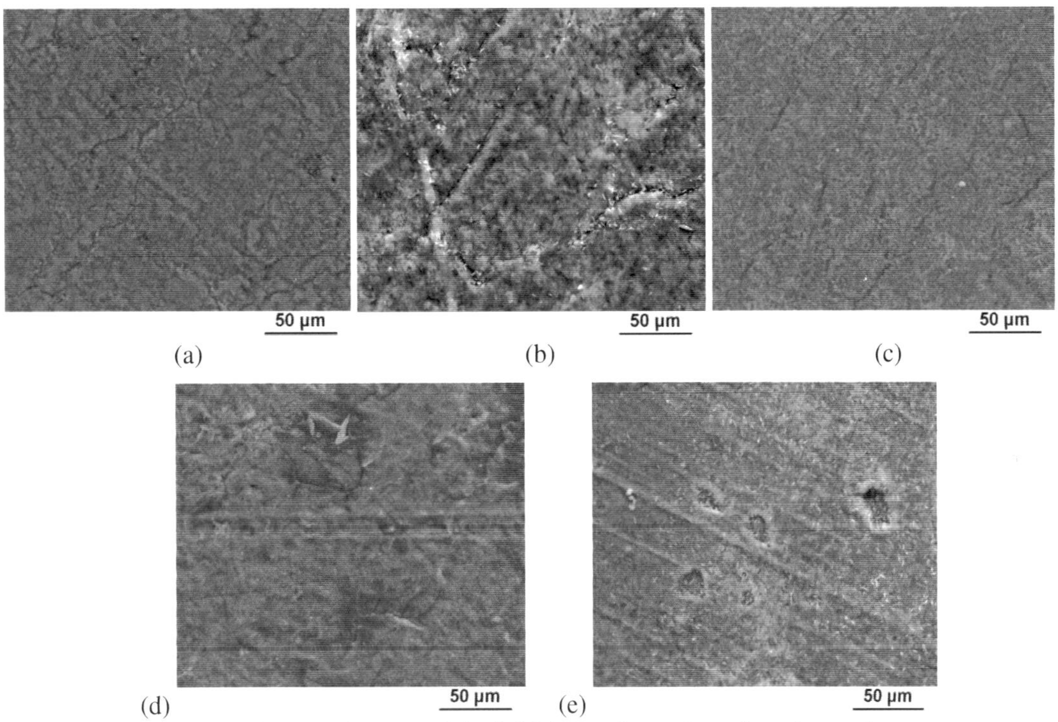

(a) (b) (c)

(d) (e)

Figure 2. SEM images of the initial sample surface. Sample number: (a) 1; (b) 2; (c) 3; (d) 4; (e) 5.

Table 2. Chemical composition of the initial surfaces of elastomers.

Sample number	Chemical composition (% of mass)										
	C	N	O	Mg	Al	Si	S	Cl	Ca	Co	Zn
1	35,28	8,58	2,79	0,46	0,37	0,10	34,26	1,84	0,25	-	16,07
2	32,22	12,00	5,14	0,58	0,59	0,01	30,30	4,49	0,06	0,11	14,50
3	37,67	11,43	6,04	1,75	1,32	0,27	22,95	4,93	0,20	0,14	13,30
4	35,64	12,49	6,84	1,12	0,89	0,04	25,2	5,13	0,12	0,12	12,33
5	40,76	11,64	7,36	0,94	0,81	0,06	22,57	4,49	0,08	0,13	11,16

It is evident that both methods give practically the same information about the topography. At the same time, the method of backscattered electrons gives a lot of information about the surface because of good visualization of various phases. In other parts of this work, the surface images were also obtained in the two above-mentioned modes and in all cases the mode of the backscattered electrons provided more information. Therefore, the results of the study of other samples, described below, will be presented only in this mode.

Figure 2 shows SEM images of the initial surface of the investigated samples 1-5, and in Table 2 the corresponding elemental composition is given. It can be seen that the initial surface of the samples is smooth and homogeneous; there is no difference in the "colour" of the phases. Samples 2-5 contain fine needle particles. The surface of sample 5 is characterized by the presence of pores of various sizes on the surface. According to the elemental composition, the samples differ insignificantly. The presence of cobalt in the chemical composition of samples 2-4 (with CNTs) can be explained by its presence in the catalytic system, used for production of nanotubes.

SEM images of the surface of elastomer samples after tribological tests (Fig. 3) show that the topography of the surface after the tests has become more developed. The phases, differing in colour and composition, also appeared at the surface. So, on the surface of samples 1 (without CNTs) and 2 (CNTs 0.3 m.p.), the lighter areas are close in composition to the original surface, while the dark ones are different. This can be explained by the nature of the interaction with the counter body: the surface of the sample is uneven and contact during friction does not occur over the whole area. In this case, the light areas correspond to the minimal contact with the counter body, while the darker areas are the contact surfaces. From the obtained results it follows that the parameters of the surface of elastomers without CNTs (or with a small number of them) vary greatly during friction. Elastomers modified with a large number of CNTs (samples 3-5) also have regions on the surface that differ in colour and topography. However, the elemental composition of these regions differs insignificantly (Table 3). Thus, it can be argued that the introduction of CNT into an elastomer stabilizes its composition.

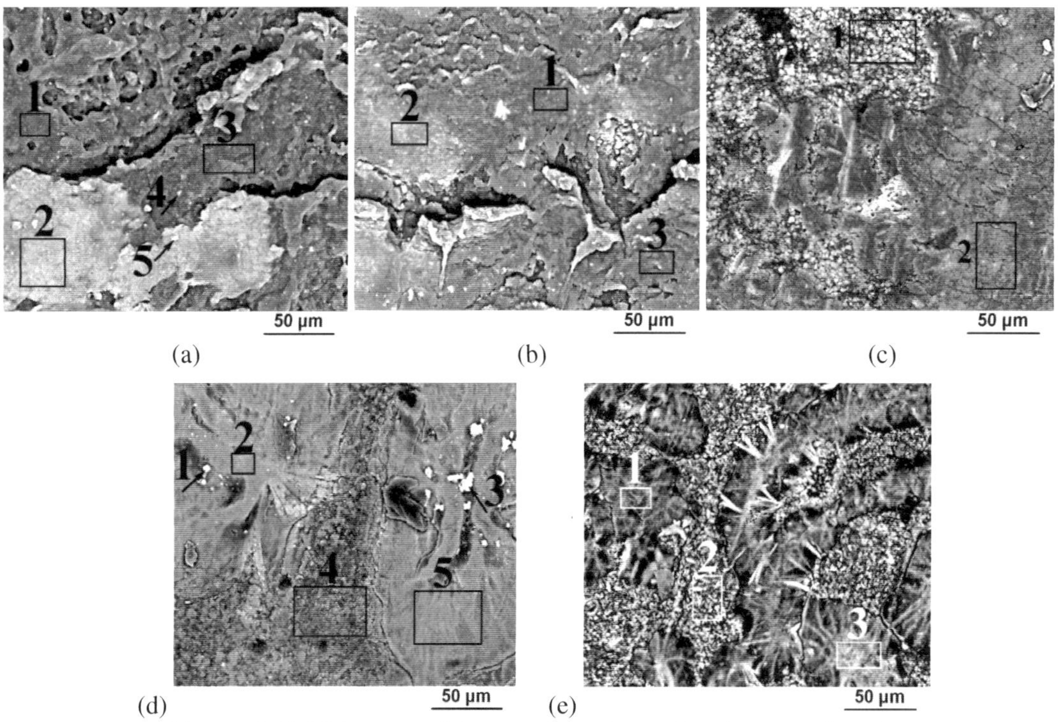

Figure 3. SEM images of the sample surface after the tests. Sample number: (a) 1; (b) 2; (c) 3; (d) 4; (e) 5.

Table 3. Chemical composition of the elastomer surfaces after the tests.

Sample number	Area number	Chemical composition (% of mass)											
		C	N	O	Mg	Al	Si	S	Cl	K	Ca	Co	Zn
1	1	73,47	3,27	9,17	1,93	1,33	0,28	4,71	2,93	-	0,32	-	2,59
	2	48,16	4,44	7,20	0,87	0,56	0,20	24,50	2,42	-	0,21	-	11,44
	3	69,23	3,51	11,67	2,16	1,37	0,27	5,12	3,78	-	0,23	-	2,66
	4	24,11	2,11	16,33	2,06	1,25	0,50	2,83	0,95	-	0,08	-	49,78
	5	21,72	2,81	37,70	1,20	0,49	0,26	7,42	0,80	-	25,51	-	2,09
	All	65,86	3,85	11,18	1,45	1,01	0,23	8,86	2,73	-	0,34	-	4,49
2	1	60,84	4,33	13,89	1,99	1,48	0,33	6,86	5,88	-	0,62	0,05	3,73
	2	48,21	5,95	10,71	1,25	1,06	0,43	17,73	5,14	-	0,28	0,09	9,15
	3	63,74	3,48	13,29	1,39	1,86	0,35	6,05	4,23	-	2,32	0,22	3,07
	All	59,76	4,68	12,78	1,32	1,17	0,29	9,75	4,36	-	0,78	0,14	4,97
3	1	38,93	10,42	7,65	1,57	1,38	0,23	22,01	4,92	-	0,12	0,09	12,68
	2	36,17	9,83	8,24	2,24	1,39	0,38	22,93	5,27	-	0,25	0,19	13,11
	All	39,06	8,87	8,48	1,79	1,31	0,30	21,78	5,08	-	0,19	0,13	13,01
4	1	19,98	3,24	8,75	0,06	0,27	0,31	6,91	27,69	29,15	0,47	0,14	3,03
	2	33,63	5,46	5,04	0,22	0,14	0,07	37,66	0,90	0,06	0,04	0,11	16,67
	3	18,36	1,82	6,64	0,00	0,02	0,07	4,21	33,86	32,09	0,38	0,09	2,46
	4	34,82	9,60	7,38	0,95	0,96	0,31	26,73	3,75	0,09	0,18	0,21	15,02
	5	31,56	5,31	4,71	0,37	0,26	0,14	38,33	1,42	0,10	0,19	0,13	17,48
	All	36,83	6,17	6,58	0,62	0,59	0,20	30,63	2,70	0,29	0,26	0,15	14,98
5	1	35,11	9,41	6,49	0,89	0,85	0,19	28,13	3,46	-	0,06	0,51	14,90
	2	41,30	8,49	8,27	0,78	0,77	0,21	24,50	2,55	-	0,29	0,17	12,67
	3	33,32	7,57	5,26	0,31	0,27	0,09	35,31	1,76	-	0,05	0,16	15,90
	All	39,15	9,39	8,83	1,03	1,00	0,40	23,51	3,39	-	0,67	0,18	12,45

For a more detailed study of the processes occurring during friction on contact surfaces, geological sections of the samples were prepared. Figure 4 shows SEM images of these sections, and in Table 4 – the corresponding elemental composition. It is easy to see a highly altered (in terms of topography and chemical composition) surface layer, but its thickness is small (about 10 μm) – therefore, its features can be estimated only at high magnification.

A study of the topography of sample sections showed that samples 3 and 4 have a more homogeneous and dense structure than samples 1, 2 and 5. Analysis of changes of the elemental composition showed that in samples with a small amount of CNTs, the surface layer significantly differs from bulk. At the same time, for samples 3, 4 and 5 (with a high content of CNTs), the element composition in depth varies to a minimal extent. Thus, it can be concluded that an increase in the number of CNTs stabilizes the composition by volume.

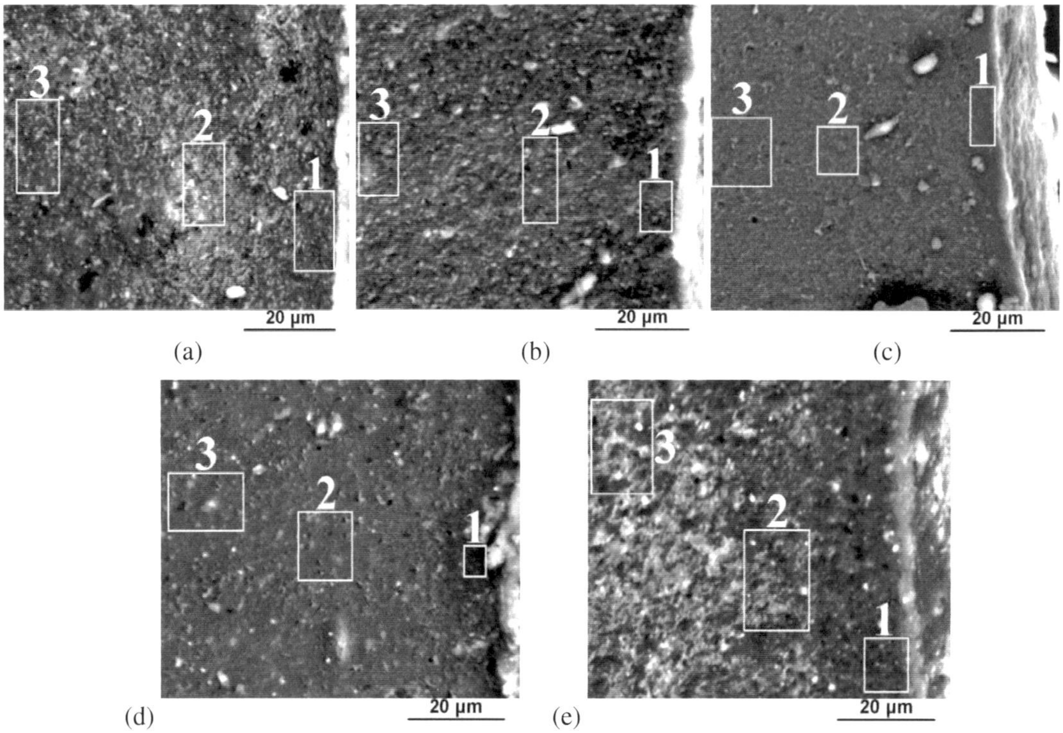

Figure 4. SEM images of the sample sections (magnification x2000). Rectangles marked areas where X-ray spectral analysis was performed. Sample number: (a) 1; (b) 2; (c) 3; (d) 4; (e) 5.

Table 4. Chemical composition of sections of samples of elastomers after testing.

Sample number	Area number	Chemical composition (% of mass)												
		C	N	O	Mg	Al	Si	S	Cl	K	Ca	Ti	Co	Zn
1	1	56,30	0,02	5,71	0,13	0,57	0,03	10,17	13,87	-	2,86	0,69	-	9,45
	3	76,30	0,60	8,75	0,87	0,79	-	3,84	5,49	-	0,13	0,12	-	3,06
2	1	62,13	0,15	7,90	1,05	1,55	0,09	7,41	11,31	0,36	0,92	0,98	0,22	5,93
	3	71,52	1,10	7,77	4,16	0,97	-	3,49	6,57	0,03	0,15	0,21	0,08	3,95
3	1	68,85	0,76	12,04	1,02	0,89	0,14	5,49	5,17	0,09	0,53	0,47	0,13	4,42
	3	72,14	1,33	10,32	1,60	1,20	0,07	4,49	5,13	0,06	0,06	0,10	0,11	3,39
4	1	71,07	0,01	3,87	0,98	0,82	0,00	5,86	9,25	0,02	0,27	0,20	0,27	7,38
	3	71,79	0,95	8,30	1,19	1,35	0,01	4,65	5,18	0,11	0,10	0,14	0,16	6,07
5	1	75,63	0,29	7,97	0,85	0,96	0,00	3,59	5,42	0,03	0,21	0,08	0,27	4,70
	3	76,16	2,10	10,14	1,15	1,03	0,06	2,93	3,78	0,06	0,10	0,07	0,14	2,28

3.2. Scanning probe microscopy

Results of measuring topography, adhesion and visco-elastic properties are given below. Figure 5 shows, as an example, the results obtained for sample 3. Similar results were obtained for other samples. The summary data for all samples are given in Table 5.

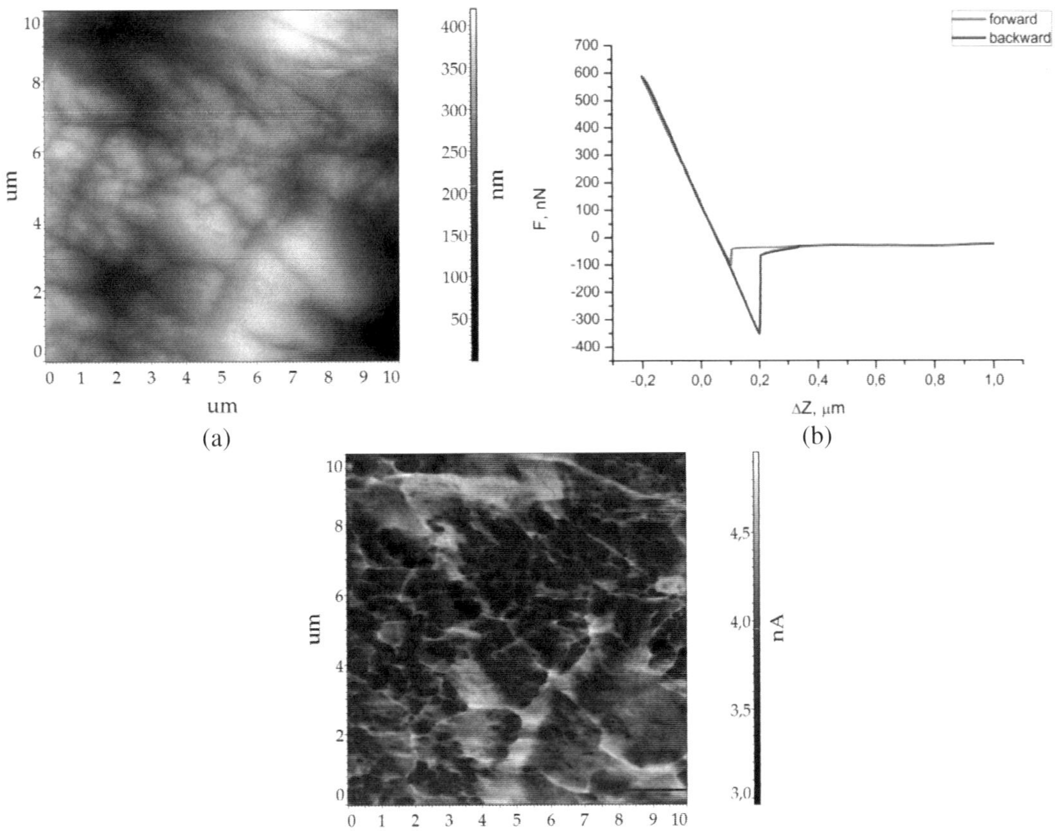

(a) (b)

(c)

Figure 5. SPM images of the sample №3: (a) the surface topography; (b) the supply-retraction curves; (c) the amplitude of the viscoelastic oscillations.

Table 5. Values of adhesion strength, Young's modulus and visco-elastic characteristics of the test samples before and after tribological tests.

№	ER+ CNTs	F_{ADH}, nN		*E*, MPa		*A*, nA	
		before	after	before	After	before	after
1	0 mp	71,8	41	92,8	82,4	2,3	4,2
2	0,3 mp	324,2	203,6	584,4	123,6	2,4	3,8
3	1,0 mp	139	322	309,7	144	3,0	3,9
4	2,0 mp	89,7	136	354,8	197,4	3,2	4,2
5	10 mp	238,3	99,8	251,9	197,1	2,6	3,3

Analysis of the obtained results allows us to make the following conclusions:

Adhesion strength. The addition of CNTs leads to an increase in adhesion: in samples without CNTs, adhesion is small enough, while adding even 0.3 parts of CNTs leads to a sharp increase in adhesion strength. However, a clear dependence of adhesion on the concentration of CNTs could not be identified. The same pattern is observed for samples after tribological tests. We note that it was not possible to reveal a general pattern for the effect of tribological tests on adhesion (after the tests the adhesion strength decreased in samples 1, 2, 5, while it increased in samples 3 and 4).

Elastic characteristics. Young's modulus was measured from the slope of the curve. For samples without CNTs and for samples with a small value of CNTs (up to 0.1 m.p.) Young's modulus is relatively small, but with an increase in the CNTs content (from 0.3 m. p), a sharp increase in the

Young's modulus occurred. For this concentration range (as for the case of adhesion forces), the dependence on the number of CNTs was not revealed. After tribological tests, the Young's modulus decreased for all samples.

Elastic characteristics. The results of the study of elasticity measured by the force modulation method suggest that the elasticity values of different samples differ little. However, it can be seen that, after tribological tests, in all cases a noticeable increase in the rigidity of the sample is observed.

In general, these investigations show that modification of the basic composition of rubber by nanotubes significantly changes the properties of rubber. At the same time, it should be noted that the conclusions obtained do not make it possible to establish an unambiguous correlation between the elastic properties expressed in terms of the Young's modulus and the elastic parameters determined by the force modulation method. Obviously, additional experiments are required here.

4. Conclusion

Unmodified samples and samples with a small amount of CNTs after tribological tests are remarkably changing both in topography and in elemental composition – the amount of carbon increases with decreasing amounts of nitrogen, sulphur, and zinc. Modification of the basic composition with CNTs significantly changes the properties of rubber. The composition of the depth becomes more stable. Samples with CNTs have a great adhesion strength and Young's modulus.

Acknowledgments

This work was supported by the Russian Scientific Foundation, project no. 14-29-00198-P (sample preparation and SEM) and the State Task № State registration AAAA-A17-117021310379-5 (SPM studies). This work was supported by the Ministry of Science and Higher Education within the State assignment FSRC «Crystallography and Photonics» RAS in part of SPM investigations (measurements of force-distance curves and force modulation).

References

[1] Diaz R, Gaarder R H and Olafsen K 2013 *Proc. of the Twenty-third International Offshore and Polar Engineering* (Anchorage: ISOPE Conference Paper) 250-6

[2] Bukhina M F and Kuryland S K 2007 *New concepts in polymer science* (Leiden, Boston: VSP) 187

[3] Eliseev O A, Chaikun A M, Buznik V M, Sokolova M D and Popov S N 2015 *J. Adv. Mater.* **11** 5-18

[4] Petrova N N, Popova A F and Fedotova E S 2002 *Kauchuk I Rezina* **3** 6-10

[5] Shashok J S and Prokopchuk N R 2014 *Application of Carbon Nanomaterials in Polymer Compositions* (Minsk: BSTU)

[6] Morozov A V, Muravyeva T I, Petrova N N, Portnyagina V V, Ammosova V N and Zagorskiy D L 2015 *Kauchuk I Rezina* **6** 22-7

[7] Stolyarova O O, Muravyeva T I, Cainutdinov R V, Morozov A V, Zagorskiy D L, Petrova N N and Portnyagina V V 2016 *J. Surf. Invest.* **9** 46-53

[8] Morozov A V and Petrova N N 2016 *J. Frict. Wear* **37** 124-8

[9] Lebedev D V, Chuklanov A P, Bucharaev A A and Druzhinina O S 2009 *ZhTF Lett.* **35** 54-61

[10] Maivald P, Butt H T, Gould S A C, Prater C B, Drake B, Gurley J A, Elings V B and Hansma P K 1991 *Nanotechnology* **2** 103-6

Influence of the domain structure on piezoelectric and dielectric properties of relaxor SBN single crystals

V A Shikhova[1], A D Ushakov[1], V V Fedorovyh[1], V A Anikin[1], A A Esin[1], V Ya Shur[1], A L Kholkin[1,2] and L I Ivleva[3]

[1] School of Natural Sciences and Mathematics, Ural Federal University, 620000, Ekaterinburg, Russia
[2] Department of Physics and CICECO – Aveiro Institute of Materials, University of Aveiro, 3810-193, Aveiro, Portugal
[3] Prokhorov General Physics Institute, Russian Academy of Sciences, 119991, Moscow, Russia

vera@urfu.ru

Abstract. We have studied the influence of initial domain structure on piezoelectric and dielectric properties of $Sr_{0.61}Ba_{0.39}Nb_2O_6$ single crystals slightly doped with Ce and Ni. Initial domain structure was created by zero-field cooling, in-field cooling, and partial switching. The difference in the frequency dependences of macroscopic piezoelectric response and temperature dependences of dielectric permittivity for various initial domain structures was revealed.

1. Introduction

The formation of stable ferroelectric micro- and nanodomain structures with proper parameters for improving piezoelectric, pyroelectric, and dielectric properties is the main direction of the domain engineering. At present, a wide class of coherent light sources based on quasi-phase matching is developed, which allowed creating a qualitatively new direction in the fabrication of electro-optical and nonlinear optical devices [1,2].

It was shown that the formation of domain structure with controlled parameters by domain engineering technique improves significantly the piezoelectric properties [3,4]. In tetragonal barium titanate crystals, the piezoelectric properties were improved due to decreasing domain sizes (increasing domain wall density) [5,6]. The reduction of the domain wall mobility in PZT ceramics was used for creation of the piezoelectric transducers, whereas its increasing was used for manufacturing of precise electromechanical drives [7]. Thus, the domain wall engineering, aimed at creating the maximum possible density of the domain walls of a certain type and orientation, presents significant importance. Based on the observed increase of the dielectric constant in barium titanate ceramics with an average domain size of about 150 nm, it was suggested that at such domain sizes the highest values of piezoelectric coefficients could be obtained [8].

Uniaxial relaxor ferroelectric strontium barium niobate ($Sr_xBa_{1-x}Nb_2O_6$, SBN) is considered as a promising material for domain engineering due to its prominent nonlinear-optical and electro-optical properties along with sufficiently high piezoelectric coefficients [9].

Two alternative theoretical approaches were used for interpretation of the diffused phase transition in relaxor ferroelectrics [10-13]. The relaxor phase is considered: (1) as a dipole-glass state or (2) as a nanoscale mixture of polar nanoregions and nonpolar matrix [14].

In this work, the relaxor phase is considered as a composite consisting of ferroelectric (polar) and paraelectric (nonpolar) nanoregions. Such an approach was supported by high-resolution transmission electron microscopy (HRTEM) [15] and piezoresponse force microscopy (PFM) [16,17]. HRTEM experiments show that the heterophase structure of the relaxor state consists of polar nanoregions with diameters down to 10 nm embedded in a nonpolar matrix [15]. Using PFM, it was observed that in SBN polar nanoregions with opposite polarization directions and nonpolar regions coexisted at the temperature close to the dielectric permittivity maximum [16,17].

It was shown that the nonpolar inclusions influenced integral characteristics of relaxor ferroelectrics. For example, in relaxor PLZT ceramics [18], PZN-PT [19] and SBN [20] single crystals below freezing temperature conventional hysteresis loops were observed, while at higher temperatures double hysteresis loops emerged. It was demonstrated that the dielectric permittivity in PZN-PT single crystals after zero-field and in-field cooling were different [19].

In this paper, we report the experimental findings of investigation of the dielectric properties and macroscopic piezoelectric response in relaxor SBN61 single crystals doped by Ce and Ni with domain structures prepared by different methods.

2. Materials and methods

The studied samples represented plates doped by Ce (0.004 wt.% CeO_2, SBN:Ce) and Ni (0.05 wt% Ni_2O_3, SBN:Ni) $Sr_{0.61}Ba_{0.39}Nb_2O_6$ single crystals grown by modified Stepanov technique [21] in the Institute of General Physics of RAS (Moscow, Russia). The 0.6-mm-thick plates were cut normally to the polar axis and carefully polished. The Cr electrodes were deposited on the sample surfaces.

Three types of the domain states created by various methods have been studied: (1) zero-field cooling (ZFC), (2) in-field cooling (FC), and (3) partial switching (PS). ZFC represent slow cooling from 200°C to the room temperature without application of electric field. The domain structure formed after ZFC consists of nanodomains with random orientation of spontaneous polarization [16,17]. FC procedure was achieved by slow cooling from 200°C to the room temperature under constant external field 600 V/mm [22]. Mostly single-domain state with residual nanodomains was formed after FC. The PS was realized in linearly increasing electric field rising for 300V/(mm·s) until the switching current maximum. The micro-size domain structure with residual nanodomains appeared after PS [23-25].

The temperature control system LINKAM THMSE-600 was used for realization of the proper heating/cooling regime. The heating/cooling rate about 1-5 K/min was used. The dielectric measurements in frequency range from 10 Hz to 10 kHz were carried out in the samples with different domain structures using LCR-meter 2B-1 (Russia) during heating from 25 to 150°C and subsequent cooling. A standard scheme of single-beam Michelson homodyne interferometer was used for measurements of the piezoelectric coefficient at room temperature in frequency range from 70 Hz to 15 kHz [26]. Detailed description of the experimental setup can be found elsewhere [26].

3. Experimental

3.1. Dielectric permittivity

The temperature dependences of the dielectric permittivity at 10 Hz measured during cooling and heating of SBN:Ce and SBN:Ni samples prepared by ZFC, FC, and PS are presented in Figure 1.

It was shown that dielectric permittivity at low temperatures was higher for samples treated by ZFC than by FC and the difference increased with heating (Fig. 2a). This difference is due to a contribution of the charged domain walls to the dielectric response. It was demonstrated earlier that the complex quasi-regular three-dimensional self-similar nanodomain labyrinth with a large number of charged domain walls formed in SBN single crystals treated by ZFC contribute significantly to the dielectric response [16,17]. On the contrary, the small domain wall area in the samples treated by FC decreased the dielectric permittivity. Decrease of depolarization fields and domain wall mobility during heating led to an apparent increase of the dielectric permittivity difference between the samples treated by ZFC and FC (Fig. 2a).

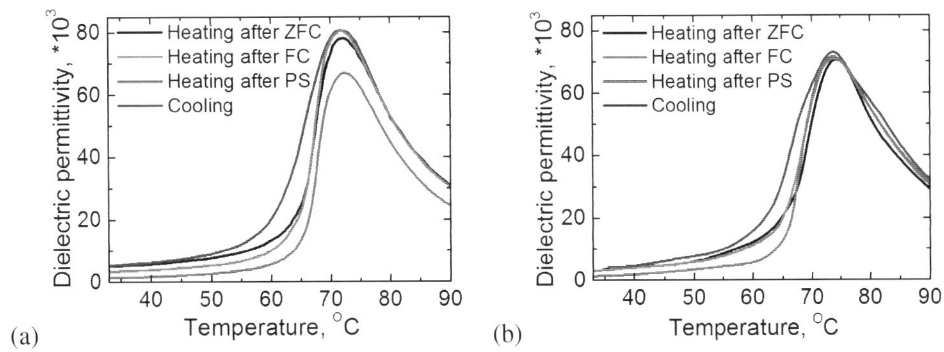

Figure 1. The temperature dependences of dielectric permittivity during cooling and heating (a) SBN:Ce and (b) SBN:Ni with different domain structures. Frequency 10 Hz.

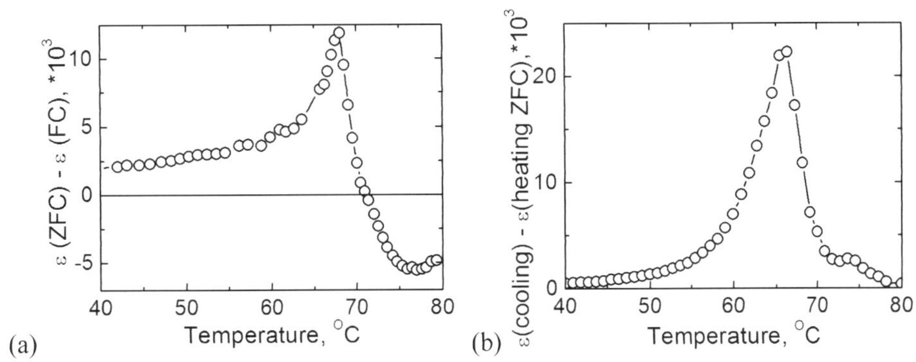

Figure 2. The temperature dependences of difference between dielectric permittivity (a) during heating after ZFC and FC and (b) during cooling and heating after ZFC. SBN:Ce. Frequency 10 Hz.

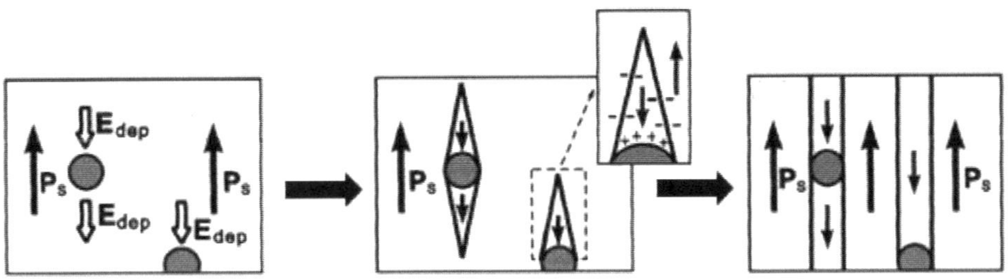

Figure 3. Schematic representation of the evolution of the domain structure samples treated by ZFC during heating. Black arrows - P_s direction, white arrows - E_{dep} direction, dark grey circles - nonpolar inclusions.

At a certain temperature, the domain contribution reached its maximum and, upon further heating, began to decrease sharply (Fig. 2a). This behaviour can be attributed to the appearance of insulated nanometre nonpolar inclusions in ferroelectric multidomain matrix [16,17]. The bound charges arising at the phase boundaries created the depolarization fields (Fig. 3) in the samples treated by FC, which could be compensated only by slow bulk screening process [27]. These fields led to a partial switching and formation of nanodomain structure with charged domain walls in the vicinity of nonpolar inclusions. Charged domain walls may contribute to the dielectric constant [27,28] due to an abnormally large concentration of steps that were mobile even in weak electric fields used for dielectric measurements.

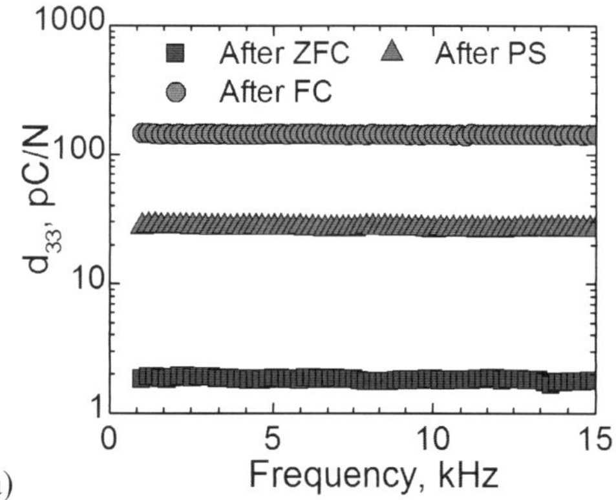

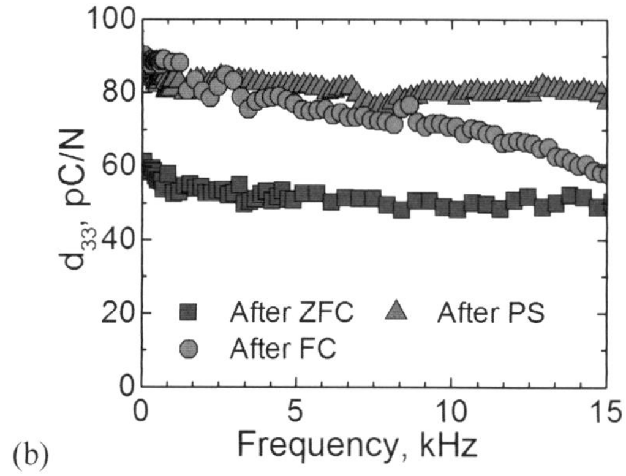

Figure 4. The frequency dependences of integrated piezoelectric response of (a) SBN:Ce and (b) SBN:Ni with different domain structures.

At low temperatures, dielectric permittivity values measured during cooling and heating for samples treated by ZFC were the same due to a small mobility of the domain walls at these temperatures. During heating, the difference between dielectric permittivity values observed (Fig. 2b). The dielectric permittivity is larger under cooling than under heating due to a higher mobility of the "fresh" domain walls formed during cooling as compared with the screened walls formed as a result of ZFC.

The dielectric permittivity for the sample prepared by PS at low temperatures is less than for those processed by other methods. It can be attributed to the lower concentration of the charged domain walls.

3.2. Piezoelectric coefficient

The frequency dependences of the integrated piezoelectric response of SBN single crystals with different domain structure measured at room temperature (below freezing temperature for the studied samples) are presented in Figure 4. Slight decrease of the macroscopic piezoelectric response with increasing frequency was seen for crystals prepared by all methods. The most pronounced decrease was obtained in SBN:Ni single crystal treated by FC.

The largest piezoelectric response was obtained in the samples treated by FC (146 pm/V (SBN:Ce) and 89 pm/V (SBN:Ni) at 1 kHz). The smallest piezoelectric response was measured in the samples treated by ZFC (1.8 pC/N (SBN:Ce) and 55 pC/N (SBN:Ni) at frequency 1 kHz). The piezoelectric response in the samples treated by PS was 28 pm/V for SBN:Ce and 80 pm/V for SBN:Ni (at frequency 1 kHz). This behaviour can be attributed to the contribution of the crystal polarization P to the piezoelectric response of the system. The polarization of the crystals treated by ZFC is rather small, while in the crystals treated by FC its value is much larger.

4. Conclusion

The influence of the initial domain structure on piezoelectric response and dielectric permittivity has been studied in SBN single crystals slightly doped by Ce and Ni. The initial domain structure was formed by three methods: (1) zero-field cooling, (2) in-field cooling, and (3) partial switching. The difference in the temperature dependences of dielectric permittivity for various initial domain structures has been attributed to the existence of charged domain walls in the samples with multi-domain structure and appearance of the non-polar inclusions during heating above the freezing temperature. The difference in the macroscopic piezoelectric response for various initial domain structures can be attributed to the contribution of the crystal polarization to the piezoelectric response.

Acknowledgements

Equipment of the Ural Centre for Shared Use "Modern nanotechnology" Ural Federal University was used. The research was made possible by Russian Foundation of Basic Research (project № 16-02-

00821-a) and state task of Ministry of education and science of the Russian Federation (No. 3.4993.2017/6.7). The work was partially supported by Government of the Russian Federation (act 211, agreement 02.A03.21.0006). Part of this work was developed within the scope of the project CICECO-Aveiro Institute of Materials, POCI-01-0145-FEDER-007679 (FCT Ref. UID /CTM /50011/2013), financed by national funds through the FCT/MEC and when appropriate co-financed by FEDER under the PT2020 Partnership Agreement.

References

[1] Shur V Ya, Rumyantsev E L, Nikolaeva E V, Shishkin E I, Batchko R G, Fejer M M and Byer R L 2001 *Ferroelectrics* **257** 191-202
[2] Shur V Ya, Akhmatkhanov A R and Baturin I S 2015 *Appl. Phys. Rev.* **2** 040604
[3] Wada S, Park S E, Cross L E and Shrout T R 1998 *J. Kor. Phys. Soc.* **32** S1290-3
[4] Wada S, Park S E, Cross L E and Shrout T R 1999 *Ferroelectrics* **221** 147-55
[5] Wada S and Tsurumi T 2004 *Br. Ceram. Trans.* **103** 93-6
[6] Wada S, Yako K, Kakemoto H, Tsurumi T and Erhart J 2004 *Key Eng. Mater.* **269** 19-22
[7] Arlt G, Hennings D and de With G 1985 *J. Appl. Phys.* **58** 1619-25
[8] Jaffe B, Cook, Jr W R and Jaffe H 1971 *Piezoelectric Ceramics* (New York: Academic Press)
[9] Prokhorov A M and Kuz'minov Yu S 1990 *Ferroelectric crystals for laser radiation control* (Bristol: Adam Hilger)
[10] Cross L E 1994 *Ferroelectrics* **151** 305-20
[11] Smolenskii G A and Agranovskaya A I 1960 *Sov. Phys. Solid State* **1** 1429-32
[12] Blinc R, Pirc R, Zalar B, Gregorovic A and Bobnar V 2004 *Ferrolectrics* **299** 1-9
[13] Shur V Ya, Lomakin G G, Kuminov V P, Pelegov D V, Beloglazov S S, Slovikovskii S V and Sorkin I L 1999 *Phys. Solid State* **41** 453-6
[14] Shur V Ya, Lomakin G G, Rumyantsev E L, Beloglazov S S, Pelegov D V, Sternberg A and Krumins A 2004 *Ferroelectrics* **299** 75-81
[15] Dai X, Xu Z and Viehland D 1994 *Phil. Mag. B* **70** 33-48
[16] Dec J, Shvartsman V V and Kleemann W 2006 *Appl. Phys. Lett.* **89** 212901
[17] Shvartsman V V, Dec J, Lukasiewicz T T, Kholkin A L and Kleemann W 2008 *Ferroelectrics* **373** 77-85
[18] Shur V Ya, Rumyantsev E L, Lomakin G G, Yakutova O V, Pelegov D V, Sternberg A and Kosec M 2005 *Ferroelectrics* **314** 245-53
[19] Shikhova V A, Shur V Ya, Pelegov DV, Rumyantsev E L and Yakutova O V 2010 *Ferroelectrics* **398** 115-26
[20] Shikhova V A, Shur V Ya, Pelegov D V and Ivleva L I 2013 *Ferroelectrics* **443** 116-23
[21] Ivleva L I 2009 *Bulletin of the Russian Academy of Sciences: Physics* **73** 1338-40
[22] Kolchina E A, Neradovskiy M M, Shikhova V A, Pelegov D V, Shur V Ya, Ivleva L I and Dec J 2016 *Ferroelectrics* **496** 149-56
[23] Shur V Ya, Pelegov D V, Shikhova V A, Kuznetsov D K, Nikolaeva E V, Rumyantsev E L, Yakutova O V and Granzow T 2010 *Phys. Solid State* **52** 346-51
[24] Shur V Ya, Shikhova V A, Pelegov D V, Ievlev A V and Ivleva L I 2011 *Phys. Solid State* **53** 2311-5
[25] Shur V, Shikhova V, Ievlev A, Zelenovskiy P, Neradovskiy M, Pelegov D and Ivleva L 2012 *J. Appl. Phys.* **112** 064117
[26] Ushakov A D, Yavo N, Mishuk E, Lubomirsky I, Shur V Ya and Kholkin A L 2016 *KnE Materials Science* **2016** 177-82
[27] Shur V Ya 1998 *Phase Transitions* **65** 49-72
[28] Shur V, Rumyantsev E, Nikolaeva E and Shishkin E 2000 *Appl. Phys. Lett.* **77** 3636-8

IOP Conf. Series: Materials Science and Engineering **443** (2018) 012032 doi:10.1088/1757-899X/443/1/012032

The limit of mass determination with an AFM cantilever-based system

I D Shumov[1], S L Kanashenko[1], V S Ziborov[2], Y D Ivanov[1], A I Archakov[1] and T O Pleshakova[1]
[1] Institute of Biomedical Chemistry, Moscow, Russia
[2] Joint Institute for High Temperatures RAS, Moscow, Russia

shum230988@mail.ru

Abstract. The application of an atomic force microscope (AFM) based microcantilever system for the determination of mass of gold nanoparticles (AuNPs) has been demonstrated. In this system, standard AFM microcantilevers for measurements in vacuum have been employed. The limit of mass determination with our AFM-based system has been determined to be of the order of 10^{-10} g. The prospects of employing AFM cantilever-based sensors for highly sensitive protein detection in proteomic studies and in diagnostics have been discussed.

1. Introduction

Nowadays, modern nanotechnology-based analytical systems employing molecular detectors, such as atomic force microscope (AFM), are being developed. Molecular detectors (AFM, nanowire sensors [1-3], etc.) are capable of detecting single target particles of nanometre size (such as nanoparticles [4], viral particles [2,5,6], and even individual biological [7,8] and synthetic [9] macromolecules). In molecular detectors, such a high sensitivity is achieved owing to the small size of the sensor element, which is comparable with that of the target molecule. Among nanotechnology-based systems, one can distinguish nanomechanical cantilever-based ones, which allow one to operate with condensed matter at the nanoscale, thus enabling determination of various characteristics of the investigated objects (mass [10], elasticity [8], etc.) at the level of single macromolecules or nanoparticles. In these systems, the target objects are adsorbed from the volume of analysed sample onto the cantilever, and the resonance frequency of the AFM cantilever (which decreases in a proportion to the added mass) is measured [11]. The adsorption of target objects increases the mass of the cantilever; this causes a shift in the resonance frequency of the cantilever, which is measured by AFM electronics. That is, in cantilever-based sensors, the mass of the particles adsorbed onto the cantilever is converted to the oscillation frequency signal. As target objects, nanoparticle-labelled single molecules of protein markers of diseases can be used [11]. The latter opens the prospects for employing AFM-based sensors as highly sensitive detectors in proteomic studies and in diagnostics [11]. In this connection, it is to be emphasized that the application of highly sensitive nanotechnology-based detectors in diagnostics is crucial, since the concentration of many clinically relevant marker proteins in the biological material is very low [12].

For the determination of mass at the nanoscale, several types of cantilever-based systems can be employed. In the paper by Naik et al. [13], the application of nanometer-sized cantilevers is discussed. Such systems exhibit an extremely high mass determination sensitivity of the order of 10^{-18} g. At the same time, the extremely small size of such cantilevers leads to long time required for the diffusion transport of target objects to the sensor surface; this fact limits the use of nanometer-sized cantilevers. In the study by Kosaka et al. [11], standard AFM cantilevers were employed. The use of standard cantilevers is much more convenient, as it allows one to perform the detection of nanoparticle-labelled protein molecules in just one hour [11]. Measurements in air, however, require additional processing of the signal to reduce the influence of noise on the determination of resonance frequency [11]. In this regard, measurements in vacuum are preferable, since the influence of noise is reduced.

In our present study, the sensitivity of a microcantilever system employing standard AFM microcantilevers for measurements in vacuum, has been estimated. We have demonstrated that such a system allows one to register ~100 gold nanoparticles (AuNPs), what corresponds to 17 μL of 10^{-17} M solution of protein labelled with these nanoparticles. This approach is convenient, because no additional equipment (such as nanobalance [10,14]) is required to measure the mass of the adsorbed protein.

2. Methods

To measure the resonance frequency shift caused by adsorption of gold nanoparticles (AuNPs) onto the cantilever, the following technique was employed. AFM cantilever (PPP-CONTR-50, Nanosensors Inc., USA; resonance frequency 6 to 21 kHz; height 450±10 μm; width 50±7.5 μm; thickness 2±1 μm; force constant 0.02 to 0.77 N/m), which resonance frequency was measured prior to the experiment, was incubated in a ~1-μL drop of solution containing 20-nm-diameter AuNPs (synthesized using a citrate reduction method [15]). The incubation time was 5 and 15 min. After the incubation, the cantilever was dried and placed into the vacuum chamber of an NTEGRA Aura AFM (NT-MDT, Zelenograd, Russia), and its resonance frequency was measured and compared with that before the incubation. All measurements have been performed in vacuum at 0.1 Torr.

The mass of the AuNPs adsorbed onto the cantilever was estimated from the resonance frequency shift according to Kosaka et al. [11]:

$$\Delta m = \frac{2m(f_1 - f)}{f},$$

where f and f_1 are the resonance frequencies of the cantilever before and after the incubation, respectively, m is the mass of the cantilever, and Δm is the mass of the adsorbed AuNPs.

Therefore, given the mass of a single AuNP, the total number of the AuNPs adsorbed onto the cantilever can be determined. The size of AuNPs and the number of AuNPs adsorbed onto the cantilever were estimated by scanning electron microscopy (SEM) visualization of the cantilever with adsorbed AuNPs employing a Hitachi S5500 electron microscope (Hitachi, Japan).

Since the resonance frequency shift for a 20-kHz cantilever, registerable with the system employed in our study, is 20 Hz (which amounts to 0.1% from 20 kHz), the mass of the adsorbed AuNP causing this shift must be 1000 times smaller than that of the cantilever itself. Let us estimate the mass of the silicon (of 2.33 g/cm^3 density) cantilever:

$$m=\rho V=2.33 \text{ g/cm}^3(450\times 50\times 1\times 10^{-12}) \text{ cm}^3=5.24\times 10^{-8} \text{ g}$$

For such a cantilever, 0.1% mass increase ($\Delta m/m$=0.001) corresponds to Δm of the order of 10^{-10} g. Since the masses of 20-nm and 100-nm AuNPs (of 19.3 g/cm^3 density) make up ~10^{-16} g and ~10^{-14} g, respectively, the number of these AuNPs causing $\Delta m=10^{-10}$ g corresponds to ~10^6 (for 20-nm AuNPs) and ~10^4 (for 100-nm AuNPs).

3. Results and discussion

Figure 1 displays the data obtained in our experiments. The resonance frequency curves shown in Figure 1a clearly indicate that adsorption of 10^6 of 20-nm AuNPs causes a 20-Hz shift in the cantilever's resonance frequency (which is in a good agreement with theoretical estimations) in contrast to the case with 10^4 of 20-nm AuNPs, when no shift in resonance frequency was registered (Fig. 1b). The number of adsorbed AuNPs was altered by varying the cantilever incubation time in AuNP-containing solution.

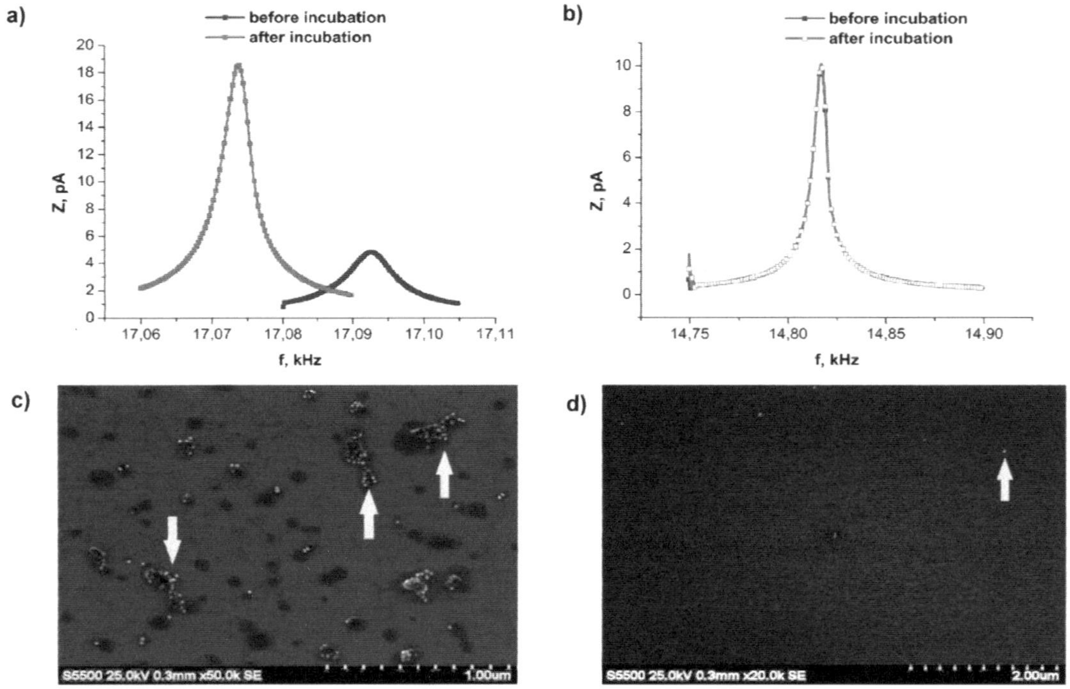

Figure 1. The resonance frequency curves of the AFM cantilever obtained before (red curves) and after (blue curves) their incubation in AuNP-containing solution, and corresponding SEM images of single AuNPs adsorbed onto the cantilever surface, in the case of adsorption of 10^6 (a, c) and 10^4 (b, d) 20-nm AuNPs.

Typical SEM images of the cantilever surface with adsorbed AuNPs presented in Figure 1c,d confirm the data obtained in the resonance frequency measurements. Indeed, in Figure 1c (illustrating the case with the adsorption of 10^6 20-nm AuNPs onto the cantilever surface) one can clearly distinguish a number of AuNPs (seen as white spots in the SEM image), which tend to form multiparticle aggregates on the cantilever surface. On the contrary, only a few (<10 particles per ~25 μm^2 image) single AuNPs are distinguishable in Figure 1d corresponding to the adsorption of 10^4 20-nm AuNPs onto the cantilever surface. In the latter case, a small amount of AuNPs adsorbed on the cantilever surface is apparently insufficient to cause a registerable shift in the cantilever's resonance frequency.

4. Conclusion

Due to high sensitivity of mass determination, the use of cantilever-based method for protein detection in biomedical applications is promising. To date, modern techniques allow labelling protein molecules with AuNPs in 1:1 ratio [12]. This fact gives an opportunity for the detection of such labelled protein molecules with high sensitivity. In this way, using 20-nm AuNPs for this purpose, the 10^{-15} M protein detection limit can be attained upon analysis of 1 mL of sample. Moreover, if 100-nm AuNPs will be used instead of 20-nm ones, the detection limit of labelled protein molecules can be shifted down to 10^{-17} M.

Acknowledgements

Experiments employing AFM were performed with financial support from Russian Science Foundation (Grant No. 14-25-00132). Electron microscopy measurements were performed using the equipment of "Human Proteome" Core Facility of the Orekhovich Institute of Biomedical Chemistry

(Russia), which is supported by Ministry of Education and Science of the Russian Federation (agreement 14.621.21.0017, unique project ID RFMEFI62117X0017) in the framework of the Program for Basic Research of State Academies of Sciences for 2013-2020. Shumov I.D. is a recipient of Russian Federation President scholarship for young scientists for 2016-2018 (project ID: SP-4280.2016.4).

References

[1] Ivanov Yu D, Pleshakova T O, Kozlov A F, Malsagova K A, Krohin N V, Shumyantseva V V, Shumov I D, Popov V P, Naumova O V, Fomin B I, Nasimov D A, Aseev A L and Archakov A I 2012 *Lab Chip* **12** 2104-11
[2] Patolsky F, Zheng G F, Hayden O, Lakadamyali M, Zhuang X W and Lieber C M 2004 *Proc. Natl. Acad. Sci. USA* **101** 14017-22
[3] Stern E, Klemic J F, Routenberg D A, Wyrembak P N, Turner-Evans D B, Hamilton A D, LaVan D A, Fahmy T M and Reed M A 2007 *Nature* **445** 519-22
[4] Georgiev P, Bojinova A, Kostova B, Momekova D, Bjornholm T and Balashev K 2013 *Colloids Surf.* A *Physicochem. Eng. Asp.* **434** 154-63
[5] Dubrovin E V, Yaminsky I V, Kirikova M N, Novikov V K and Drygin Y F. 2004 *Colloid J.* **66** 673-8
[6] Ivanov Yu D, Kaysheva A L, Frantsuzov P A, Pleshakova T O, Krohin N V, Izotov A A, Shumov I D, Uchaikin V F, Konev V A, Ziborov V S and Archakov A I 2015 *Int. J. Nanomedicine* **10** 1597-608
[7] Limanskaya L A and Limanskii A P 2006 *Russ. J. Bioorgan. Chem.* **32** 444-59
[8] Ivanov Yu D, Bukharina N S, Pleshakova T O, Frantsuzov P A, Krokhin N V, Ziborov V S and Archakov A I 2011 *Biophysics* **56** 892-6
[9] Kumaki J, Kawauchi T and Yashima E 2006 *Macromolecules* **39** 1209-15
[10] Kiselev G A and Yaminsky I V 2007 *Nanotechnologies in Russia* **2** 112-7
[11] Kosaka P M, Pini V, Ruz J J, da Silva R A, González M U, Ramos D, Calleja M and Tamayo J 2014 *Nat. Nanotechnol.* **9** 1047-53
[12] Rissin D M, Kan C W, Campbell T G, Howes S C, Fournier D R, Song L, Piech T, Patel P P, Chang L, Rivnak A J, Ferrell E P, Randall J D, Provuncher G K, Walt D R and Duffy D C 2010 *Nat. Biotechnol.* **28** 595-99
[13] Naik A K, Hanay M S, Hiebert W K, Feng X L and Roukes M L 2009 *Nat. Nanotechnol.* **4** 445-50
[14] Ukraintsev E V, Kiselev G A, Bagrov D V, Gorelkin P V, Kudrinsky A A, Lisichkin G V and Yaminsky I V 2007 *Sens. Syst.* **1** 18-21
[15] Kimling J, Maier M, Okenve B, Kotaidis V, Ballot H and Plech A 2006 *J. Phys. Chem.* B **110** 15700-7

IOP Conf. Series: Materials Science and Engineering **443** (2018) 012033 doi:10.1088/1757-899X/443/1/012033

Pronounced light trapping effect and enhanced photo-electrochemical property of type (II) aligned graphitic-C_3N_4 with embedded 1-D ZnO nanostructures

S Singh[1] and M Khanuja[2]

[1] Shaheed Rajguru College of Applied Sciences for Women, University of Delhi, New Delhi 110096, India
[2] Centre for Nanoscience and Nanotechnology, Jamia Millia Islamia (A Central University), New Delhi 110025, India

sonal.singh0811@gmail.com, manikakhanuja@gmail.com

Abstract. Graphitic carbon nitride (g-C_3N_4) with embedded 1D-ZnO nanostructures was prepared by simple two step thermal decomposition process. Photocatalyst exhibited higher optical absorption than pure ZnO due to π-π conjugations of C_3N_4, which thus allow greater adsorption on its surface. Conduction band (CB) edge of g-C_3N_4 being more negative than ZnO and Valence band (VB) of ZnO is more positive than VB of g-C_3N_4; this alignment of Type (II) resulted in transfer of excited electrons in g-C_3N_4 to the CB of ZnO. This charge transfer created the internal electrostatic potential, which aided in reducing the recombination rate of charge carriers. Finally, samples were tested for photo-electrochemical activity measured by generated photocurrent density with respect to voltage and time under illumination conditions.

1. Introduction

Graphitic Carbon Nitride (g-C_3N_4) has already attracted much attention in the field of organic nanomaterials owing to its unusual excellent properties towards environmental protection and remediation through artificial photocatalysis, photo-electrochemical and other solar energy conversion related methods [1,2]. g-C_3N_4 is metal-free catalyst with a band gap of 2.7 eV, which is smaller than commercially used and much advanced inorganic TiO_2. It is regarded as the most stable phase of C_3N_4 at ambient conditions andwith controlled structure and morphology at the nanoscale, it can possess properties such as high hardness, low friction coefficient, and reliable chemical inertness, which classify it as a suitable photoactive material. Moreover, it is highly sustainable being composed of nitrogen and carbon only [3,4]. However, the quantum yield of the system still remains low and the photocatalytic efficiency of pristine g-C_3N_4 is still not satisfactory due to factors like high recombination rate of photogenerated electron-hole pairs. Therefore, it is required to assist its activity in some sort. Several routes like chemical doping, non-metallic doping, and composite formation by coupling g-C_3N_4 with other semiconductors have been adopted and there is an effective strategy to modify the electronic structure as well as surface properties of semiconductors [3,5].

Although, various studies related to formation of ZnO nanostructures, such as nanosheets, nanotubes, and nanoparticles coupled with g-C_3N_4 as a composite have been reported for their photocatalytic activity so far, photoelectrochemical (PEC) properties of 1D-ZnO nanostructures, i.e. nanorods with embedded g-C_3N_4 are still poorly studied [6-9]. Studies reveal that individual materials, here ZnO, of different morphology (particles, rods, sheets, etc.) exhibit different optical and electronic properties under the same conditions [10-14]. ZnO of 1D morphology, such as nanorods and nanotubes, has shown improved photoactivity by light trapping effect, wherein it achieves improved electron transport due to its internal electric field. So, we have purposely chosen this particular morphology and attempted to embed these nanostructures into a narrow band gap material, g-C_3N_4. The ability of 1D nanostructure to trap photons theoretically remains very high as compared to flat surface, which thus increases light absorption [15]. Coupling a wide band gap material with a narrow one is a promising approach to achieve high photoactivity, as it helps overcome the demerits of individual components and instigate a synergistic effect, such as improved photostability, enhanced harvesting of light, and efficient photo-excited charge separation [16,17]. Further, due to sp_2 hybridization, formation of the delocalized conjugated π structure of g-C_3N_4 brings effective charge separation in the photocatalysis. Adopting such strategy may help the material to overcome some of its existing drawbacks, such as (1) typical confinement below 470 nm wavelength due to insufficient sunlight absorption, (2) moderately low specific surface area, (3) low efficiency with low charge mobility, and (4) inability to oxidize water due to its inappropriate valance band (VB) position at about 1.4 V vs. NHE [18].

Herein, we attempted to prepare g-C_3N_4 nanofilms with ZnO nanorods embedded to it via simple two step thermal decomposition process and tested its PEC properties.

2. Experimental

All chemicals were of analytical grade and were used in experiments without any further purification. Pure ZnO nanorods were prepared by mechanical thermal decomposition method using [$Zn(CH_3COO)_2$ $2H_2O$], zinc acetate dehydrate as a precursor. 2 g of powder were kept in alumina crucible after mechanical grinding in mortar pestle for 45 min. To thermally decompose the powder, it was placed in programmable furnace at 500°C with ramp rate 4°C/min for 4 h. The obtained powder sample was washed several times with distilled water and then dried in oven for 8 h at 100°C. Graphitic carbon nitrite sheets were synthesized via simple heat treatment method as reported in Ref. [19]. To prepare zinc oxide embedded g-C_3N_4, the as prepared ZnO nanostructures were added to the g-C_3N_4 precursors during its synthesis. In brief, the light-yellow powder of melamine (5 g) was obtained after annealing and kept for cooling overnight under ambient temperature. The sample was annealed with ramp rate of 4°C per min for 4 h at 550°C. The sample was treated with HCl (50 mL/g of g-C_3N_4, 18.5 wt. %) for protonation followed by addition of ZnO nanostructures to the solution at room temperature for 4 h. This suspension was diluted, filtered, and washed several times with de-ionized water and dried at 80°C for 6 h.

Transmission electron microscopy (TEM, FEI, Technai G2, 200kV) was used to study the morphologies and crystallographic structure of the as-prepared samples. To investigate the crystalline structure of pure g-C_3N_4, pure ZnO and ZnO embedded g-C_3N_4 samples, X-ray diffractometer (XRD) (Bruker, D2-Phaser) using CuKα (λ= 1.5403Å) X-ray was used. XRD spectrum was recorded from 2θ = 5 to 80° operating at 10 KV and 30 mA. XRD data peaks were verified with the help of JCPDS files. The photo luminescence (PL) emission spectra were examined using photo luminescence spectroscopy (RF 5301 PC, Shimadzu) with 325 nm excitation wavelength.

PEC performance of pristine ZnO and ZnO-embedded-C_3N_4 samples was evaluated and compared using three-electrode-electrochemical system. Chrono-amperometry and linear sweep voltammetry (LSV) measurements were carried out under illumination at bias potential of 0 V vs Ag/AgCl. The samples were prepared over conducting glass and used for PEC as working electrode keeping Pt and Ag/AgCl electrode as counter electrode and reference electrode, respectively. The electrolyte used was 1 M NaOH at pH 14. The film was spin coated on Fluorine-doped tin oxide (FTO) substrate/glass at 5000 rpm for 50 seconds and annealed at 350°C for 1 hr.

3. Results and discussion

Figure 1 shows the XRD spectra for both pure ZnO and ZnO-embedded-C_3N_4 sample. The XRD characteristic peaks were verified with the help of JCPDS card no (06-2151), which confirmed the successful formation of the samples. The as-calculated lattice parameters values for ZnO (a=3.25 Å, c=5.2 Å) were consistent with the standard values for wurtzite hexagonal structure confirming its structure. XRD pattern for ZnO-embedded-C_3N_4 exhibited characteristic peaks of both ZnO and C_3N_4. However, the strongest peak for C_3N_4 at (002) is not distinctly visible in the graph, as the intensity of the peak is proportional to its concentration compared to counter material in the composite. As the concentration of the material in the composite increases, the XRD peak intensifies. In our case, amount of C_3N_4 is less compared to ZnO. No phase change was observed for both the materials in nanocomposite rather change in value of peak intensities indicated a change in crystallinity. In case of C_3N_4-ZnO, a decline in peak intensity values was observed for planes (100), (002), and (101) as compared to pure ZnO as shown in Table 1.

Figure 2 shows the transmission electron microscope micrographs for pure ZnO and C_3N_4-ZnO samples. Figure 2a reveals the clear formation of 1-Dimensional morphology for pure ZnO in the form of nanorods. From Figure 2b, it is clearly observed that ZnO nanorods are completely incorporated in C_3N_4 nanosheets. The images of C_3N_4-ZnO confirm the huge dispersion of ZnO nanorods in C_3N_4 nanostructures and thus lead to an establishment of close contact of latter with the former. This sets for prompt transfer of charges between the two.

Table 1. Data showing peak intensities for ZnO and C_3N_4-ZnO at various planes.

Sample	(100)	(002)	(101)
ZnO	8525	6424	14160
C_3N_4-ZnO	4733	3789	8426

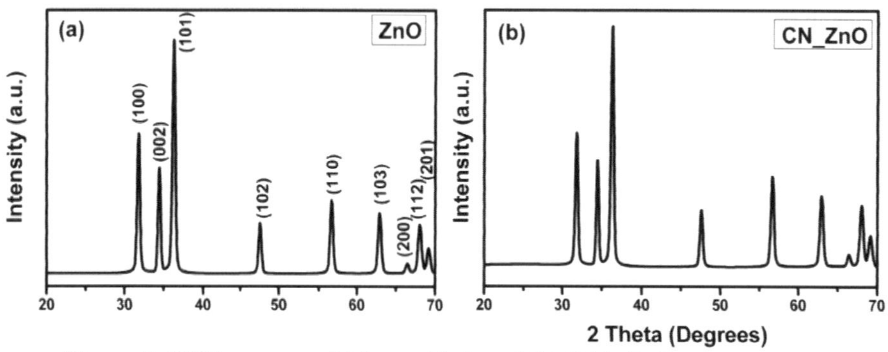

Figure 1. XRD patterns of (a) pure ZnO and (b) C_3N_4-ZnO nanostructure.

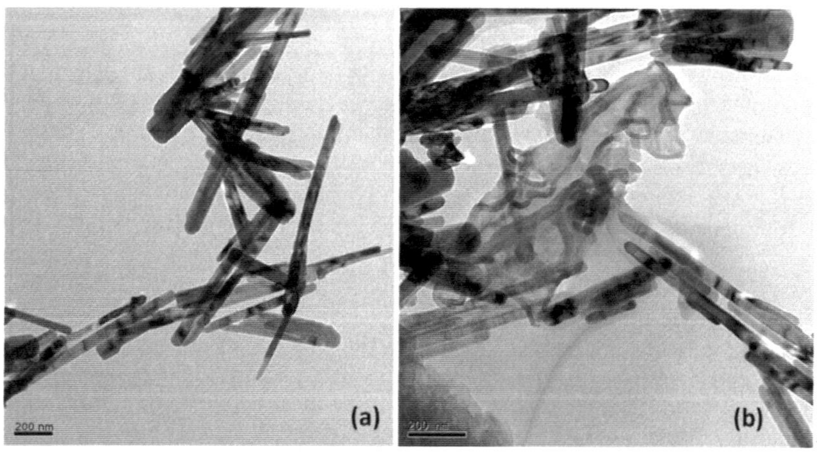

Figure 2. TEM micrograph of synthesized (a) ZnO nanorods and (b) C_3N_4-ZnO nanostructures. Scale-bar 200 nm.

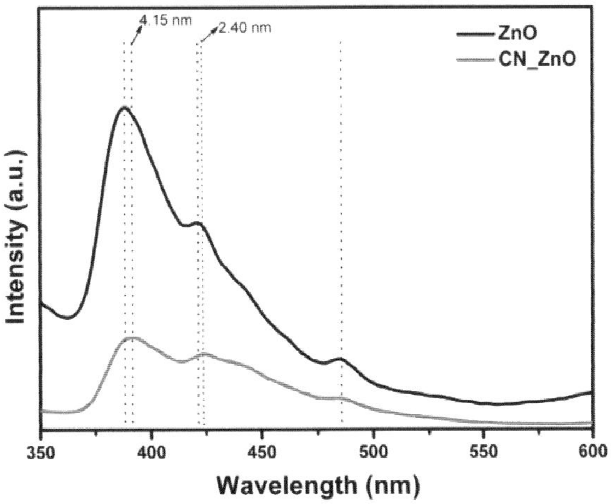

Figure 3. PL spectra for ZnO and C_3N_4-ZnO nanostructure.

To understand the high rate of charge carrier migration and get an insight into the mechanism process in case of close contact formation in composite, PL analysis was carried out. Figure 3 shows the PL spectra for pure ZnO and C_3N_4-ZnO with 325 nm excitation wavelength. Analysis of PL spectra is based on two primary factors: (a) peaks intensity and (b) peaks shift. As shown in Figure 3, major emission peak for pure ZnO sample corresponds to the light energy being almost equal to the band gap of ZnO. After adding ZnO to C_3N_4, it is clearly observed that there is a significant decrease in intensity of peaks for C_3N_4-ZnO PL spectra, along with the slight red-shift of the spectra. It is well known that decrease in intensity of peaks indicates reduced recombination rate of photogenerated charge carriers, which is the case in our C_3N_4-ZnO sample. This fact of prolonged electron-hole separation in the material contributes significantly towards improved charge transfer between ZnO and C_3N_4 and overall enhanced PEC activity of the material. Further, the shift in the spectra of C_3N_4-ZnO confirms the existence of close contact between the two materials, which is consistent with the results obtained for TEM micrographs. The shift of spectra towards red region of visible region also indicates a reduction in band gap as compared to pristine sample, which further aids in the superior photoactivity of the composite material.

PEC performance of pristine ZnO and C_3N_4-ZnO samples with respect to voltage and time is shown in Figure 4. LSV measurements under illumination conditions clearly show an increase in current linearly of +4.6 mA for C_3N_4-ZnO as compared to +2.0 mA for ZnO and almost 0 mA current for ZnO in dark. It is clearly visible that C_3N_4-ZnO shows the enhanced current and maximum current density with respect to pure ZnO, which does not produce sufficiently high current. The results obtained for LSV were found to be consistent with those of current density vs. time curves. C_3N_4 showed maximum photocurrent density and thus was the most stable with respect to time as compared to other samples.

Based on the PEC results obtained in this work, mechanism for boosted photoactivity is proposed in Figure 5. With the PEC results, it was clearly observed that ZnO alone was not able to contribute to the photoactivity as it could not be excited in solar visible region due to large band gap. Whereas, C_3N_4-ZnO exhibited remarkable photo electrochemical activity. The energy potentials for Conduction Band (CB) and Valence Band (VB) of C_3N_4 lies at -1.12 eV and 1.57 eV, respectively. The conduction band edge potential of ZnO is 0.5 eV, which is more positive than that of C_3N_4. The photoinduced electrons reaching the CB of C_3N_4 from its VB, can now easily transfer or inject to the CB of ZnO due to well-developed close contact between the two materials. The CB of ZnO now has excess of electrons reaching via VB of ZnO and CB of C_3N_4. This route of direct injection of charges from C_3N_4 to ZnO reduces the amount of energy required for the excitation of the material, thus making the material active in visible region of solar spectrum. This promotes the photo-electrochemical reactions of the composite material. Also, the reduced recombination rate of electron-hole pairs results in prolonged separation of charges further enhancing its photoactivity.

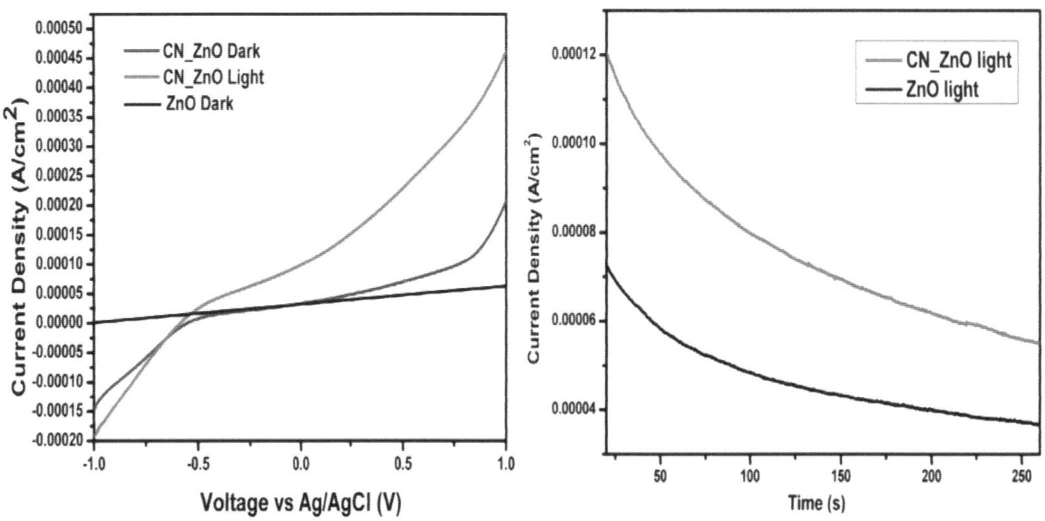

Figure 4. PEC performance of ZnO and C_3N_4-ZnO samples with respect to voltage and time.

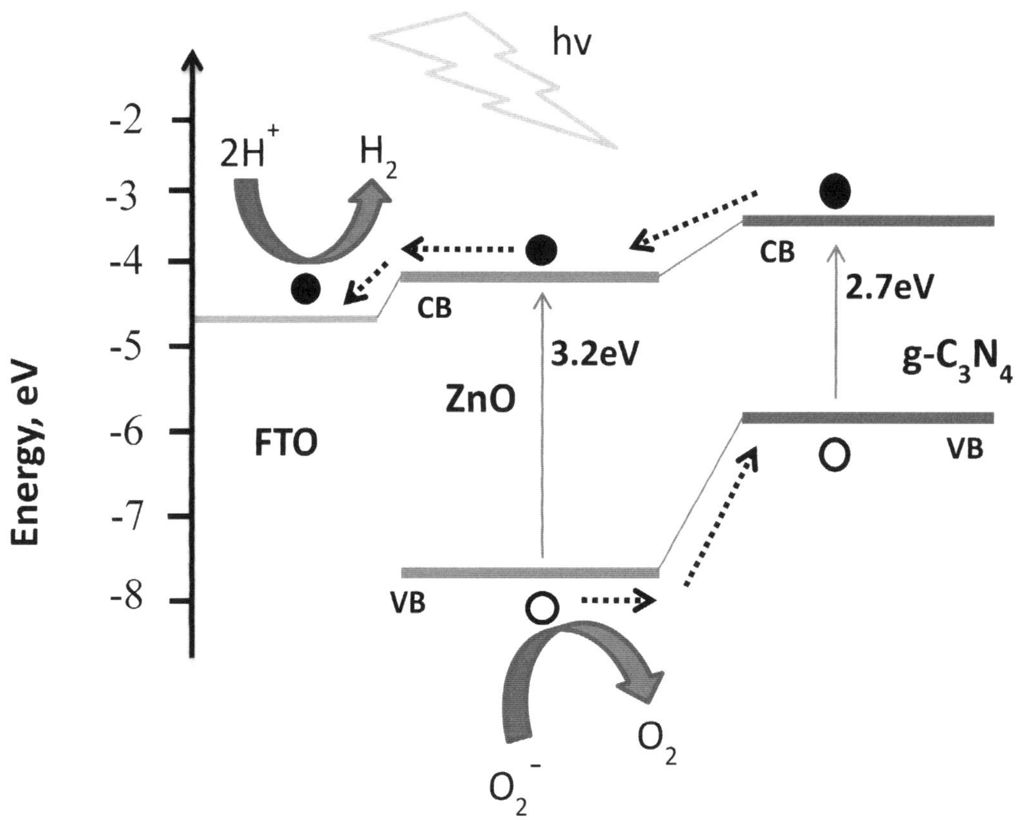

Figure 5. Schematic diagram of PEC action of C_3N_4-ZnO nanorods.

4. Conclusion

In summary, graphitic-C_3N_4 nanosheets embedded with 1D-ZnO nanorod structures were successfully formed by two step thermal decomposition process to test their photoelectrochemical activity and compare the results with that of pure ZnO. It was found that ZnO-embedded-C_3N_4 sample showed an overall increase of +2.6 mA current density as compared to pristine ZnO, which was attributed to the fact that close contact and type (II) alignment between the two materials promoted prompt and easy charge transfer. Further, 1D structure of ZnO contributed towards improved light trapping effect, which reduced the photon losses and increased average lifetime of charge carriers. Reduction in band gap of ZnO and increased electron-hole pair separation assisting to its pronounced photoelectrochemical activity is proposed with possible action mechanism.

Acknowledgements

The authors are thankful to Dr. Aadesh P. Singh from Department of Physics, Indian Institute of Technology, Delhi, India for help in preparation of ZnO-nanorods with embedded graphitic-C_3N_4 nanosheets samples.

References

[1] Wen J, Xie J, Chen X and Li X 2017 *Appl. Surf. Sci.* **391** 72-123
[2] Cao S and Yu J 2014 *J. Phys. Chem. Lett.* **5** 2101-7
[3] Zheng Y, Liu J, Liang J, Jaroniecc M and Qiao S Z 2012 *Energy Environ. Sci.* **5** 6717-31
[4] Li Y, Zhang J, Wang Q, Jin Y, Huang D, Cui Q and Zou G 2010 *J. Phys. Chem. B* **114** 9429-34
[5] Yan H 2012 *Chem. Commun.* **48** 3430-2
[6] Liu W, Wang M, Xu C, Chen S and Fu X 2013 *J. Mol. Catal. A* **9** 368-9
[7] Wang Y J, Rui S, Jie L and Zhu Y F 2011 *Energy Environ Sci.* **4** 2922-9
[8] Fu M, Pi J, Dong F, Duan Q and Guo H 2013 *Int. J. Photoenergy* **2013** 158496
[9] Lv H, Ji G, Yang Z, Liu Y, Zhang X, Liu W and Zhang H 2015 *J. Colloid Interf Sci.* **450** 381-7
[10] Xu L, Hu Y-L, Pelligra C, Chen C-H, Jin L, Huang H, Sithambaram S, Aindow M, Joesten R and Suib S L 2009 *Chem. Mater.* **21** 2875-85
[11] Mclaren A, Valdes-Solis T, Li G and Tsang S C 2009 *J. Am. Chem. Soc.* **131** 12540-1
[12] Baral A, Khanuja M, Islam S S, Sharma R and Mehta B R 2017 *J. Lumin.* **183** 383-90
[13] Bhuyan T and Khanuja M 2015 *J. Nanoparticle Res.* **17** 1-11
[14] Singh S, Pendurthi R, Khanuja M, Islam S S, Rajput S and Shivaprasad S M 2017 *Appl. Phys. A* **123** 184
[15] Moniz S J A, Zhu J and Tang J 2014 *Adv. Energy Mater.* **4** 1-8
[16] Singh S, Sharma R, Joshi G and Pandey J K 2017 *Korean J. Chem. Eng.* **34** 500-10
[17] Chai B, Peng T, Mao J, Li K and Zan L 2012 *Phys. Chem. Chem. Phys.* **14** 16745-52
[18] Zhang J, Sun J, Maeda K, Domen K, Liu P, Antonietti M, Fu X and Wang X 2011 *Energy Environ. Sci.* **4** 675-8
[19] Singh A P, Arora P, Basu S and Mehta B R 2016 *Int. J. Hydrog. Energy* **41** 5617-28

IOP Conf. Series: Materials Science and Engineering **443** (2018) 012034 doi:10.1088/1757-899X/443/1/012034

Electrical properties of irradiated individual multi-walled carbon nanotubes after gas adsorption

D V Sokolov[1], N A Davletkildeev[1,2], V V Bolotov[1] and I A Lobov[1]

[1] Omsk Scientific Centre SB RAS, 644024, Omsk, Russia
[2] F.M. Dostoevsky Omsk State University, 644077, Omsk, Russia

classicsub-zero@mail.ru

Abstract. The electrical parameters of individual nitrogen-doped carbon nanotubes (irradiated by electrons, protons, and argon ions) were determined during the ammonia and nitrogen dioxide exposure using the methods of scanning probe microscopy.

1. Introduction

Carbon nanotubes (CNTs) present excellent properties when used as sensing materials, supercapacitors, and others [1]. The contribution of nitrogen doping for CNTs enhances electrical conductivity, and nitrogen-containing defects serve as anchoring sites for specific gas molecules. In order to change electrical properties and create adsorption centres in nitrogen-doped CNTs, it is necessary to functionalize them, for example, by electron or ion irradiations [2]. Irradiation of doped nanotubes leads to the formation of adsorption centres that makes it possible to use them in sensory nanoelectronic.

Individual CNTs are ideal materials for nanodevices manufacturing but electrical measurements of a single nanotube are extremely difficult. Scanning probe microscopies techniques allows to get information about electrical properties individual nanoobjects with high resolution.

In the present work, electrostatic force microscopy (EFM) and conductive atomic force microscopy (C-AFM) have been successfully used for determining the free carrier concentration and the electrical conductivity in individual CNTs (irradiated by electrons, protons, and argon ions) during adsorption of oxidative and reductive gases.

2. Material and methods

Synthesis of CNTs was described in Ref. [3]. CNTs were irradiated by various beams: electrons (with energy 3 MeV and doses 10^{16}cm^{-2}), protons (140 keV, 10^{14} cm^{-2}), and argon ions (15 keV, 10^{16} cm^{-2}). The samples for the investigation were single CNTs deposited on structures with gold electrodes [3].

Conductivity and carrier concentration measurements were performed on the atomic force microscope MFP-3D SA (Asylum Research) using C-AFM and EFM methods [3]. The conductivity of CNTs was calculated according to the lengths and outer diameters of CNTs obtained from first pass EFM-images.

Scanning probe microscopy (SPM) measurements of individual CNTs were carried out in nitrogen atmosphere with reducing and oxidizing gases. Changes in the electrical properties of CNTs during ammonia and nitrogen dioxide adsorption at a concentration of 1000 ppm were studied.

3. Results and discussion

Figure 1 presents typical values and range of the conductivity of as-grown and irradiated CNTs in dry nitrogen atmosphere and after ammonia and nitrogen dioxide adsorption. Inset on Figure 1 presents a first pass of EFM image visualizing CNT, which lays between two neighbouring electrodes.

Figure 2 shows the free carrier concentration of as-grown and irradiated CNTs in dry nitrogen atmosphere and after exposure to ammonia and nitrogen dioxide. Ammonia molecules (reductive gas) adsorbed on CNTs act as electron donors to the nanotubes, which lead to increase of conductivity and electron concentration. With adsorption of acceptor molecules (nitrogen dioxide), conductivity and electrons concentration of doped CNTs decrease.

The type of irradiation particles effects on conductivity and carrier concentration in individual CNTs. Irradiated CNTs have a variety of charge carriers due to the kind and concentration of irradiation-induced defects in CNT walls. A change in conductivity of irradiated CNTs correlates with a change in charge carrier concentration in CNTs under adsorption of ammonia and nitrogen dioxide molecules.

Exposure to nitrogen dioxide increases the hole concentration in CNTs irradiated by protons and argon ions, because the extracted electrons by oxidizing gas molecules result in the generation holes in the valence band. Conductivity of these CNTs increases because the majority charge carriers are holes. The conductivity of the p-type CNTs under exposure to reducing gas (ammonia) decreases because generated electrons recombine with holes.

In CNTs irradiated by electrons, the majority charge carriers are electrons. Adsorption of nitrogen dioxide leads to a change of the majority charge carriers. As a result, holes concentration is more than the electron concentration in the CNTs before to nitrogen dioxide adsorption, so the conductivity increases. Exposure to reductive gas (ammonia) increases the electrons concentration in CNTs irradiated by electrons, and therefore the conductivity increases. Since the Fermi level shift during adsorption of the oxidative and reductive gases slightly changes, the concentrations of the majority charge carriers are practically equal, but they are opposite in sign. Therefore, C-AFM results show that the electrical conductivity of CNTs irradiated by electrons increases during adsorption in both gases.

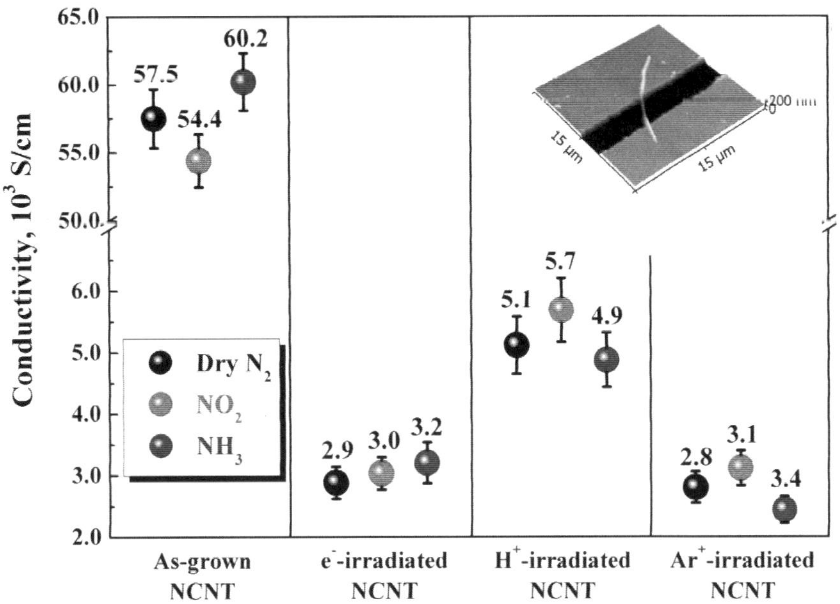

Figure 1. Mean values and range of electrical conductivity of individual nitrogen-doped CNTs (as-grown and irradiated by electrons, protons, and argon ions) during adsorption of ammonia and nitrogen dioxide molecules. Inset: 3D first pass of EFM image of individual CNT on gold electrodes.

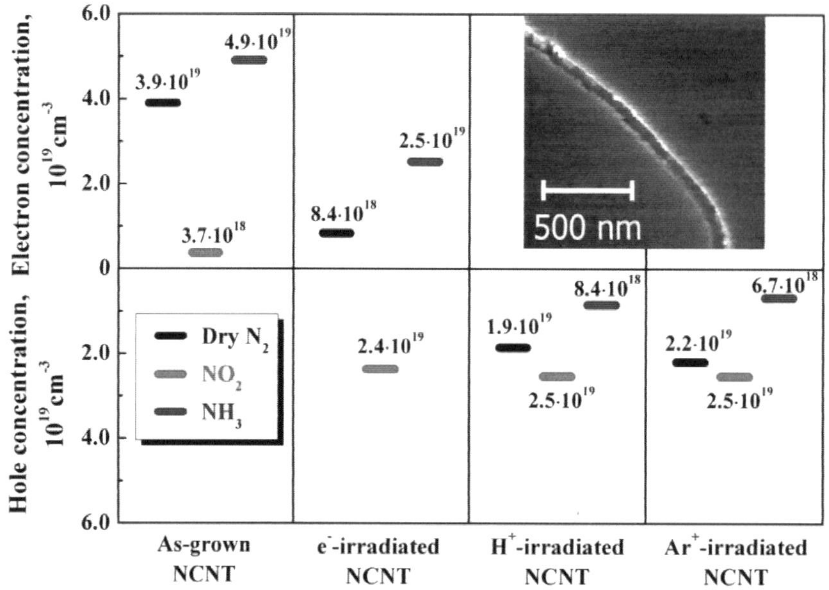

Figure 2. Mean values of free carrier concentration in individual doped CNTs (as-grown and irradiated by electrons, protons, and argon ions) during adsorption of ammonia and nitrogen dioxide molecules. Inset: EFM image of individual CNT at tip bias +5V.

4. Conclusion

Combination of two techniques of scanning probe microscopy (C-AFM and EFM) allows getting values of conductivity and carriers concentration of individual CNTs after gas exposition. The dependence of the change in conductivity of irradiated CNTs is due to the type of majority charge carriers. Using an independent measurement of conductivity and charge carrier concentration, it is possible to determine the change in mobility in CNTs after irradiation.

Irradiation of CNTs produces defects that decreases conductivity (by an order of magnitude) and creates molecular centres of adsorption. These centres can be used in composite structures for binding with additional components.

Acknowledgments

The work was carried out according to the state task of the OSC SB RAS (project registration number AAAA-A17-117041210227-8). This work was partially supported by the Russian Foundation for Basic Research (Grant 16-08-00763a).

References

[1] Williams T S, Orloff N D, Baker J S, Miller S G, Natarajan B, Obrzut J, McCorkle L S, Lebron-Colon M, Gaier J, Meador M A and Liddle J A 2016 *ACS Appl. Mater. Interfaces* **8** 9327–34

[2] Adjizian J-J, Leghrib R, Koos A A Suarez-Martinez I, Crossley A, Wagner P, Grobert N, Llobet E and Ewels P C 2014 *Carbon* **66** 662-73

[3] Sokolov D V, Davletkildeev N A, Bolotov and V V, Lobov I A 2017 *IOP Conf. Ser.: Mater. Sci. Eng.* **256** 012018

IOP Conf. Series: Materials Science and Engineering **443** (2018) 012035 doi:10.1088/1757-899X/443/1/012035

Investigation of opal nanostructures using scanning probe microscopy

A B Syritskii[1] and E V Panfilova[2]

[1] Department of Metrology and Interchangeability, Moscow State Technical University Bauman, Moscow, 105005, Russia
[2] Department of Electronic Technologies in Mechanical Engineering, Moscow State Technical University Bauman, Moscow, 105005, Russia

syritsky@bmstu.ru, panfilova.e.v@bmstu.ru

Abstract. The paper presents the research results on the creation of multilayer structures based on opal films using the methods of atomic force microscopy, tunnelling microscopy, and current spectroscopy. It is found that tunnelling microscopy methods are applicable for studying chromium-opal-gold-carbon layered structures. Surface images and current-voltage characteristics of each layer are shown. It is shown that the formation of film structures on the surface of opal matrix starts with "islands" on tops of silica spheres. It is found that the deposition of carbon films on the surface of the chrome-opal-gold structure leads to increase in tunnel currents in the probe-sample gap. The presented results may be used for the formation of a variety of layered structures and particle arrays on the surface of opal matrix, e.g., photonic devices, sensitive elements, and emission devices.

1. Introduction

Possible application areas of opal films are not only photonics, but also plasmonics, sensor technology, emission, laser and microwave technology, medicine, and a number of other ones. Development of technologies for the formation of appropriate structures requires a significant number of different studies. Therefore, it is necessary to elaborate the equipment, methods, and techniques for comprehensive studies of the formed structures [1,2].

One method of research is scanning probe microscopy (SPM), using techniques of atomic force (AFM) and scanning tunnelling microscopy (STM). SPM can be easily used because of the ability to place the sample in the atmosphere. The advantage of AFM is the ability to study almost any materials, and this method is widely used in the study of opal structures [3]. The advantage of the STM method is ultra-high resolution: up to 10^{-3} nm normally to the surface (vertical) and up to 10^{-2} nm along the axes of parallel surfaces (lateral). Modern equipment allows studying the topography and morphology of the surfaces, moving individual atoms, carrying out local chemical reactions, manipulating individual molecules, investigating the direction of the atom spins [4], displaying the density of states and the work function, obtaining the dependences I(U), I(Z), dI/dU and dI/dZ, where I – tunnel current, U – voltage in the gap, Z – the gap, and determining the chemical type of bond between the atoms of the object surface and the chemical composition of the surface layer of the object (current spectroscopy). The tunneling current recorded during scanning is sufficiently small (0.5 pA...50 nA), which makes it possible to study samples with low conductivity [5].

This paper presents the results of a layer-by-layer study by AFM and STM methods of a metal-dielectric film composition created on the basis of an opal film (opal sublayer). Such structures can be used to create giant Raman scattering substrates, emission and sensor devices.

2. Samples description

The choice of the films compositions was due to the prospect of creating emission structures based on the opal matrix. A schematic representation of the sample structure under study is shown in Figure 1. The opal film with the sizes of silica spheres 220 ... 250 nm and thickness of about 700 nm was formed by vertical pulling from colloidal liquid on the chromium sublayer deposited on the glass substrate by magnetron sputtering. The gold film with the thickness from 20 to 200 nm was applied to the opal film by magnetron sputtering. A carbon layer α-CH with the thickness from 10 to 100 nm was deposited on the gold surface by chemical deposition from the gas phase. The gold and carbon films thicknesses were controlled on a witness sample – a glass substrate. The chromium film provided the conductivity for the study on STM and acceptable adhesion of opal to the substrate. Opal layer served as a matrix for formation of the upper layers' topology.

The gold film was a catalyst for the growth of carbon nanostructures. The choice of layer thickness was determined by the results of previous studies [3]. When analysing STM images of opal films, it was found that acceptable image quality could be obtained only for a film with small thickness, consisting of one to three globular layers (200...750 nm). The choice of the thickness of gold films was determined by the revealed dependence of the gold layer surface topology deposited on the surface of the opal matrix on the thickness of the gold film measured on a smooth witness sample. Previously, it was shown that at the initial stage of film growth the metal was deposited on the tops of the opal matrix of silica spheres, then the film grew and the relief began to smooth, but to thicknesses of about 200 nm with a diameter of spheres 200...250 nm its relief repeated the relief of the matrix.

3. Results

The samples were studied using scanning probe microscope Solver P-47 (NT-MDT, Russia) by means of AFM in semi-contact mode and STM in topographic mode by direct current method and current spectroscopy mode. Minimum step STM-scanning was 0,006 nm. Previously it was found that the image of the relief of opal films on the metal sublayer depended on the tunnel voltage generated in the gap. This effect was most likely not related to mass transfer and was explained by the presence of emission processes. During scanning the voltage was from 0.1 to 1.0 V.

The results of the study indicate the possibility of studying metal-dielectric multilayer opal structures using STM. To be able to scan, the sum of the thickness of the opal, gold, and carbon layers should not exceed 1 μm. Moreover, it is possible to obtain both direct images of the sample surface and current-voltage characteristics.

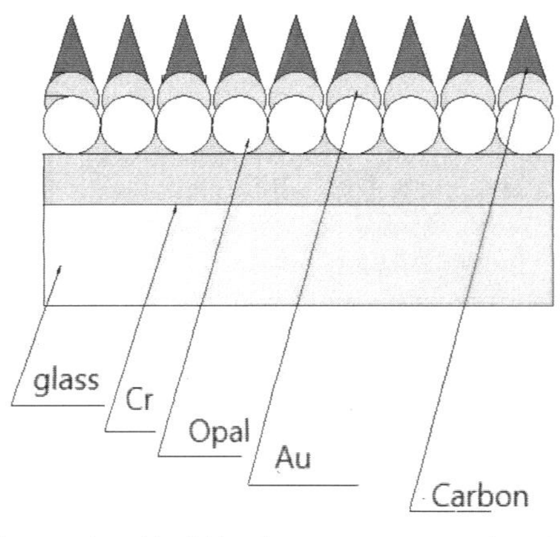

Figure 1. Chrome/opal/gold/carbon structure on the glass substrate.

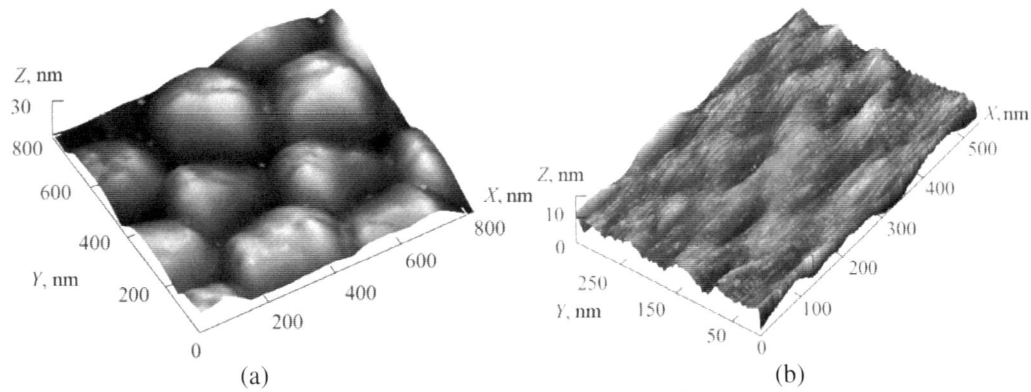

Figure 2. Three-dimensional view of multilayer structure relief obtained by (a) AFM, (b) STM.

Image quality depended on the thickness of individual films. It was found that for the compositions of chromium-opal-gold STM method was available at the gold films thickness from 100 to 200 nm. This fact is confirmed by the mechanism of formation of metal films on the opal matrix presented in Ref. [3]. When the thickness reached 100 nm, interspherical voids began to be filled with the deposited material, reaching at a small thickness of the opal film of the chromium sublayer; the metal "islands" formed at the tops of the spheres grew and merged, starting to form a continuous film. Tunnel currents increased, and the sample became available for study by STM. The structure and, hence, conductivity of the carbon layer also depended on the thickness. Thus, for the samples under consideration, the quality of the images of the surface of the composition chromium-opal-gold-carbon with a thickness of the upper carbon film of 100 nm was better than that of the thinner films and close to the quality of the image obtained by AFM. For comparison, Figures 2 and 3 present three-dimensional views, surface images, and profiles of such a sample obtained by AFM and STM.

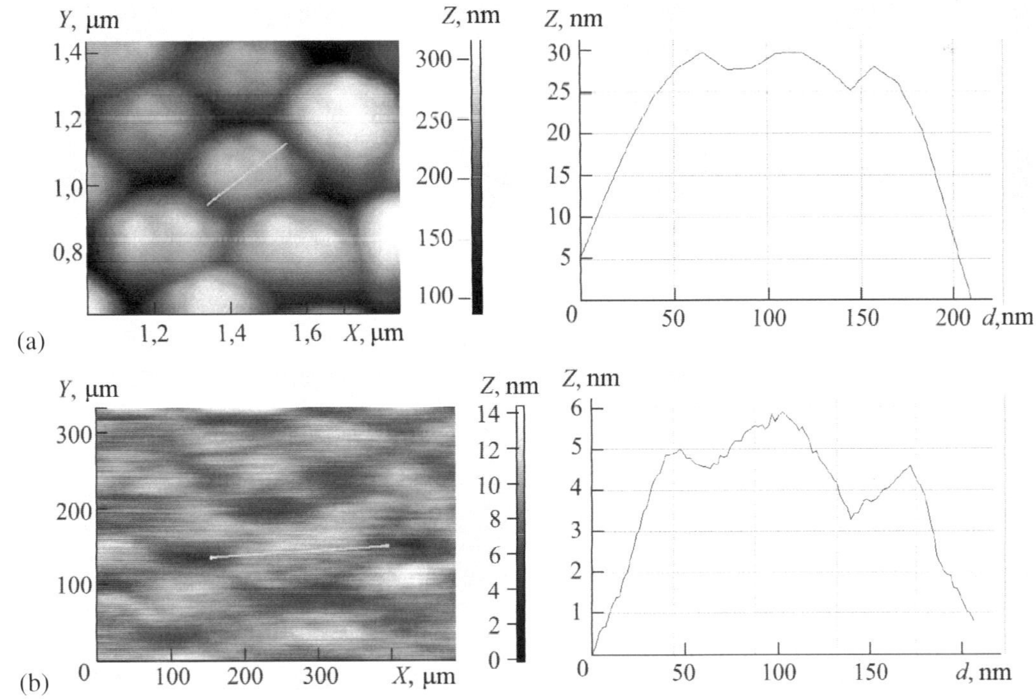

Figure 3. Surface images and profiles of carbon structures obtained by (a) AFM, (b) STM.

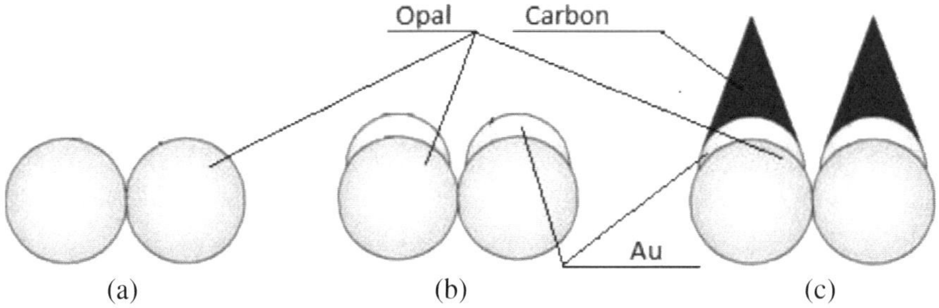

Figure 4. Scheme of the layers' formation of the film composition. The sample deposited on the chromium sublayer: (a) opal film, (b) gold film formed on the opal film, (c) carbon film formed on the gold film.

4. Discussion

The schematic representation of the studied structures and the systematized results of the layer-by-layer study of the samples are presented in Figures 4-6.

Comparison of the surface films formed of opal and the layer of gold once again confirmed previously identified mechanism of growth of metal films on the surface of opal matrix [3]. The study of images and profiles of the surface of the carbon deposited on the gold layer showed that carbon structures were formed mainly in areas located above the tops of the silica spheres on the "islands" of gold, which led to an increase in the elevation of these areas. The carbon layer has a "bumpy" surface, which is noticeable on the surface images (Figs. 2a, 3a) and on relief profiles (Figs. 3, 5c). At the same time, the relief of the peaks became more developed. Moreover, the application of carbon not from the gas phase, but by magnetron method, performed for comparison on some samples, led to the formation of a surface with a similar nature of the relief.

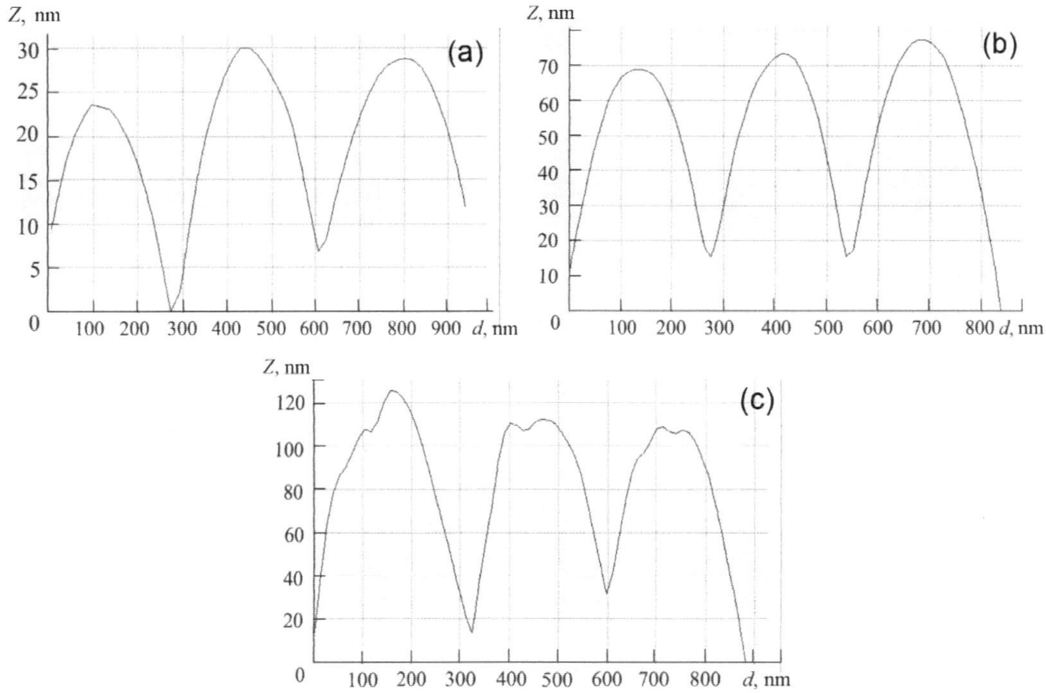

Figure 5. AFM-profiles of the surface topography: (a) opal film, (b) gold film formed on the opal film, (c) chromium film formed on the gold film.

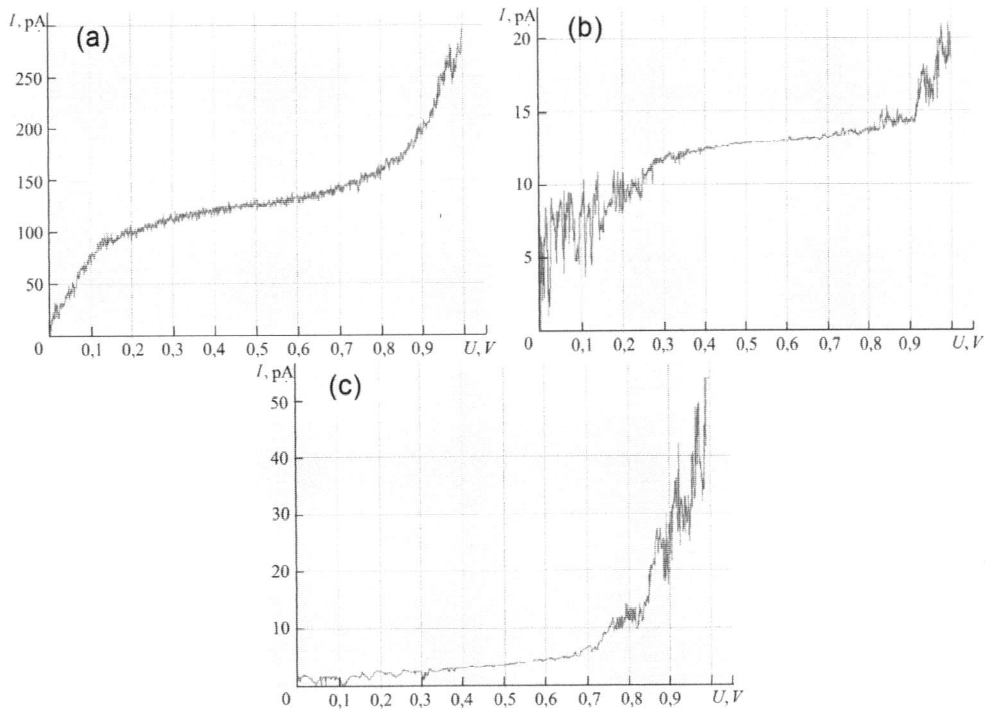

Figure 6. I-V curves of the gaps:(a) tip-opal film, (b) tip-gold film, (c) tip-chromium film.

When analysing current-voltage characteristics of the samples and the film compositions of the individual layers, it was established that I-V curve demonstrated hysteresis. Appearance of a small hysteresis can be attributed to the nonlinearity of piezoceramics. I-V curve analysis confirms the dependencies found in topographic studies. For opal films, the I-V curve has a nonlinear symmetric form typical for the metal-dielectric contact. The deposition of gold with a thickness of 100 nm or more on the surface of the opal film retains the nature of the dependence, but leads to an increase in the magnitude of the currents. Subsequent carbon deposition entails a change in the nature of the U(I) dependence, expressed more strongly for samples with a greater thickness of the carbon film in the studied thickness range. Quantitative assessment of the amplification of the electric field on the samples with carbon films with a thickness of 100 nm on the witness, made in the coordinates of the Fowler-Nordheim, testifies that at a voltage of 1.0 V for a number of points, the gain field in the gap between tip and substrate reached values of 400...420.

5. Conclusion

The results of the work indicate the prospects provided by scanning probe microscopy for the study of topography and electrophysical properties of structures based on opal films. The advantage of the method is the ability to conduct comprehensive studies of the formed structures. The authors see the development of the work in the creation and study of emission structures formed on the opal sublayer.

References

[1] Yankovskii G M,Komarov A V, Puz'ko R S, Baryshev A V, Afanas'ev K N, Boginskaya I A, Bykov I V, Merzlikin A M, Rodionov I A and Ryzhikov I 2016 *Phys. Solid State* **58** 2503-10
[2] Shakhnov V A, Vlasov A I and Tokarev S V 2016 *IOP Conf. Ser.: Mater. Sci. Eng.* **151** 012005
[3] PanfilovY V 2017 *MATEC Web of Conference* **129** 02019
[4] Deulin EA et al. 2010 *Mechanics and Physics of Precise Vacuum Mechanisms. Fluid Mechanics and its Applications* **91** (Dordrecht: Springer) p 234
[5] Komshin A S and Orlova S R 2016 *Meas. Tech.* **59** 589-94

IOP Conf. Series: Materials Science and Engineering **443** (2018) 012036 doi:10.1088/1757-899X/443/1/012036

Formation of ZnO memristor structures by scratching probe nanolithography

R V Tominov, V A Smirnov, V I Avilov, A A Fedotov, V S Klimin and N E Chernenko

Institute of Nanotechnologies, Electronics and Equipment Engineering, Southern Federal University, Taganrog, 347928, Russia

roman.tominov@gmail.com

Abstract. This work presents the results of the formation and investigation of Al_2O_3/ZnO:In/ZnO/Ti memristor structures. It is shown that using Ti layer can improve memristive effect. Resistive switching from high resistance state (HRS) to low resistance state (LRS) occurred at 2.1±0.3 V, and from LRS to HRS at -1.5±0.3 V. Endurance test showed that HRS was 5.51±0.13 GΩ and LRS was 0.25±0.05 GΩ. HRS/LRS coefficient equalled to 22. The results can be useful for micro- and nanoelectronics elements manufacturing, as well as micro- and nanosystem engineering using probe nanotechnologies and for ZnO- based RRAM fabrication.

1. Introduction

Development of the next generation computer memory is the promising direction of electronics. One of the promising types of memory is non-volatile resistive random access memory (RRAM) based on metal oxide memristor structures; it has high-density integration, low-power consumption, and fast write/read operations. Zinc oxide (ZnO) is the promising material, which is widely used in electronic element developments, sensors, and microsystem technology. ZnO demonstrates memristor effect and is compatible with semiconductor technology [1-6]. To fabricate ZnO-based RRAM, it is necessary to carry out prototyping and investigations of ZnO memristor structures. Thus, it is necessary to develop new nanolithography techniques, which allow fabricating the structure of RRAM elements at lower expenses and within shorter periods than conventional methods of photo- and electron-beam lithography [7,8]. One of the promising methods for the formation of nanoscale structures is scratching probe nanolithography (SPN) of atomic force microscope (AFM) [9-11]. The SPN method involves the modification of thin polymer films by the formation of profiled nanosized structures using the tip of the AFM probe. Through the profiled nanosized structures, various technological operations (deposition and etching) could be performed. The advantages of SPN include high resolution and the absence of physical templates. Various structures of the elements of nanoelectronics and microsystem engineering were formed using SPM [12-15].

The aim of this work is fabrication of ZnO thin film-based memristor structures using scratching probe nanolithography, and investigation of memristor effect on them.

2. Experiment details

ZnO thin film was grown using pulsed laser deposition technique (Fig. 1a). Al_2O_3/ZnO:In was used as a wafer. Deposition was performed under the following conditions: wafer temperature 400°C, target-wafer distance 50 mm, O_2 pressure 1 mTorr, and pulse energy 300 mJ. The solution of photoresist/thinner (FP-383/RPF383F) at volume ratios of 1:10 was transferred onto ZnO using the centrifugal method at the rotation speed of a Laurell WS-400B-6NPP centrifuge 5000 rpm (Fig. 1b). After the deposition of the film, the photoresist/thinner film was dried at a temperature of 90°C for 25 min. Thickness of the photoresist/thinner film was equal to 75.1±3.3 nm.

Scratching probe nanolithography (SPN) on the photoresist/thinner film was performed using a Solver P47 Pro scanning probe microscope (NT-MDT, Russia) (Fig. 1c). Thus, array of the 9 squared nanostructure-grooves was formed (Fig. 1d). Then thin Ti film was deposited using BOC Edwards Auto 500 system (Fig. 1d). After that lift-off process was applied using dimethylformamide (Fig. 1e).

Electric measurements of the Al_2O_3/ZnO:In/ZnO/Ti structure was carried out using Probe Nanolaboratory Ntegra (NT-MDT, Russia). ZnO:In film was grounded during measurements. W_2C AFM probe was used as a top electrode (Fig. 1f). AFM images were obtained in semicontact mode.

3. Results and discussion

Figure 2 shows experimental investigation of ZnO films morphology. It is shown that ZnO film surface has a granular structure with 0.35±0.26 nm roughness (Fig. 1b). The ZnO film thickness was investigated using AFM by scanning bottom electrode/ZnO film boundary and was equal to 10.2±3.4 nm.

Figure 3 shows experimental investigation of Ti structures morphology. AFM profiles of all 9 Ti structures were obtained (Fig. 3b) and it was shown that thickness of Ti structures was 2.5±0.4 nm. Ripped edges of Ti structures are the result of lift-off process.

Figure 4 shows current-voltage characteristic (CVC) of Al_2O_3/ZnO:In/ZnO/Ti/W_2C structure. at –3 to +3 voltage sweep. It was shown, that Al_2O_3/ZnO:In/ZnO/Ti/W_2C structure had nonlinear bipolar behaviour when the electric potential gradient was the dominant parameter of memristive effect [16]. Resistive switching from high resistance state (HRS) to low resistance state (LRS) occurred at -3.0 V (SET), and from LRS to HRS – at 2.1 V (RESET) (Fig. 4a).

Investigation of memristive effect uniformity of Al_2O_3/ZnO:In/ZnO/Ti/W_2C structure in a single point (endurance test) showed that HRS was 5.51±0.13 GΩ and LRS was 0.25±0.05 GΩ (Fig. 4b). It was shown that HRS/LRS coefficient was equal to 22 at –3 to +3V voltage sweep. Read voltage was -2V. HRS/LRS = 22 is sufficient for RRAM fabrication [17].

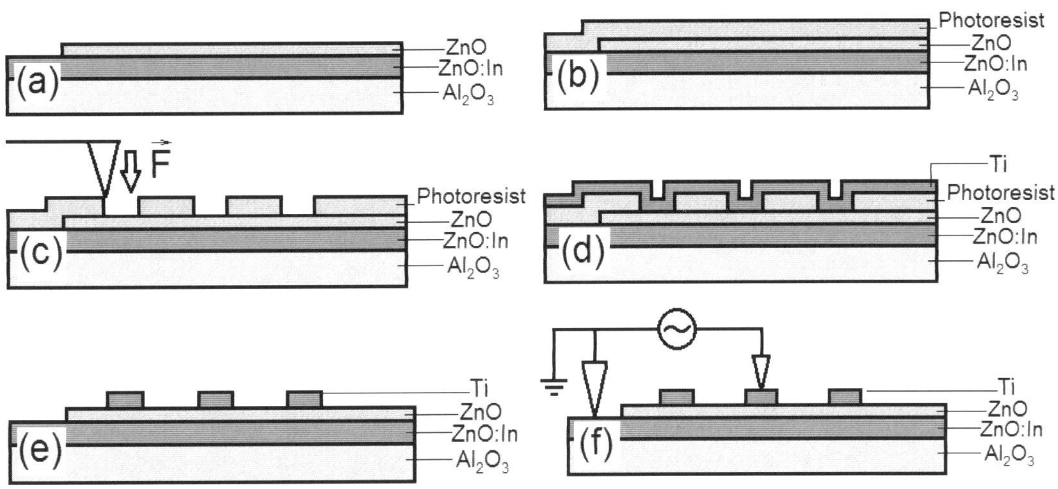

Figure 1. Process diagram of formation and investigations of Al_2O_3/ZnO:In/ZnO/Ti memristor structures: (a) ZnO film growth; (b) deposition of photoresist to the ZnO film surface; (c) nanostructures formation on the photoresist surface using scratching probe nanolithography; (d) Ti film deposition; (e) lift-off process; (f) electric measurements.

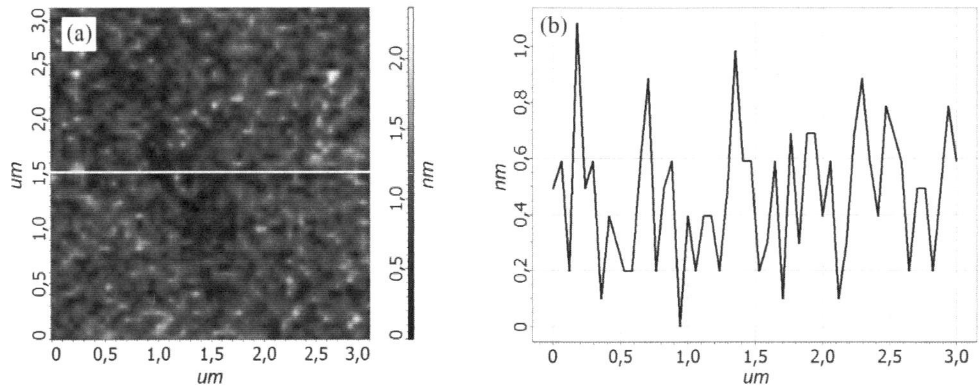

Figure 2. Photoresist surface morphology: (a) AFM-image; (b) AFM profile along line on (a).

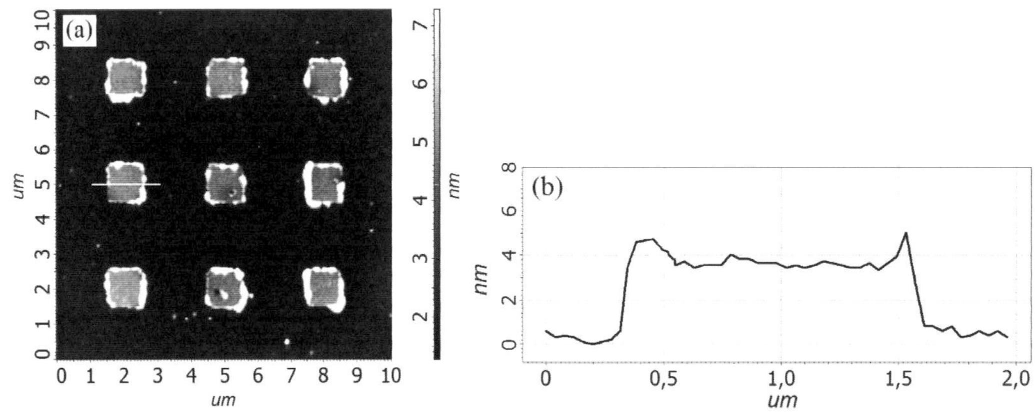

Figure 3. 3×3 array of Ti structures: (a) AFM-image; (b) AFM profile along line on (a).

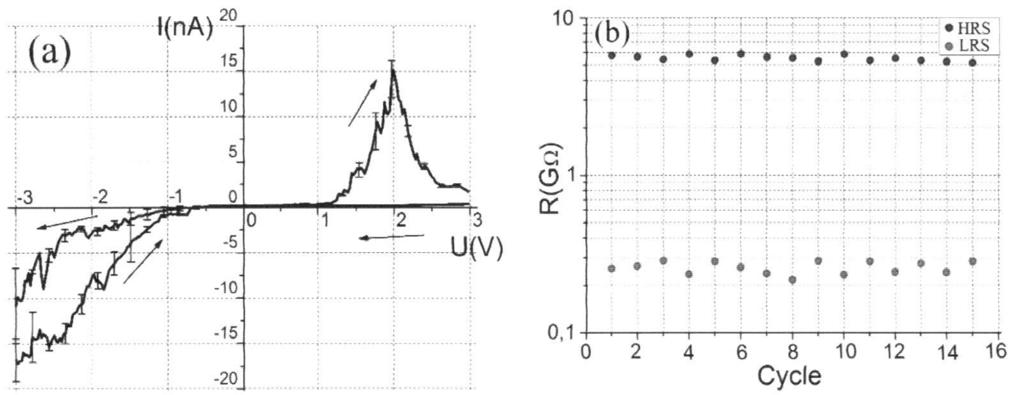

Figure 4. Electric measurements of Al_2O_3/ZnO:In/ZnO/Ti/W_2C structure: (a) current-voltage characteristic at –3 to +3V voltage sweep; (b) endurance test.

To evaluate Ti film impact to memristor effect in Al_2O_3/ZnO:In/ZnO/Ti/W_2C structure, electric measurements in Al_2O_3/ZnO:In/ZnO/W_2C structure were carried out. Endurance test of Al_2O_3/ZnO:In/ZnO/W_2C structure showed that HRS was 2.16±0.17 GΩ and LRS was 0.31±0.06 GΩ. Resistive switching from HRS to LRS occurred at -3.0 V, and from LRS to HRS – at 2.5 V. HRS/LRS coefficient was equal to 7 at –3 to +3V voltage sweep. Read voltage was also -2 V. HRS/LRS = 7 is insufficient for RRAM fabrication [17].

Analysis of the results shows that the addition of Ti layer in the Al_2O_3/ZnO:In/ZnO/Ti/W_2C structure leads to the increase in the HRS/LRS coefficient and to the decrease in RESET voltage.

The increase in the HRS/LRS coefficient may be due to the increase in the additional concentration of oxygen ions at the Ti/ZnO interface, which leads to an increase in the quantity of recombining pairs "oxygen vacancy-oxygen ion" during the transition of the film from LRS to HRS, which in turn leads to an increase in the length of the destroyed section of the nanosized conduction channel.

4. Conclusion

In summary, we fabricated and investigated Al_2O_3/ZnO:In/ZnO/Ti memristor structures. It was shown that using the Ti layer could improve memristive effect. Resistive switching from high resistance state (HRS) to low resistance state (LRS) occurred at -3.0 V, and from LRS to HRS – at 2.1 V. Endurance test showed that HRS was 5.51±0.13 GΩ and LRS was 0.25±0.05 GΩ. HRS/LRS coefficient was equal to 22. The results can be useful for micro- and nanoelectronics elements manufacturing, as well as micro- and nanosystem engineering using probe nanotechnologies, and for ZnO- based RRAM fabrication.

Acknowledgments

This work was supported by Grant of the President of the Russian Federation No.MK-2721.2018.8, by RFBR according to the research project № 18-37-00299 and the Southern Federal University (grant VnGr-07/2017-26). The results were obtained using the equipment of the Research and Education Centre and Centre for Collective Use "Nanotechnologies" of Southern Federal University.

References

[1] Ageev O A, Smirnov V A, Solodovnik M S, Rukomoikin A V and Avilov V I 2012 *Semiconductors* **46** 1616-21
[2] Ageev O A, Konoplev B G, Rubashkina M V, Rukomoikin A V, Smirnov V A and Solodovnik M S 2013 *Nanotechnologies in Russia* **8** 23-8
[3] Avilov V I, Ageev O A, Smirnov V A, Solodovnik M S and Tsukanova O G 2015 *Nanotechnologies in Russia* **10** 214-9
[4] Ageev O A, Alyab'eva N I, Konoplev B G, Polyakov V V and Smirnov V A 2010 *Semiconductors* **44** 1703-8
[5] Ageev O A, Klimin V S, Solodovnik M S, Eskov A V and Krasnoborodko S Y 2016 *J. Phys.: Conf. Ser.* **741** 012178
[6] Avilov V I, Ageev O A, Kolomiitsev A S, Konoplev B G, Smirnov V A and Tsukanova O G 2013 *Semiconductors* **106** 50-7
[7] Khakhulin D A, Vakulov Z E, Smirnov V A, Tominov R V, Yoon J-G and Ageev O A 2017 *J Phys.: Conf. Ser.* **9** 917
[8] Avilov V I, Polupanov N V, Tominov R V, Smirnov V A and Ageev O A 2017 *IOP Conf. Ser.: Mat. Sci. Eng.* **256** 012001
[9] Bizyaev D A, Bukharaev A A, Lebedev D V, Nurgazizov N I and Khanipov T F 2012 *Tech. Phys. Lett.* **7** 645-8
[10] Smirnov V A, Tominov R V, Chernenko N E and Ageev O A 2017 *Nanotechnologies in Russia* **12** 650-7
[11] Akhavan O and Abdolaha M 2009 *Appl. Surf. Sci.* **255** 3513-7
[12] Tang Q,Shi S Q and Zhou L 2004 *J. Nanosci. Nanotechnol.* **4** 948-63
[13] Garcia R, Armin W and Riedo E 2014 *Nat. Nanotech.* **9** 577-87
[14] Sung I H and Kim D E 2005 *Appl. Surf. Sci.* **239** 209-21
[15] Chen Y J, Hsu J H and Lin H N 2005 *Nanotechnology* **16** 1112-5
[16] Yang J J and Pickett M D 2008 *Nat. Nanotech.* **3** 429-33
[17] Waser R, Dittmann R, Staikov G and Szot K 2009 *Advanced materials* **21** 2632-63

IOP Conf. Series: Materials Science and Engineering **443** (2018) 012037 doi:10.1088/1757-899X/443/1/012037

Microstructure of $(Ba_{0.75},Sr_{0.25})TiO_3$ based glass-ceramics doped by Mn

A P Turygin[1], A S Abramov[1], D O Alikin[1], D S Chezganov[1], A A Esin[1], I S Baturin[1], X Song[2], T Zhang[2], Y Zhang[2], K Hu[2], Z Zhao[2] and V Ya Shur[1]

[1] School of Natural Sciences and Mathematics, Ural Federal University, Ekaterinburg, 620000, Russia

[2] Beijing Key Laboratory of Fine Ceramics, State Key Laboratory of New Ceramics and Fine Processing, Institute of Nuclear and New Energy Technology, Tsinghua University, Beijing 100084, PR China

anton.turygin@urfu.ru

Abstract. We studied microstructure, surface morphology, and domain structure of $(Ba_{0.75},Sr_{0.25})TiO_3$ based glass-ceramics with different amount of Mn additive (from 0 to 0.5%) prepared from melted and quenched mixed powders using several microscopic methods. The as-quenched sample was annealed and subjected to a controlled crystallization in air for 2h at temperatures 850 and 950°C. It was shown that the addition of the Mn up to 0.5% had no impact on morphology and domain structure. The dendrite structure was revealed in the samples crystallized at 850°C. The geometry of the clusters was analysed in terms of fractal approach. The faceted grains with size about 150 nm were found in ceramics annealed at 950°C. Individual grains demonstrated non-uniform piezoresponse, which could be attributed to existence of domain structure.

1. Introduction

Glass-ceramics representing ferroelectric grains embedded in a glass matrix are of significant interest for power electronic applications, especially for energy storage devices [1,2]. It has been shown that the application of the glass-ceramics based on the ferroelectric materials can improve energy storage properties by the utilization of a high dielectric constant of ferroelectrics and relatively high breakdown voltage of glass [2]. It is known that the aluminosilicate glass-ceramics based on $(Ba,Sr)TiO_3$ (BST) are one of the most promising materials for the energy storage applications [3]. An addition of SrO is used for tuning the Curie temperature and increasing the dielectric constant at room temperature.

It has been shown that the energy storage properties of the BST glass-ceramics depend on their microstructure [4,5]. Different additives (AlF_3, La_2O_5, MnO_2, etc.) were used to control the microstructure, thus optimizing the dielectric properties [5].

The effect of space charges (interfacial polarization) at the interphase boundaries between crystalline grains and glass on the breakdown voltage and the dielectric properties has been discussed [6-8]. At the same time, properties of the ferroelectric materials strongly depend on the domain structure, but little is known about the domain structure of the BST glass-ceramics. The purpose of the present paper is to study the domain structure and its relation with the microstructure of the BST glass-ceramics with different sintering conditions and concentrations of Mn additive.

2. Experimental

The samples of the BST based glass-ceramics with different concentrations of Mn additive were prepared from melted and quenched mixed powders. The compositions of the studied glass-ceramics are presented by a chemical formula (27.75-0.75x) $BaCO_3$-(9.25-0.25x) $SrCO_3$-29 TiO_2-20 SiO_2-14 Al_2O_3 - x MnO_2 (mol %) with x = 0.1-0.5. The compositions were prepared from well-mixed powders of $BaCO_3$, $SrCO_3$, TiO_2, SiO_2, Al_2O_3, and MnO_2. All raw materials were prepared by a ball milling in ethanol media using zirconia balls in polyethylene jars. After drying, the powder mixture was melted in a platinum crucible at 1550ºC for 3 h. The melt was poured into a preheated Ø30 mm×20 mm Cu mould. The casted glass was immediately annealed at 650ºC for 3 h to relieve residual stresses. Finally, prepared samples with different Mn concentrations were subjected to a controlled crystallization in air at 850 and 950°C for 2h with the heating rate of 5°C/min to convert the glass into the glass-ceramics. A detailed description of the BST glass-ceramics with Mn additives preparation and its dielectric properties was published [5].

The sample surface was polished by diamond paste with decreasing of grain sizes from 6 to 0.25 μm. The final polishing was done by colloidal silica (SF1 Polishing Suspension, Logitech, UK). The sample thickness after polishing was about 400 μm.

The microstructure of the glass-ceramic samples was studied by a scanning electron microscope (SEM, Merlin, Carl Zeiss, Germany) with field-emission using a secondary electron imaging with charge compensation.

P-E hysteresis loops were measured using FE module of AixACCT TF 2000 system (aixACCT Systems GmbH, Germany). Bipolar triangular voltage pulses were applied to the sample and the current was measured using a virtual ground transimpedance amplifier. For measurement, silver paste electrodes with area about 10 mm^2 were used.

The visualization of the domain structure with high spatial resolution was performed using piezoresponse force microscopy (PFM). The samples were mounted by conductive silver paste to metal disks. PFM measurements were realized by a scanning probe microscope NTEGRA Aura (NT-MDT, Russia). The silicon HA_NC/W2C tips (ScanSens, Russia) with a tungsten carbide conductive coating with a radius of curvature R = 35 nm, resonance frequency f = 140 kHz, and spring constant k = 3.5 N/m were used. Piezoresponse measurements were realized by the application 5 V AC far from contact resonance frequency to the SPM tip.

Polarization reversal was done using NI-6251 multifunction data acquisition board and high-voltage amplifier Trek-677B by application of rectangular DC pulses. The local switching was realized at the ambient conditions with relative humidity about 40% [9,10].

3. Results and discussion

Investigation of the BST glass-ceramics microstructure by SEM allowed revealing dendrite-like clusters of crystal phase in the samples annealed at 850°C (Fig. 1a,b). The clearly visible cluster boundaries resemble Voronoi diagram. The average cluster size about 5 μm (Fig. 1c) does not depend on doping level. The formation of dendrite structures in glass-ceramics with addition of AlF_3 and MnO_2 were reported [5,11], but not clearly identified. In contrast, the faceted grains with size about 150 nm were found in samples annealed at 950°C. The structure type and grain sizes did not demonstrate significant dependence on the doping level (Fig. 1d-e).

Detailed study of the samples annealed at 850°C allowed revealing the individual branches in clusters with width below 100 nm (down to 30 nm) (Fig. 2a,b). The structure of the crystal phase clusters is similar to the famous clusters formed by diffusion limited aggregation (DLA). The number of cluster particles on radius *r* from the centre is [11]:

$$N(r) = N_0 \left(\frac{r}{R_g}\right)^D f\left(\frac{r}{R_g}\right), \tag{1}$$

where N_0 is total number of particles , D is fractal dimension, $R_g = \left({N_0}^{-1} \sum_i r_i^2\right)^{1/2}$ is gyration radius, and $f(x) = \{1\ for\ x < 1, x^{-D}\ for\ x > 1\}$ is transition function.

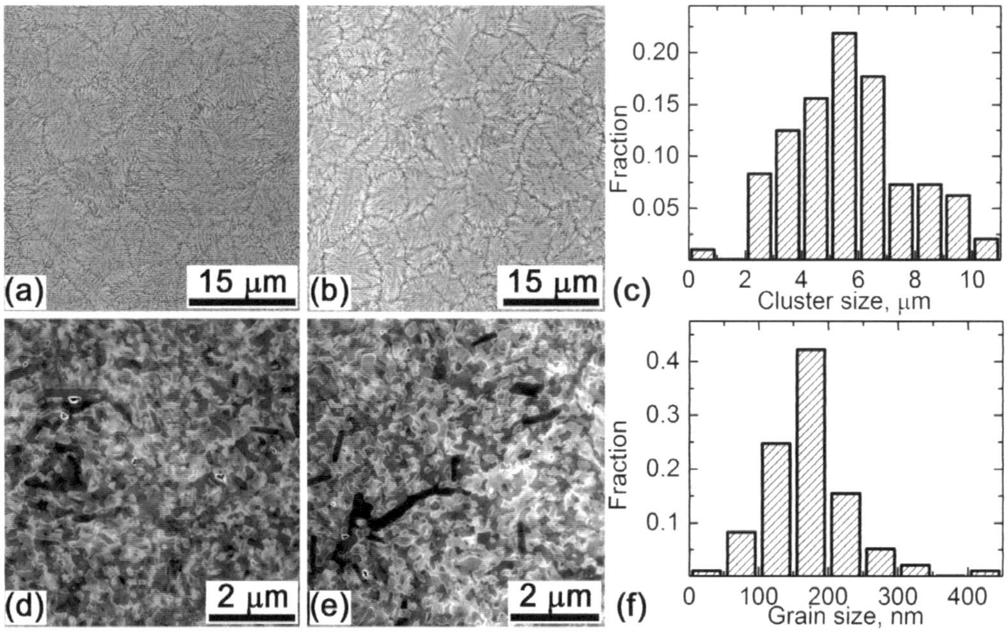

Figure 1. SEM images of BST glass ceramics doped by Mn annealed at 850°C: (a) 0.1%, (b) 0.5%, (c) histogram of cluster size distribution; at 950°C: (d) 0.1%, (e) 0.5%, (f) histogram of grain size distribution.

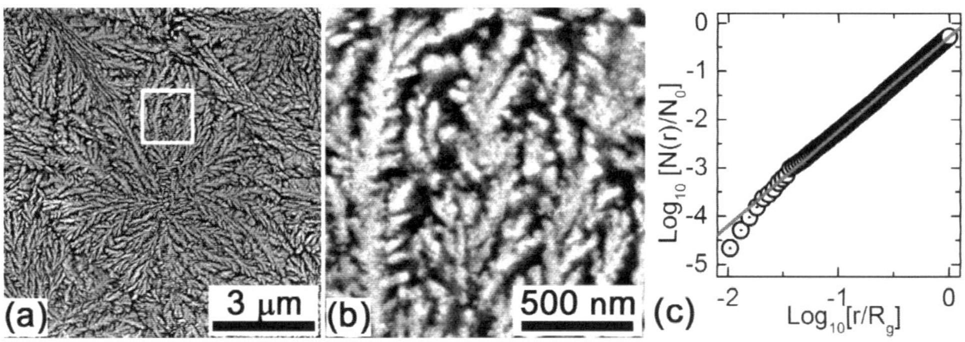

Figure 2. High resolution SEM images of BST ceramics doped by 0.1% Mn annealed at 850°C: (a) individual cluster, (b) zoomed region, (c) dependence of number of particles in cluster on radius.

Although dendrite clusters are three-dimensional, only the surface two-dimensional SEM image can be analysed (Fig. 2a). The analysis has been done after binarization of the image and fractal dimension was calculated using Eq. (1) (Fig. 2c). The obtained D = 1.95 is sufficiently larger than typical fractal dimension for DLA clusters (D = 1.70) [12]. The analysis of the same image by Box Counting [13] method gave the similar result (D = 1.94).

The P-E loops in the studied samples demonstrated different behaviour for different annealing temperatures (Fig. 3). The value of remnant polarization (P_r) increased significantly with annealing temperature. Hysteresis loop for sample annealed at 850°C was narrow with P_r = 0.3 C/cm^2, whereas in sample annealed at 950°C the loop was wider with P_r = 0.8 C/cm^2. The lower value of P_r in the sample annealed at 850°C can be attributed to lower fraction of the ferroelectric phase in this sample.

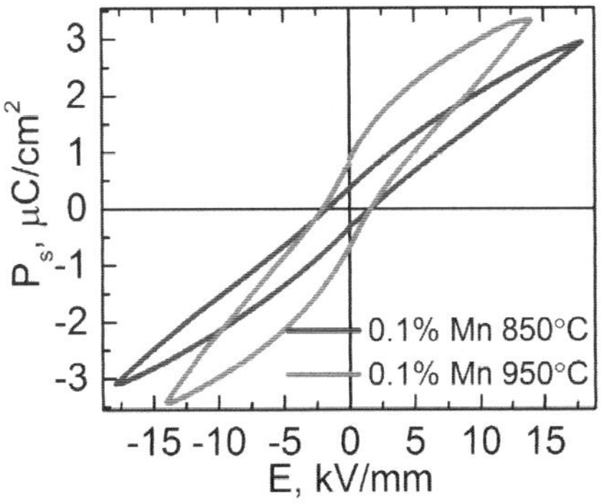

Figure 3. P-E hysteresis loops for the BST glass ceramics with different annealing temperatures. The measurements were made at 1 Hz.

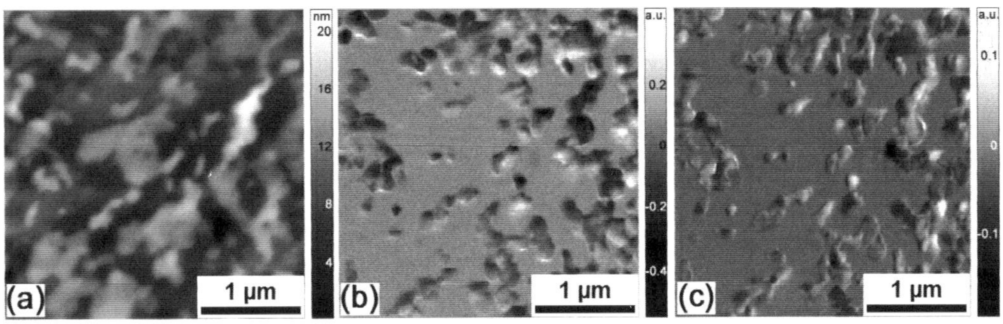

Figure 4. PFM images ($R \cdot \cos \Theta$) of domain structure in BST glass-ceramics, annealed at 950°C: (a) topography, (b) out of plane, (c) in plane.

The PFM contrast was revealed only in the samples annealed at 950°C (Fig. 4). The non-polar glass matrix and polar grains associated with ferroelectric phase can be clearly distinguished on PFM images (Fig. 4). The individual grains demonstrated spatially non-uniform piezoresponse, which can be attributed to existence of the domain structure. The calculated fraction of the non polar phase [14] averaged over several images was about 0.5.

No significant piezoresponse was measured in the samples annealed at 850°C. The obtained signals can be associated only with the surface relief (topography). Absence of piezoresponse contrast correlates with narrow P-E loops in these samples. Application of the single voltage pulse with amplitude up to ±200 V and duration 60 s did not allow realizing local polarization reversal. This fact can be attributed to effective bulk screening [15] or domain wall clamping by glass matrix [16,17].

4. Conclusion

The statistical information about grain sizes was obtained for the BST glass-ceramics with different Mn concentrations. It was shown that the increase of the Mn concentration up to 0.5% did not lead to visible changes in morphology or domain structure. It was found that crystallization temperature had a critical role on ceramics morphology. The dendrite-like clusters with extremely high value of fractal dimension were found in glass ceramics crystallized at 850°C. The obtained information about the domain structure can be used for further improvement of the dielectric properties of BST glass-ceramics.

Acknowledgements
The equipment of the Ural Centre for Shared Use “Modern Nanotechnology” Ural Federal University was used. The research was made possible by Russian Foundation for Basic Research (grant 18-52-53032) and state task of Ministry of education and science of the Russian Federation (No. 3.4993.2017/6.7). The work was supported partially by Government of the Russian Federation (act 211, agreement 02.A03.21.0006).

References
[1] Gorzkowski E P, Pan M, Bender B and Wu C C M 2007 *J. Electroceramics* **18** 269-76
[2] Herczog A 1973 *IEEE Trans. Parts, Hybrids, Packag.* **9** 247-56
[3] Fletcher N H, Hilton A D and Ricketts B W 1996 *J. Phys. D. Appl. Phys.* **29** 253-8
[4] Chen J, Zhang Y, Deng C, Dai X and Li L 2009 *J. Am. Ceram. Soc.* **92** 1350-3
[5] Chen J, Zhang Y, Deng C and Dai X 2009 *J. Am. Ceram. Soc.* **92** 1863-6
[6] Zhang Y, Ma T, Wang X, Yuan Z and Zhang Q 2011 *J. Appl. Phys.* **109** 084115
[7] Huang J, Zhang Y, Ma T, Li H and Zhang L 2010 *Appl. Phys. Lett.* **96** 042902
[8] Liebault J, Vallayer J, Goeuriot D, Treheux D and Thevenot F 2001 *J. Eur. Ceram. Soc.* **21** 389-97
[9] Ievlev A V, Morozovska A N, Shur V Ya and Kalinin S V 2014 *Appl. Phys. Lett.* **104** 092908
[10] Shur V Ya, Ievlev A V, Nikolaeva E V, Shishkin E I and Neradovskiy M M 2011 *J. Appl. Phys.* **110** 052017
[11] Wang X, Zhang Y, Ma T, Yuan Z and Zhang Q 2013 *Int. J. Appl. Ceram. Technol.* **10** 301-6
[12] Meakin P 1991 *Rev. Geophys.* **91** 317-54
[13] Feder J 1988 *Fractals* (Plenum Press, NY, 1988)
[14] Alikin D O, Turygin A P, Walker J, Rojac T, Shvartsman V V, Shur V Ya and Kholkin A L 2015 *J. Appl. Phys.* **118** 072004
[15] Shur V Ya 2006 *J. Mater. Sci.* **41** 199-210
[16] Robels U and Arlt G 1993 *J. Appl. Phys.* **73** 3454-60
[17] Patel S, Chauhan A, Kundu S, Madhar N A, Ilahi B, Vaish R and Varma K B R 2015 *AIP Adv.* **5** 087145

IOP Conf. Series: Materials Science and Engineering **443** (2018) 012038 doi:10.1088/1757-899X/443/1/012038

Domain structure imaging in PMN-PT crystals using channelling-contrast backscattered electron microscopy

E O Vlasov[1], D S Chezganov[1], L V Gimadeeva[1], A D Ushakov[1], Q Hu[2], X Wei[2] and V Ya Shur[1]

[1] School of Natural Sciences and Mathematics, Ural Federal University, 620000, Ekaterinburg, Russia

[2] Electronic Materials Research Laboratory, Key Laboratory of the Ministry of Education & International Centre for Dielectric Research, Xi'an Jiaotong University, 710049, Xi'an, China

evgeny.vlasov@urfu.ru

Abstract. We demonstrate the abilities of various SEM techniques for domain imaging in PMN-PT crystals with different compositions. It is shown that the imaging of the chemically etched relief is limited by its destructivity and potential contrast – by low resolution and contrast instability. The optimal parameters of the imaging by backscattered electron channelling allow obtaining the high-resolution domain imaging in PMN-PT crystals. The domain structure change as a result of the phase transition are imaged. The correlation between the images of the surface domain structure obtained by scanning electron microscopy and piezoresponse force microscopy is demonstrated.

1. Introduction

The lead magnesium niobate-lead titanate (1-x)Pb($Mg_{1/3}Nb_{2/3}$)O_3-xPbTiO_3 (PMN-PT) crystals are well known for their outstanding physical properties, such as high electromechanical and piezoelectric coefficients, high strain, and low dielectric losses [1,2]. It makes PMN-PT crystals promising for various applications including ultrasonic transducers, piezoelectric actuators, etc. Moreover, PMN-PT is a potential candidate for non-linear optics and photonics.

The domain imaging by several microscopic techniques with a high spatial resolution is important for studying the domain structure produced by various treatments [3-5]. The domain imaging by scanning electron microscopy (SEM) possesses high resolution and comparatively fast image acquisition. The method can be used in wide temperature range and allows visualizing slow domain evolution [6]. SEM was used for imaging of the domain structure of bulk ferroelectric single crystals [7-10], ceramics [11,12], and thin films [13]. The imaging was realized via several techniques, such as visualization of etched domains [10,14], secondary electron imaging (potential contrast) [7,15], electron beam stimulated polarization currents [9], electron backscatter diffraction [16], and secondary and backscattered electron imaging using electron channelling [6,11,17-19].

The backscattered electron (BSE) mode via electron channelling is a relatively simple technique with resolution below 10 nm. BSE represents primary electrons, which after experiencing one or multiple scatterings in the sample, eventually emerge from the surface. Two types of BSE are distinguished: (1) high angle BSE as a result of Rutherford scattering and (2) low angle BSE realized by Mott scattering. The first one carries information about atomic number, while the second one is sensitive to the surface orientation. The domain contrast is based on the channelling of incident electron beam within domains of different orientations, which modulates the BSE emission and provides a contrast caused by crystallographic orientation. The method was applied for studying the domain structure change under mechanical stress [6,17] and electric field [18,19] in wide temperature range. Moreover, the electron channelling contrast is a powerful technique for imaging crystal defects, such as dislocations, stacking faults, twins, and grain boundaries [20].

In this paper, we demonstrate the first successive application of the BSE channelling mode for the domain imaging in PMN-PT crystals.

2. Experimental

We used the field emission scanning electron microscope (FE-SEM) Merlin (Carl Zeiss NTS GmbH) with Schottky field emitter. The domain structures were visualized using four quadrant angular selective BSE detectors with accelerating voltage 1-20 kV and electron beam current from 100 pA to 2 nA.

We studied the PMN-PT crystals grown by modified Bridgman technique of various compositions with different phases at the room temperature: rhombohedral (R), tetragonal (T), and morphotropic phase boundary (MPB) [21]. The following procedure was used for sample preparation: (1) polishing by water suspension of diamond particles with sizes down to 1 μm and (2) mechanochemical polishing with silica compound SF1 suspension to remove residual deformation and amorphous surface layer. The samples were mounted on a holder by a conductive sticky tape. In order to remove excess charge induced by electron beam, the samples were covered by 6-nm-thick carbon layer.

3. Results and discussion

We checked the abilities of the most common methods of domain imaging for PMN-PT. It is well known that the combination of the selective chemical etching and SEM allows visualizing the domain structures with a resolution down to 2 nm. We applied this method to PMN-PT single crystals. Thus, we obtained the topography corresponding to the domain structure revealed after mechanochemical polishing with a colloidal silica (Fig. 1a,b). The technique revealed the maze domain structures with a characteristic width of 0.5-2 μm (Fig. 1a) and needle-like structures with a width down to 150 nm (Fig. 1b).

We used the potential contrast in order to visualize domain structure and observed mentioned above obstacles, such as low resolution and stability (Fig. 1c). Moreover, surface charging caused by electron beam scanning led to disappearance of the surface potential contrast due to screening.

In order to define optimal parameters for domain imaging by channelling contrast, we studied the dependence of domain contrast and resolution on accelerating voltage (primary electron beam energy), electron beam current, and working distance (distance from the pole piece of the electron column to the sample surface). It was shown that decrease of working distance increased the resolution and domain contrast due to increase of the solid angle of the electron beam, and BSE detection resulted in prevailed detection of the low angle BSE. The increase of the accelerating voltage also increased the

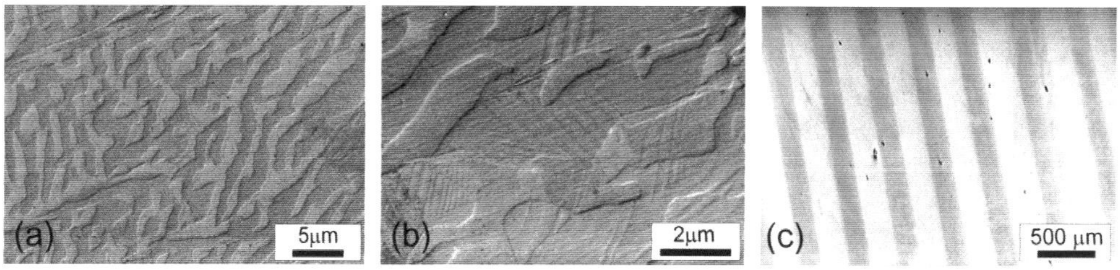

Figure 1. SEM images of the domain structures in PMN-PT crystals: (a,b) domains revealed by selective chemical etching during mechanochemical polishing, (c) potential contrast of *a*-domains.

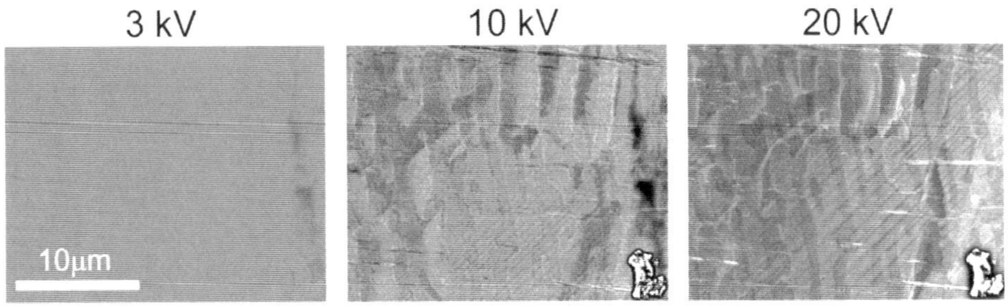

Figure 2. SEM images of the domain structures in PMN-PT crystals at various accelerating voltages in electron channelling mode.

resolution and the domain contrast (Fig. 2) due to reduction of the electron probe size and increase of the interaction volume. It should be noted that the presence of the damaged surface layer formed during polishing, hampered the domain imaging at low electron energies. It was demonstrated that rise of the beam current enhanced the contribution of the topography contrast due to the increase of the SE quantity, which could be detected at the same time by BSE detector. Thus, the optimal parameters were revealed (Table 1).

It was shown that the electron beam scanning during the image acquisition might lead to the change of the initial domain state (Fig. 3) caused by generation of the electric field induced by the charge accumulation. This issue can be overcome by deposition of the surface conductive layer. Deposition of the 6-nm-thick carbon layer allowed removing the charging effect and keeping the domain contrast.

Table 1. Optimal parameters for domain imaging using channelling contrast in PMN-PT crystals.

Parameter	Value
Working distance (mm)	below 4
Accelerating voltage (kV)	above 20
E-beam current (pA)	about 0.3

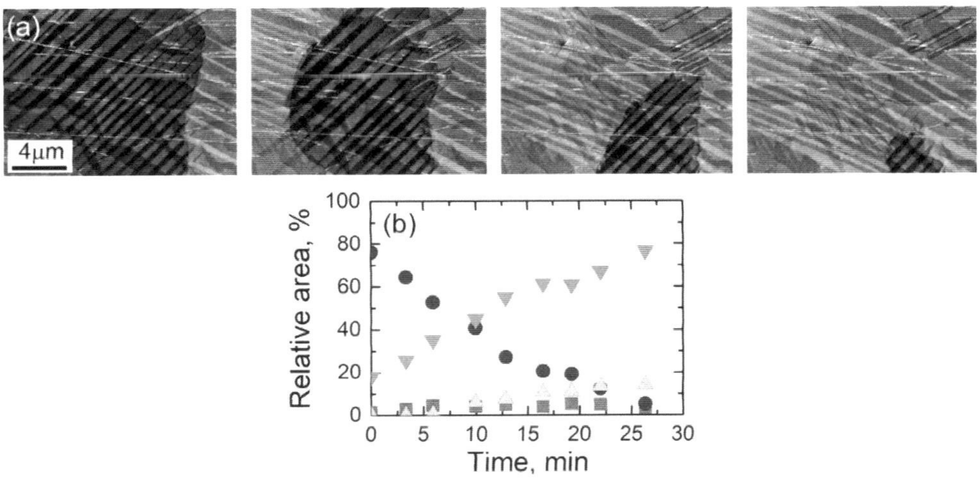

Figure 3. (a) SEM images of the domain structures in PMN-PT crystals obtained by sequential scanning. Scanning time is 3 min, (b) time dependence of the relative area for each domain type.

IOP Conf. Series: Materials Science and Engineering **443** (2018) 012038 doi:10.1088/1757-899X/443/1/012038

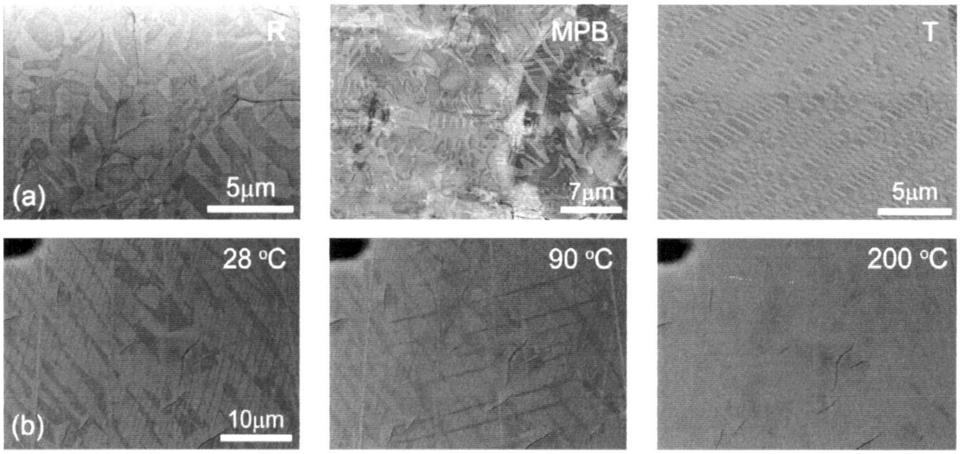

Figure 4. SEM images of the domain structures in PMN-PT crystals obtained: (a) for various compositions, (b) at the different temperatures.

We used the obtained information to visualize the nanodomain structures in PMN- PT in rhombohedral and tetragonal phases and the composition corresponding to the MPB region (Fig. 4a). Moreover, we demonstrated the ability to image the domain structure change as a result of the phase transition during the temperature treatment in 0.67PMN-0.33PT crystals (Fig. 4b). Two types of as-grown domain structures: quasi-regular patterns of pin-stripe domains and the laminar domains were visualized at the room temperature. The heating up to 70°C led to disappearing of the pin-stripe domains and formation of the long needle-like domains. Such behaviour can be attributed to the transition from monoclinic to tetragonal phase. Further heating above 200°C resulted in domain disappearance due to transition to the paraelectric phase.

The good correlation between domain images obtained in MPB phase of PMN-PT crystals by piezoresponse force microscopy (PFM) and SEM in channelling mode (Fig. 5) was demonstrated. The better quality of SEM images can be attributed to the difference in depth of the obtained signal: tens of nanometres for SEM and above micron for PFM. It should be noted that, the image acquisition time for SEM is essentially shorter. It makes the SEM channelling mode more attractive as compared with PFM.

4. Conclusion

The abilities of various SEM techniques for domain imaging were demonstrated in PMN-PT crystals with different compositions. It was shown that the imaging of the chemically etched relief was limited by its destructivity and potential contrast – by low resolution and contrast instability. The channelling

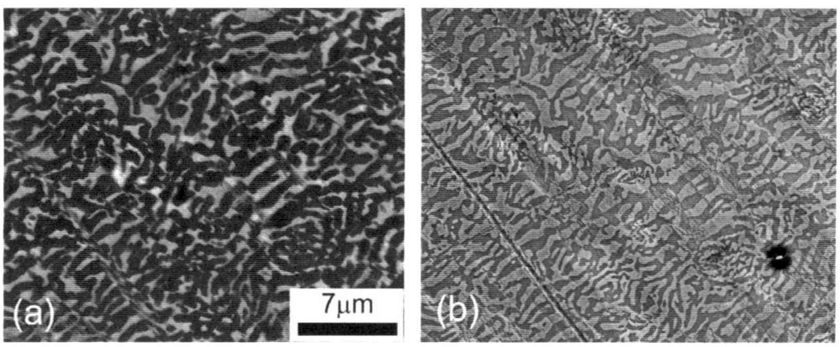

Figure 5. Domain images of the MPB PMN-PT crystals obtained by: (a) PFM, (b) SEM.

mode BSE detection looks preferable. The optimal parameters of the BSE channelling imaging were obtained. The high-resolution domain imaging in PMN-PT crystals in rhombohedral and tetragonal phases and the composition corresponding to the morphotropic phase boundary region were shown. The domain structure change as a result of the phase transition was imaged. The comparison of SEM and PFM domain images showed good correlation.

Acknowledgments
The equipment of the Ural Centre for Shared Use "Modern Nanotechnology" Ural Federal University has been used. The research was made possible by RFBR (grant 17-52-80116-BRICS_a) and state task of Ministry of education and science of the Russian Federation (No. 3.4993.2017/6.7). The work was partially supported by Government of the Russian Federation (Act 211, Agreement 02.A03.21.0006).

References

[1] Park S-E and Shrout T R 1997 *J. Appl. Phys.* **82** 1804-11
[2] Zhang R, Jiang B and Cao W 2001 *J. Appl. Phys.* **90** 3471-5
[3] Soergel E 2005 *Appl. Phys. B* **81** 729-51
[4] Tagantsev A K, Cross L E and Fousek J 2010 *Domains in Ferroic Crystals and Thin Films* (New York: Springer)
[5] Potnis P, Tsou N-T and Huber J 2011 *Materials* **4** 417-47
[6] Reichmann A, Zankel A, Reingruber H, Pölt P and Reichmann K 2011 *J. Eur. Ceram. Soc.* **31** 2939-42
[7] Le Bihan R 1989 *Ferroelectrics* **97** 19-46
[8] Zhu S and Cao W 1997 *Phys. Rev. Lett.* **79** 2558-61
[9] Sogr A A and Kopylova I B 1997 *Ferroelectrics* **191** 193-8
[10] Kuznetsov D K, Chezganov D S, Mingaliev E A, Kosobokov M S and Shur V Y 2016 *Ferroelectrics* **503** 60-7
[11] Gruner D and Shen Z 2010 *J. Am. Ceram. Soc.* **93** 48-50
[12] Omori M, Mishima T and Fujimoto T 2011 *Jpn. J. Appl. Phys.* **50** 09NC03
[13] Ihlefeld J F, Michael J R, McKenzie B B, Scrymgeour D A, Maria J-P, Paisley E A and Kitahara A R 2017 *J. Mater. Sci.* **52** 1071-81
[14] Robinson G Y and White R M 1967 *Appl. Phys. Lett.* **10** 320-3
[15] Kokhanchik L S 2018 *Phys. Solid State* **60** 1778-85
[16] Howell J A, Vaudin M D and Cook R F 2014 *J. Mater. Sci.* **49** 2213-24
[17] Karlsen M, Einarsrud M-A, Lein H L, Grande T, Hjelen J and Vullum P E 2009 *J. Am. Ceram. Soc.* **92** 732-7
[18] Leach C, Ali N K and Hall D A 2011 *Ceram. Int.* **37** 2185-91
[19] Otonicar M, Reichmann A and Reichmann K 2016 *J. Eur. Ceram. Soc.* **36** 2495-504
[20] Zaefferer S and Elhami N-N 2014 *Acta Mater.* **75** 20-50
[21] Wang R, Xu H, Yang B, Luo Z, Sun E, Zhao J, Zheng L, Dong Y, Zhou H, Ren Y, Gao C and Cao W 2016 *Appl. Phys. Lett.* **108** 152905

IOP Conf. Series: Materials Science and Engineering **443** (2018) 012039 doi:10.1088/1757-899X/443/1/012039

Applications of atomic force microscope manipulator operating in hybrid mode

A A Zhukov

Institute of Solid State Physics, Russian Academy of Science, Chernogolovka, 142432, Russia

azhukov@issp.ac.ru

Abstract. The topography images by atomic force microscope (AFM) operating in a new hybrid mode with nanometre scale noise are obtained. Using the improved feedback of the AFM hybrid mode, the possibility of the manipulations of the microdrops by microscope tip is shown. Additionally, we introduce a method of an extra gentle movement of nanowires by liquid flow organized by the AFM tip.

1. Introduction

Since introducing the atomic force microscopy (AFM) [1], numerous modes of operations of the microscopes have been developed [2]. Three basic modes such as static contact tapping [3], dynamic tapping [4,5], and non-contact [6] modes are the standard ones realized in most of AFM devices on the market. Each of the basic modes allows obtaining topography image of the solid sample with atomic resolution at certain conditions [7-9].

Essential improvement of the AFM electronics allowed assembling a huge amount of data during scanning. A number of more advanced AFM modes were introduced, such as peak-force mode, where the force curve is measured, stored and analysed for each point of the scan. Peak-force mode, as well as the multi-frequencies modes not only allow the easy extraction of an additional information about mechanical properties of the surface under investigation such as viscosity, elasticity etc., but they are also essential important in investigations of the biological objects [2].

AFM probes are widely applied to manipulate both solid objects [10] and drops of the liquids [11]. Usually standard static contact mode is successfully used for this purpose [12,13], but the dynamic tapping mode may be applied for nanodrops deposition as well [14].

Current report is dedicated to improvement and application of the new hybrid mode introduced in Ref. [15]. We demonstrate the essential improvement in topography mapping as well as application of the AFM tip for microdrops manipulations. Finally, we introduce the novel method to replace the InAs nanowires without touching the wire by the tip, but using the liquid flow organized by the AFM tip only.

2. Topography mapping using AFM in hybrid mode

All manipulations and scanning presented in this report were done by home-made two-probe manipulator. Detailed description of this manipulator can be found elsewhere [15]. It is worth noting that both AFMs of this manipulator are operated in so called hybrid mode. The main idea of this mode is to unite the permanent pressing of the tip probe to the sample surface of the contact mode and the amplitude oscillating of the tuning fork as the feed-back signal similar to dynamic modes. The schematic of realization of this mode is shown in Figure 1. The tip is mounted not vertically but with a slope of 20 to 30 degrees to the sample surface. So, the tip of the probe stays on the sample surface due to the bending of the tip, while the base of the probe mounted on the tuning fork prong oscillates (Fig. 1).

Several examples of this mode application, such as measurements of adhesive force and electronic transport can be found in Ref. [15]. Tungsten tips for manipulations and PtIr for electronic transport measurements were used. Due to bulk PtIr tips, the current up to 1 μA can be safely applied during the measurement of electronic transport with ohmic contact resistance (less than 6 kOhm) and 100\% of duty ratio [15]. But the first attempts of the topography mapping had too large noise in vertical direction to apply the obtained image for allocation of the nano-objects on the substrate [15]. So, the scanning electron microscopy image has to be used instead of AFM mapping to get the final result of the nanowires manipulations in case when optical microscopy has insufficient resolution [15].

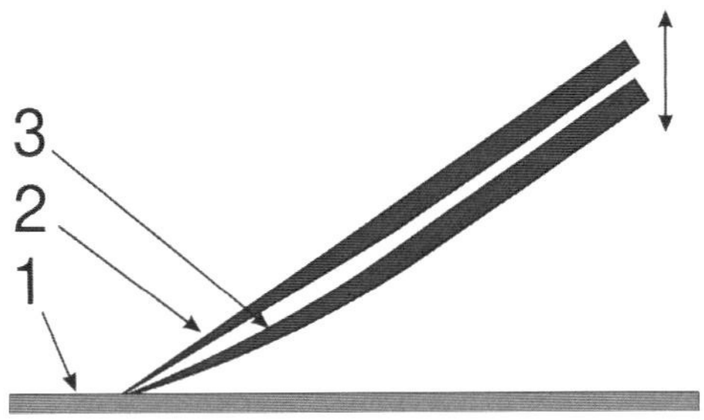

Figure 1. The sketch of the hybrid mode. 1 - substrate, 2 - tip in the topmost position, 3 - tip in the bottommost position. The tip of the probe attached to the oscillating tuning fork touches the surface even at the topmost position. Two-arrows vertical line – direction of oscillations of the tuning fork prong.

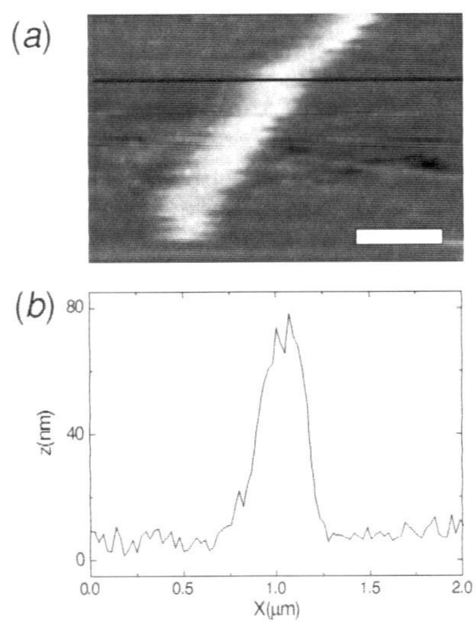

Figure 2. (a) AFM topography image of InAs nanowire obtained using hybrid mode with artificially damped to quality factor equal to 22 of the tuning fork assembly. (b) cross-section at black line of (a). Vertical noise of less than 10 nm is clearly visible. The length of the scale bar is 0.5 μm.

The main problem to obtain the better AFM image, besides the lack of resolution in direction parallel to the tip, was the overregulation of the AFM feedback. The standard AFM electronics was used, while the response from the piezo-actuator was more than an order of magnitude larger than the typical value [15]. One of the solutions for such kind of problem is the decreasing of the quality factor of the tuning fork and the tip assembly. It can be done both electronically [16] and mechanically. The second method was chosen. Length of the tip was increased to 3 mm and the thin stripe of Varnish glue on the top of the prong was added, if necessary, to damp the quality factor of the tuning fork assembly to 20. It was found that the optimal quality factor for topography mapping was from 20 to 30. Overdamping results in the unacceptable signal to noise ratio, so the noise in the detecting amplitude of the tuning fork oscillations determines the topography mapping quality.

The example of the topography image in optimally tuned system is shown in Figure 2a. The quality factor was of 22 and resonance frequency of f_0=16800 Hz. The cross section of the image (Fig. 2b) demonstrates the vertical noise of essentially less than 10 nm. This more than an order of magnitude improvement of the noise level allows using AFM mapping images gathered in hybrid mode to allocate nanoobjects, such as nanowires on the substrate with precision of nanometre scale.

It is worth noting that an additional wire glued to the side of the prong to contact PtIr tip for electronic transport measurements additionally damping the quality factor serves not only for galvanic isolation of the tip, but works additionally toward more stable feed-back, so more reliable and stable ohmic contact between the AFM tip and measured object can be achieved.

3. Manipulations of microdrops

Essential improvement of the feed-back control allows extending the manipulator application to liquid microdrops replacements. This is again quite similar to applications of the standard static contact AFM mode, additionally to conductance measurements and measurements of the adhesive forces.

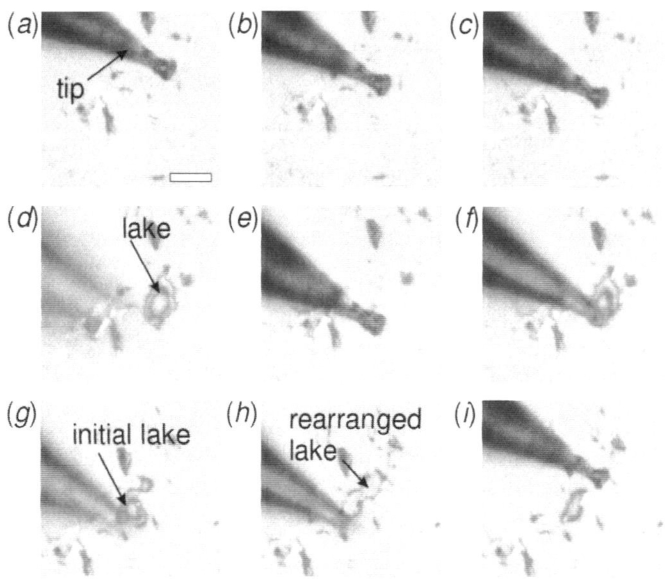

Figure 3. (a)-(c) Replacement of the drop beneath the tip. (d) Initial lake, (e) the lake shrunk by the landing tip and (f) the lake released by the tip jumped slightly up. (g)-(i) Step by step the process of the division of the large lake in two parts: (g) initial large lake, (h) the lake rearranged using the landing tip sweeping over the surface, (i) the landing tip assembling the part of the lake liquid, the division of the initial lake is completed. The length of the scale bar is 10 μm the same for all figures.

For demonstration, a Si wafer with 100 nm SiO_2 thermally grown surface layer was used as a substrate. The liquid was the heavy fractions remained after the evaporation of the isopropanol (Component reaktiv, TU 6-09-07-1718-91). Detailed chemical composition of such fractions are not important in the current investigation, because we are not interested in the resulted drops shapes, but the main issue is the demonstration of the potential of the hybrid AFM mode only.

Figures 3a-c show the replacement of the drop beneath the tip. There is no visible trace of the liquid remained after the AFM tip movement, so the volume of the moved drops presumably keeps almost constant. Figures 3d-f show the initial lake, the lake shrunk by the landing tip, and the lake released by the jumped up tip, correspondently. The shape and the size of the lake is similar before tip landing and after tip taking off. Figures 3h-i demonstrate the division of the large lake into two parts. Figure 3g shows the initial large lake assembled from the several smaller drops. Then, the lake is rearranged using the landing tip sweeping over the surface, wetting the substrate and, so, defining new oblong shape of the lake. The lake with a new shape is shown in Figure 3h. Landing tip assembling the part of the lake liquid is shown in Figure 3i. So, the division of the initial lake is completed.

All the images are obtained using direct short focus optical microscope, which allows adequate and comfort live checking the manipulations results. For drops manipulations, standard tungsten tips were used. Wetting of the tip was enough to perform all presented manipulations with drops. Additionally, we checked that the size of the drop with diameter approximately of 7 μm (Figs. 3d-f) had no visible changes of size, shape and, presumably, the volume after landing and taking off of the tip performed ten times. So, it is possible to conclude that the wetting of the tip still does not change the size of the drop after the single manipulation for more than 5 %.

Let us discuss in detail the landing of the tip, assembling of the drop and taking off of the AFM tip (Figs. 3a-c as illustrations from the top view). Figure 4a shows the drop on the substrate surface with initial shape and the tip placed over it. Figure 4b shows the position, when the tip just touches the surface. Drop starts to wet the tip. Bended tip shown in Figure 4c assembles the liquid drop beneath itself. Bending of the tip essentially helps to replace the drop. When the tip is taking off, it initially gets straight, this returns the drop to its initial shape. Finally tip lefts the substrate surface. It is worth noting that the size of tip surface touches the liquid in drop could be approximately an order of magnitude larger, while tip is bending comparing to the straight one. This helps to relocate drops keeping their volumes almost constant.

Realization of the hybrid mode in liquid environment, which is of the essential importance for biological applications, is still under construction.

4. Application of the liquid flow for nanowire replacing

Easy and well-controlled manipulation of the liquid drops with AFM tip allows realizing a new extremely gentle nanowire replacement method without touching the nanowire with the AFM tip at all. Replacement is performed by liquid flow arranged by the tip.

Experimental measurements of the adhesion forces of the nanowires demonstrate quite a large error of measured values, namely more than an order of magnitude [10]. Presented method is applicable to nanowires, which stick to the substrate not too strong.

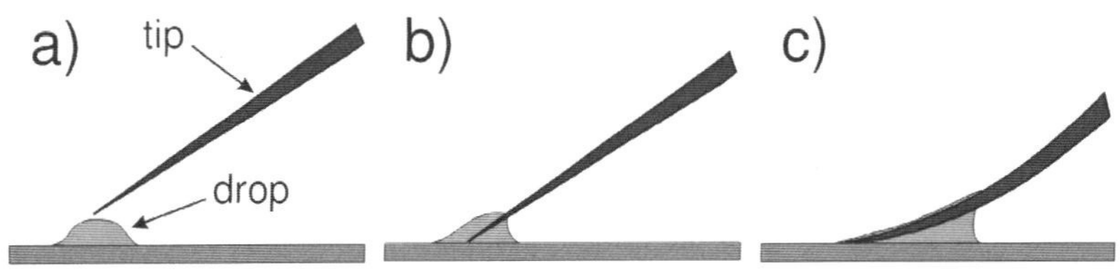

Figure 4. (a) The drop on the substrate surface with initial shape tip placed over it. (b) The tip just touches the surface. (c) Bended tip assembling the liquid drop beneath the tip.

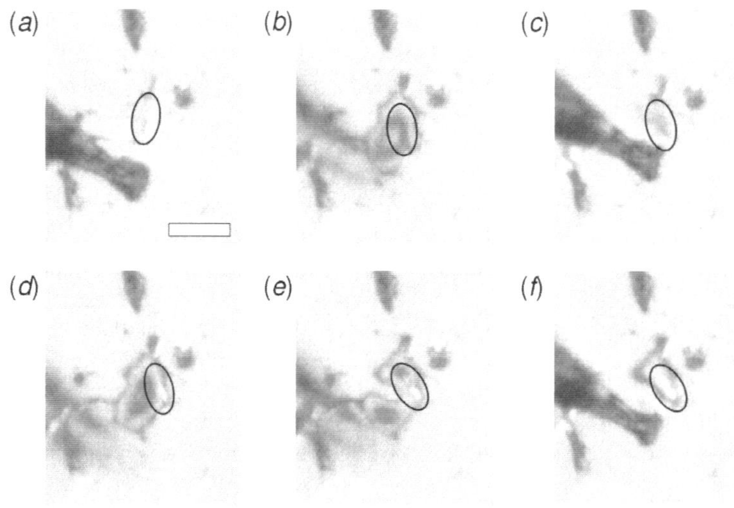

Figure 5. Replacing of the InAs nanowire by a liquid flow. (a), (c), (f) AFM tip assembling the large lake of liquid. (b), (d), (e) Lake released by the AFM tip. Ovals mark the InAs nanowire replaced by the liquid flow. The length of the scale bar is 10 μm same for all figures.

Figure 5 shows replacing of the InAs nanowire by a liquid flow. AFM tip assembles (Figs. 5a,c,f) and releases (Figs. 5b,d,e) the large lake resulting in InAs nanowire (marked by oval) replacement over the SiO_2 substrate. The replacement of the wire is not well controlled by now. The direction of the movement is generally defined by the flowing of the liquid only. Additional efforts must be made to improve this "nanoflood" or "nanorafting" method.

The improved realization of this method should be performed provisionally in the following way and might include several steps. The first step is to pick up the drop and to drive the tip along the path of the future channel wetting the SiO_2 substrate, similar to lake reshaping mentioned in the previous section. The nanowire needed to be moved must be on the way of the channel. The second step is the formation of the large lake of the liquid (Fig. 5). The third step is to connect the lake to the channel. Some tilt of the substrate for 4 to 5 degree might help nanowire movement by the liquid flow. The fourth step (if necessary) is liquid transferring drop-by-drop with AFM tip to add the volume of the liquid to the top (source) lake. It is worth noting that the design of the AFM manipulator we use fits well to work on the tilted surfaces, because the working range in all directions including vertical is 50 microns.

5. Conclusion

In this report, we demonstrate an order of magnitude improvement of decreasing the vertical (z-directional) noise in topography mapping in hybrid AFM mode. The resulting images are acceptable for allocation of the nanoobjects with nanometre lateral resolution. Examples of microdrops manipulations using AFM tip operating in hybrid mode are presented. Finally, the new method of extreme gentle manipulations of the nanowire with the liquid flow organized by AFM tip is introduced.

Acknowledgments

We would like to acknowledge for the InAs samples to H. Hardtdegen and Th. Schaepers. This work is supported by the Russian Foundation for Basic Research (Grants 14-02-00192-a, 17-02-00454-a, and 13-02-12127-ofi-m), programs of the Russian Academy of Science, the Program for Support of Leading Scientific Schools, and the International Bureau of the German Federal Ministry of Education and Research within the project RUS 09/052.

IOP Conf. Series: Materials Science and Engineering **443** (2018) 012039 doi:10.1088/1757-899X/443/1/012039

References

[1] Binnig G, Quate C F and Gerber C 1986 *Phys. Rev. Lett.* **56** 930-3
[2] Dufrêne Y F, Ando T, Garcia R, Alsteens D, Martinez-Martin D, Engel A, Gerber Ch and Müller D J 2017 *Nat. Nanotechnol.* **12** 295-307
[3] Marti O, Colchero J and Mlynek J 1990 *Nanotechnology* **1** 141-4
[4] Garcia R and Pérez R 2002 *Surf. Sci. Rep.* **47** 197-301
[5] Zhong Q, Inniss D, Kjoller K and Elings V 1993 *Surf. Sci. Lett.* **290** L688
[6] Albrecht T R, Grütter P, Horne D and Rugar D 1991 *J. Appl. Phys.* **69** 668-73
[7] Binnig G, Gerber Ch, Stoll E, Albrecht T R and Quate C F 1987 *Europys. Lett.* **3** 1281-6
[8] Giessibl F J 2005 *Materials Today* **8** 32-41
[9] Binnig G and Rohrer H 1999 *Rev. Mod. Phys.* **71** S324
[10] Conache G, Gray S, Bordag M, Ribayrol A, Fröberg L E, Samuelson L, Pettersson H and Montelius L 2008 *J. Phys.: Conf. Ser.* **100** 052051
[11] O'Connell C D, Higgins M J, Marusic D, Moulton S E and Wallace G G 2014 *Langmuir* **30** 2712-21
[12] Piner R D, Zhu J, Xu F, Hong S and Mirkin Ch A 1999 *Science* **283** 661-3
[13] Ginger D S, Zhang H and Mirkin C A 2004 *Angewandte Chemie International Edition* **43** 30-45
[14] Fang A, Dujardin E and Ondarcuhu Th 2006 *Nano Letters* **6** 2368-74
[15] Zhukov A A, Stoliarov V S and Kononenko O V 2017 *Rev. Sci. Instrum.* **88** 063701
[16] Ctistis G, Frater E H, Huisman S R, Korterik J P, Herek J L, Vos W L and Pinkse P W H 2011 *J. Phys. D: Appl. Phys.* **44** 375502

AUTHOR INDEX

AUTHOR INDEX